Toxicity Testing

Strategies to Determine Needs and Priorities

Steering Committee on Identification of Toxic and Potentially Toxic Chemicals for Consideration by the National Toxicology Program

Board on Toxicology and Environmental Health Hazards

Commission on Life Sciences

National Research Council

NATIONAL ACADEMY PRESS
Washington, D. C.   1984

NATIONAL ACADEMY PRESS   2101 CONSTITUTION AVE., NW   WASHINGTON, DC 20418

The work on which this publication is based was performed pursuant to Contract NO1-ES-0-0008 with the National Toxicology Program.

Library of Congress Catalog Card Number 84-60095
International Standard Book Number 0-309-03433-7

Printed in the United States of America

COMMITTEE ON CHARACTERIZATION OF STATUS OF TOXICITY
DATA ELEMENTS FOR A SELECT UNIVERSE OF COMPOUNDS
(Committee on Toxicity Data Elements)

John Doull, <u>Co-Chairman</u>
Department of Pharmacology
University of Kansas Medical Center
Kansas City, Kansas

Emil A. Pfitzer, <u>Co-Chairman</u>
Department of Toxicology and Pathology
Hoffmann-LaRoche, Inc.
Nutley, New Jersey

Eula Bingham
Kettering Laboratory
University of Cincinnati Medical Center
Cincinnati, Ohio

David J. Brusick
Department of Genetics
Litton Bionetics Research Laboratory
Kensington, Maryland

George T. Bryan
Department of Human Oncology
Clinical Science Center
Madison, Wisconsin

Robert T. Drew
Medical Department
Brookhaven National Laboratory
Upton, New York

E. Marshall Johnson
College of Medicine
Thomas Jefferson University
Philadelphia, Pennsylvania

Morton Lippmann
Aerosol and Inhalation Research Laboratory
New York University
New York, New York

<u>Staff</u>

Scott R. Baker
Project Director

Gordon W. Newell
Project Director (until October 1981)

Barbara B. Jaffe
Senior Staff Officer (until July 1983)

Resha M. Putzrath
Staff Officer (until September 1983)

Thomas Mack
Cancer Surveillance Program
University of Southern California
  School of Medicine
Los Angeles, California

Gilbert J. Mannering
Department of Pharmacology
University of Minnesota
Minneapolis, Minnesota

Donald E. McMillan
Department of Pharmacology
University of Arkansas
Little Rock, Arkansas

Robert A. Neal
Chemical Industry Institute
  of Toxicology
Research Triangle Park, North Carolina

Edward O. Oswald
Department of Environmental
  Health Sciences
University of South Carolina
Columbia, South Carolina

Marvin A. Schneiderman
Clement Associates
Washington, D.C.

Carrol S. Weil
Carnegie-Mellon Institute
  of Research
Carnegie-Mellon University
Pittsburgh, Pennsylvania

Hanspeter R. Witschi
Biology Division
Oak Ridge National Laboratory
Oak Ridge, Tennessee

Leslye B. Giese
Research Assistant

Ronnie M. Good
Bibliographic Assistant

Michele W. Zinn
Administrative Assistant

Shirley A. Perry
Secretary

COMMITTEE ON RESEARCH OF AGENTS
POTENTIALLY HAZARDOUS TO HUMAN HEALTH
(Committee on Priority Mechanisms)

Arthur C. Upton, <u>Chairman</u>
Institute of Environmental Medicine
New York University Medical Center
New York, New York

Bernard D. Astill
Health, Safety and Human Factors
   Laboratory
Eastman Kodak Company
Rochester, New York

Stephen L. Brown
Center for Health and
   Environmental Research
SRI International
Menlo Park, California

Patricia A. Buffler
School of Public Health
University of Texas
Houston, Texas

Richard M. Cooper
Williams and Connolly
Washington, D.C.

Baruch Fischhoff
Decision Research, Inc.
Eugene, Oregon

Corwin H. Hansch
Department of Chemistry
Pomona College
Claremont, California

Sheldon D. Murphy
Department of Pharmacology
University of Texas Medical School
Houston, Texas

Michael R. Overcash
Department of Chemical Engineering
North Carolina State University
Raleigh, North Carolina

Talbot R. Page
Environmental Quality Laboratory
California Institute of Technology
Pasadena, California

Verne A. Ray
Medical Research Laboratory
Pfizer, Inc.
Groton, Connecticut

Harold R. Ward
Center for Environmental
   Studies
Brown University
Providence, Rhode Island

<u>Staff</u>

Samuel B. McKee
Project Director

Walter G. Rosen
Project Director (until July 1983)

Azora L. Irby
Secretary

Roger O. McClellan
Lovelace Inhalation Toxicology
  Research Institute
Albuquerque, New Mexico

Daniel Menzel
Departments of Pharmacology and Medicine
Duke University Medical Center
Durham, North Carolina

Norton Nelson
Institute of Environmental Medicine
New York University Medical Center
New York, New York

Staff

Devra Lee Davis, Executive Director

Jacqueline K. Prince, Staff Assistant

Shirley A. Ash, Administrative Secretary

PREFACE

In September 1980, the National Toxicology Program (NTP), through the National Institute of Environmental Health Sciences, contracted with the National Research Council (NRC) and the National Academy of Sciences for a study with two principal charges:

(1) To characterize the toxicity-testing needs for substances to which there is known or anticipated human exposure, so that federal agencies responsible for the protection of public health will have the appropriate information needed to assess the toxicity of such substances.

(2) To develop and validate uniformly applicable and wide-ranging criteria by which to set priorities for research on substances with potentially adverse public-health impact.

The study, titled "Identification of Toxic and Potentially Toxic Chemicals for Consideration by the National Toxicology Program," was established in the Board on Toxicology and Environmental Health Hazards of the NRC's Commission on Life Sciences.

A Steering Committee was formed to address the charges and coordinate the efforts of three operating committees. The Committee on Statistical Sampling Methods ("Committee on Sampling Strategies") and the Committee on Characterization of the Status of Toxicity Data Elements ("Committee on Toxicity Data Elements") addressed the first charge. The Committee on Research of Agents Potentially Hazardous to Human Health ("Committee on Priority Mechanisms") addressed the second charge.

The two charges of the study involved concepts with many similarities, but they also required different approaches. The first charge--to characterize the status of toxicity-testing information on substances--required a detailed examination of existing toxicity data based on available criteria for qualitative and quantitative characteristics of adequacy. The approach required judgments on the adequacy of testing of specific substances, independently of whether the substances were potent agents or relatively inert substances with low toxicity. The second charge--to develop criteria for setting research priorities--required a detailed examination of methods of selecting potentially toxic substances from a large universe of substances on which there was little or no toxicity information. The approach required judgments on the predictability of testing procedures, the nature and severity of potential toxic effects, and the systemic modeling of steps in decision-making.

This is the final report of the four committees. It represents the work of their members, NRC staff, a large number of consultants, and others who provided invaluable assistance to the committees. This report is designed to be a stand-alone document, summarizing and synthesizing the full scope of work first described at two interim stages of evolution in documents entitled Strategies to Determine Needs and Priorities for Toxicity Testing--Volume 1: Design and Volume 2: Development. These earlier reports were published at the end of the first and second years of the study.

The activities to address the two charges were conducted with substantial interchange of concepts and practices, but they were recognized to be sufficiently different in approach to warrant presentation as two distinct parts. The specific differences are presented in a final appendix to Part 2. The reader should be aware of this intent when reading Parts 1 and 2 of this report.

We of the Steering Committee thank all the participants, who have contributed so much to this study. The amount of work performed indicates the magnitude of the task that faces NTP. In particular, we wish to recognize the enormous effort dedicated to this study by the three operating committees and the NRC staff. The dossiers containing the analyses of all toxicity data that were available on a subsample of 100 substances constitute approximately 2 cubic meters of decision-making information. The collection, analysis, and interpretation of information on just this small number of substances in general use required thousands of hours of the time of volunteer committee members, the staff, and other participants over a 3-year period. The thoughtfulness and computer work needed to select the study samples and model the criteria for priority-setting required no less intensive an effort.

The large scope of work addressed by the committees could not have been accomplished without the assistance of the skilled and dedicated staff of the Board on Toxicology and Environmental Health Hazards. We wish to thank Robert G. Tardiff, who was project director for this study during his tenure with the Board, and Alvin G. Lazen, who was associated with the study in its last stages. We also thank Scott R. Baker for his important contributions to the "needs" section of this report and Samuel B. McKee and Walter G. Rosen for their work on the "priorities" section; Lamar B. Dale, Leslye R. Giese, Ronnie M. Good, Norman Grossblatt, Cheryl J. Haily, Veronica C. Harris, Azora L. Irby, Barbara B. Jaffe, Paula H. Morris, Gordon W. Newell, Shirley A. Perry, Frances M. Peter, Jacqueline K. Prince, Resha M. Putzrath, Joyce A. Russell, Patricia A. Sterling, Robert J. Thomas, and Michele W. Zinn for their professional and administrative contributions; and Andrew M. Pope, Gerald M. Rosen, and Jeannee K. Yermakoff for their contributions while they were serving as National Research Council Fellows.

We are particularly grateful for the assistance provided by Raymond E. Shapiro, who served as the study's project officer for NTP.

James L. Whittenberger, Chairman
John C. Bailar
John Doull
Emil A. Pfitzer
Arthur C. Upton

ACKNOWLEDGMENTS

Valuable contributions to this study were made by Steven Wilhelm and Yvonne Hales, who served as technical consultants, and by Lynn Jones, Nora Riley, and Janice Miller, who assisted in the acquisition of documents.

The following colleagues in both the public and private sectors generously shared information, resource material, and expertise: Wellman Bachtel, Susan Bloodworth, Miles Bogle, Keith Booman, Edward Brooks, Dorothy Canter, Michael Conners, Paul Craig, Sonia Crisp, Denny Daniels, Karen Dickerson, David Disbennett, Ronald Dunn, Donald Dunnom, Robert Elder, Curt Enslein, Norman Estrin, Theodore Farber, Kenneth Fisher, Gary Flamm, Richard Ford, Vasilios Frankos, John Froines, Vera Glocklin, John Gordon, Martin Greif, Allen Heim, Joseph Highland, Louis Hodes, David Hoel, Maurice Hubert, Vera Hudson, Julius Johnson, Judy Jones, Henry Kissman, Mary Rose Kornreich, Raymond Kukol, Ann McCann, Arthur McCreesch, Jerald McEwan, Carolyn McHale, Joseph Merenda, Mary Lou Miller, William Milne, Victor Morgenroth, Therese Murtaugh, Ian Nisbet, William Olson, Donald Opdyke, Norbert Page, William Payne, Alan Rulis, Andrew Sage, Virginia Salzman, Phillip Sartwell, Takashi Sugimura, John Tinker, Justine Welch, and Karen Wetterhahn.

CONTENTS

EXECUTIVE SUMMARY

Abstract: A "select universe" of 65,725 substances that are
of possible concern to the National Toxicology Program (NTP)
because of their potential for human exposure was
identified. Through a random sampling process, 675
substances covering seven major intended-use categories were
selected. From this sample, a subsample of 100 substances
was selected by screening for the presence of at least some
toxicity information. In-depth examination of this subsample
led to the conclusion that enough toxicity and exposure
information is available for a complete health-hazard
assessment to be conducted on only a small fraction of the
subsample. On the great majority of the substances, data
considered to be essential for conducting a health-hazard
assessment are lacking. By inference, similar conclusions
were made for the select universe from which the sample and
the subsample were drawn. This report presents criteria for
selecting substances and determining toxicity-testing needs,
provides estimates of those needs, and describes some useful
criteria for assigning priorities for toxicity testing.

The potential public-health impacts of chemicals lead society to seek
information for determining the probability and magnitude of such
impacts. Such information is based primarily on predictions from results
of toxicity studies. The development of a strategy for obtaining
appropriate information requires an estimation of the quantity and
quality of available toxicity data applicable to the assessment of human
health hazard, as well as knowledge of the number of substances on which
necessary experimental data are not yet available. A characterization of
the magnitude of needed testing would be valuable to those who allocate
resources for such testing. However, because resources for developing
sound scientific bases for identifying public-health hazards are limited,
it is important to establish priorities among chemicals and to select
those known or expected to have the greatest impact on human health.

A major function of the National Toxicology Program (NTP) is the
selection and testing of chemicals for toxicity. NTP has under
continuing review candidate chemicals for testing, as they are nominated
by the federal agencies served by the program, by state and local
governments, and by academic, industrial, and labor groups. Such
candidates are of interest to NTP because of their potential for human
exposure and public-health impact.

In September 1980, NTP, through the National Institute of
Environmental Health Sciences, contracted with the National Research
Council (NRC) and the National Academy of Sciences for a study. This
study was undertaken because NTP recognizes that the number of
substances, both natural and man-made, in the human environment is very
large and is increasing, with no clear indication of the nature and
amount of toxicity information that might be needed on these substances

to ascertain their potential for adverse effects on human health. It is useful for NTP to know as precisely as practicable the toxicity-testing needs for substances to which humans are potentially exposed.

NTP asked NRC to address these matters in two major objectives of the study:

(1) To characterize the toxicity-testing needs for substances to which there is known or anticipated human exposure, so that federal agencies responsible for the protection of public health will have the appropriate information needed to anticipate the extent of testing needs.

(2) To develop and validate uniformly applicable and wide-ranging criteria by which to set priorities for research on substances with potentially adverse public-health impact.

The study, titled "Identification of Toxic and Potentially Toxic Chemicals for Consideration by the National Toxicology Program," was established in the Board on Toxicology and Environmental Health Hazards of the NRC Commission on Life Sciences.

In this report, the Committee on Sampling Strategies and the Committee on Toxicity Data Elements describe in detail the criteria and procedures they used to determine the nature and extent of toxicity testing and their collective judgment on the testing needs for a "select universe" of chemical substances. The underlying strategy for the characterization of toxicity-testing needs involved four major steps:

(1) Definition of the select universe of substances that might be of interest to NTP because of their potential for human exposure.

(2) Drawing of a random sample of representative substances from the select universe.

(3) Statistical analysis of the sample to determine the quantity and quality of available information and detailed description of testing needs for the sample.

(4) Predictions, based on the sample analysis, of the testing needs for the select universe.

The Committee on Priority Mechanisms presents criteria and a decision-making framework that could be used to set priorities for research on substances with a potential for adverse public-health impact.

SELECT UNIVERSE OF SUBSTANCES

According to an estimate based on the Chemical Abstracts Service (CAS) Registry, the universe of known chemicals consists of over 5 million entities. To define toxicity-testing needs for substances in the human environment, it was necessary to select a manageable subset of the universe that would include most of the substances to which humans are likely to be exposed in the United States. The construction of this "select universe" of substances--a core that would be the reference for the study--relied on a search for lists of substances preselected for human exposure potential and computerized for reasonably easy access. A search for such lists revealed several that could be assembled to form the select universe, provided that most duplications of substances on the combined lists could be identified to permit statistical adjustments. The lists used included the Toxic Substances Control Act (TSCA) Inventory of 48,523 chemical substances in commerce; a list of 3,350 pesticides (active and inert ingredients) registered for use by the Environmental Protection Agency (EPA); a list of 1,815 prescription and nonprescription drugs approved by the Food and Drug Administration (FDA) and excipients used in drug formulations; a list of 8,627 food additives, including those approved for use by FDA; and a list of 3,410 cosmetic ingredients from the Cosmetic, Toiletry and Fragrance Association. This select universe did not systematically include environmental decomposition products, manufacturing contaminants, or natural substances (e.g., plant pollens and foods). The sum of the above, 65,725 entries from the lists, was taken as the select universe for the purposes of this study. Statistical adjustment for duplications indicated that the select universe contained about 53,500 distinct entities. The Committee on Toxicity Data Elements and the Committee on Sampling Strategies regarded the contents of the select universe as closely approximating the expected universe of interest to NTP.

CHARACTERIZING TOXICITY-TESTING NEEDS

During the planning stages of the study, it was recognized that the Committee on Toxicity Data Elements would not be able to examine the information on all 53,500 substances in the select universe, because of limitations of available resources. Therefore, the Committee on Sampling Strategies developed a method for drawing from the select universe a sample that was (1) small enough to be thoroughly examined for completeness, quality, and utility of information within the limitations of available resources and (2) designed to permit an extension of the committees' findings from the sample back to the select universe from which it was drawn. With a stratified random process, 675 substances were selected from the 65,725 listings. A random subsample of 100 substances with at least minimal toxicity information (described in Chapter 2 of Part 1) as prescribed by the Committee on Toxicity Data Elements was then selected from the random sample. The select universe, the sample, and the subsample contained representatives of seven categories of substances defined by the lists that make up the select universe: (1) pesticides and inert ingredients of pesticide formulations, (2) cosmetic ingredients, (3) drugs and excipients in drug formulations, (4) food additives, and chemicals in commerce, which were divided into (5) those with 1977 production of 1 million pounds or more, (6) those with 1977 production of less than 1 million pounds, and (7) those whose 1977 production was unknown or inaccessible because of manufacturers' claims of confidentiality. The sizes of each category in the select universe, the sample, and the subsample are presented in Figure 1.

The lists of substances making up the select universe were compiled on the basis of intended use, rather than toxicity. The intended-use theme was preserved throughout sample selection, data analysis, and inference-making, and it is reflected in the conclusions. Some structural classes of chemicals might be "overrepresented" in the subsample (e.g., cottonseed oil, linseed oil, and peanut oil). However, the Committee on Toxicity Data Elements made no attempt to define the select universe by chemical-structure classes. Rather, it relied on the probabilities inherent in small random samples to choose an appropriate number of substances from each major chemical grouping.

It is important to recognize that the subsample of 100 is drawn from the seven categories defined above and that these categories often include "inert" substances that are used to formulate "active" substances. Furthermore, these categories are not defined by chemical structure, so structurally similar substances in the subsample of 100 should not be combined for inference to the select universe. Similarly, inferences about a category should not be limited to the "active" substances in the category (e.g., drugs), but rather should be applied to all the substances (e.g., drugs and excipients in drug formulations) in that category. Nor is the subsample representative of substances involved in specially publicized episodes of toxic effect, such as those associated with thalidomide or dioxins. These kinds of substances are often selected for toxicity testing, because there is a particular interest in them--e.g., because some toxic effect has been observed. Thus, selection of substances that have already had some testing does not necessarily constitute random sampling of all possible substances in the select universe.

The Committee on Toxicity Data Elements developed a well-structured, stepwise approach to the determination of toxicity-testing needs for substances in the select universe. This required agreement on a strategy for judging the adequacy of toxicity data, establishment of guidelines for assessing the quality of individual toxicity studies, and creation of a decision-making system for reviewing and evaluating the total data base on the hazard of a substance--its toxicity, its exposure potential for humans, and its chemical and physical characteristics.

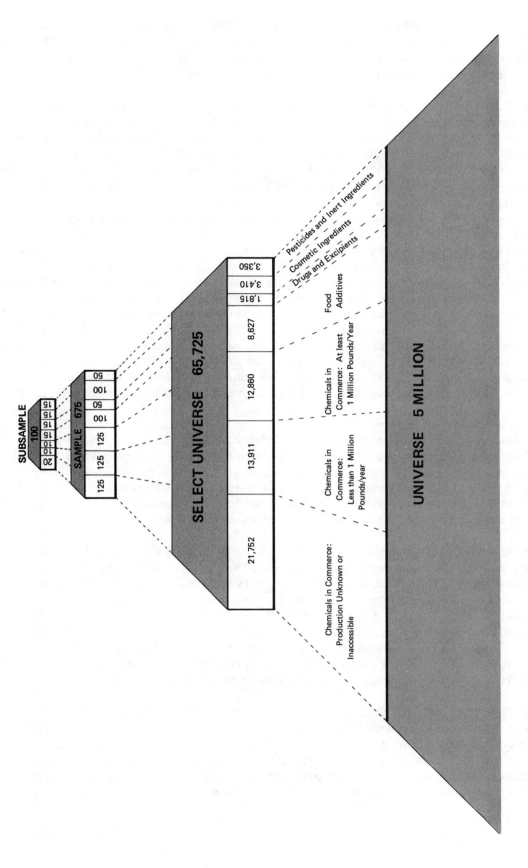

FIGURE 1 Content and size of each category of substances in the select universe, sample, and subsample.

5

Results of these efforts were applied to the subsample to establish the extent of additional toxicity testing that might be needed. Data on the sample and the subsample were then used by both committees to estimate toxicity-testing needs for the entire select universe.

The Committee on Toxicity Data Elements judged the quality and completeness of the toxicity data base on each substance in the subsample. To ensure quality, available information was checked against established reference guidelines for toxicity-testing protocols (e.g., those of the Organisation for Economic Co-operation and Development) that have been widely reviewed and generally accepted. The committee also relied on the accumulated experience and expertise of its members, whose combined judgment was used to determine the adequacy of an individual study if it did not meet the standards of the reference protocol guidelines. The committee's determination of the adequacy of toxicity testing for conducting a health-hazard assessment was based on information derived from experiments performed according to the reference protocol guidelines and other information that met the committee's own basic criteria for evaluating scientific methods (described in Chapter 4 of Part 1). Using this combination, the committee assessed the adequacy of the toxicity-testing protocols and the need for further toxicity tests in detailed evaluations, tabulations, and analyses for all substances in the subsample. The committees recognize that regulatory agencies' standards and requirements for testing may differ from those used in this study.

As analysis of the data bases proceeded, the Committee on Toxicity Data Elements established a working document with a standardized format, content, and method of reporting for each of the 100 substances in the subsample as the focal point for all document control efforts and all evaluations of testing adequacy. The working document or dossier became the unit of record for all committee decisions and actions.

The approach developed to collect data on each substance included searches of open literature (primary sources) through automated, on-line data retrieval files; secondary-source literature, such as reference manuals and textbooks; government technical reports; files of U.S. regulatory agencies; and files provided by some chemical manufacturers and trade associations. The data obtained from searches of the primary and secondary open literature constituted the bulk of the information in the dossiers. Search strategies were carefully developed to ensure the most efficient screening of the data bases selected.

The findings of the Committee on Sampling Strategies and the Committee on Toxicity Data Elements are based on analyses of the sample of 675 substances randomly chosen from the select universe and the subsample of 100 randomly chosen from the sample that had at least what the latter committee defined as prescribed minimal toxicity information. Some specific analyses are derived solely from the sample or the subsample. Others are derived from combined information on both the sample and the subsample. Confidence limits are given for the results of analyses in Chapter 5 of Part 1. In some cases, the confidence limits are wide.

The committees recognize that, despite extensive efforts to obtain all information, they might not have had access to results of some toxicity tests. Toxicity-testing information on the subsample of 100 substances was sought from industries and other interested parties via a Federal Register notice and by direct contact with manufacturers and importers of sampled chemicals in commerce, but some industrial information probably remained unavailable to the committees. Similarly, the committees were not able to examine toxicity, physical, and chemical information on cosmetic ingredients, drugs, excipients in drug formulations, and food additives that may be in the files of FDA, except in the case of food additives listed as substances generally regarded as safe (GRAS).

The documentation for decisions about the quality of tests that have been conducted and about toxicity-testing needs lends particular strength to this study. Scientists have varied opinions about protocol guidelines for toxicity tests, about testing needs for specific uses of substances, and about grounds for claims of adequacy or inadequacy of a particular test as it was performed. Where such varied opinions are important, they can become an integral part of the decision-making process to provide new estimates of testing needs. In the context of this study, scientific judgments used by the committees were recorded and subjected to peer review in a flexible study framework that would accommodate changes in estimates brought about by the presentation of new data.

QUANTITY AND NATURE OF TESTING

It was recognized from the beginning that the quantity and nature of testing needs were such that they could never be fulfilled adequately only by the use of specific testing regimens. Although tests of substances will always be needed, a better understanding of the "how" and "why" of toxic injury itself at the subcellular, cellular, organ, and whole-animal levels will be necessary in the future to fulfill the needs in the most efficient and economical manner. The Committee on Toxicity Data Elements used a battery of toxicity tests as the basic "measuring stick" for quantitation of testing needs. At the same time, it rejected the concept that every substance in the select universe required the adequate performance of a complete battery of toxicity tests for a human-health hazard assessment, even if that were practical. Thus, other criteria, including data from human exposures, were also used for judgments about testing adequacy. The Committee on Toxicity Data Elements recognizes that meeting the testing needs will require the establishment of priorities for the tests and the substances needing them.

In the seven categories of the sample of 675 substances, testing for acute and subchronic effects was generally present more frequently than testing for chronic, mutagenic, or reproductive and developmental effects (see Table 7 in Chapter 5 of Part 1). On the basis of an analysis of the randomly selected sample of 675 substances, 75% of the drugs and excipients in drug formulations in the select universe have at least some information on acute toxicity and 62% have information on subchronic testing. For pesticides and inert ingredients of pesticide formulations, these values are 59% and 51%, respectively. Testing was absent most frequently for substances on the TSCA list of chemicals in commerce, particularly for chronic, reproductive, and developmental effects.

More specifically, substances in the subsample of 100 were most frequently tested with acute oral rodent studies and acute parenteral studies (see Table 8 in Chapter 5 of Part 1). Except for drugs and excipients in drug formulations, the next most commonly conducted test was for genetic toxicity. Dermal and eye irritation studies had often been done with substances in the three production categories of chemicals in commerce.

QUALITY OF TESTING

The Committee on Toxicity Data Elements tabulated the quality ratings from evaluations of a total of 664 tests of the 100 substances in the subsample, without regard for either intended-use category or type of test conducted (see Appendix H of Part 1). When judged against currently accepted standards for toxicity testing, only 8% of the tests in the subsample met the standards of the reference protocol guidelines and another 19% of tests performed were judged to be adequate by the committee's standards. The percentages are based on the one study of highest quality when two or more studies of the same type were done.

The quality of design, execution, and reporting of toxicity studies was not uniform among the various types of experiments. Some test types (acute oral administration in rodents, acute dermal application, acute eye irritation and corrosivity, guinea pig skin sensitization, and subchronic dermal application for 90 days) were deemed not to require repetition in most cases where they had been conducted.

Four acute tests of substances in the select universe were often of acceptable quality: acute oral administration in rodents (83%), acute dermal application (87%), acute dermal irritation and corrosivity (81%), and acute eye irritation and corrosivity (76%). Fewer chronic test types were of acceptable quality; these included multigeneration reproduction in rodents (33%), carcinogenicity in rodents (52%), chronic toxicity (38%), and combined carcinogenicity and chronic toxicity in rodents (50%). Overall, more testing is needed for chronic toxicity than for acute toxicity. These findings should be viewed in perspective: the comparison is of simple acute tests with more complex chronic tests;

far fewer chronic tests were performed than acute tests; and, although the percentages themselves are high, they are derived, on the whole, from small numbers of evaluated tests, particularly in the case of chronic studies.

Evaluation of individual study protocols by the Committee on Toxicity Data Elements was always accompanied by documentation of the reasons for the particular ratings given to the studies. For the most part, these reasons were statements of specific adequacies or inadequacies in a testing protocol. They were collectively tabulated for analysis to assess which deficiencies were most prevalent and what values should be placed on the deficiencies when the overall value of a study was assessed.

Some of the more common deviations from reference protocol guidelines that nevertheless resulted in a test's being rated as adequate included the use of too few animals per dosage group, the use of too few or improper doses, and the absence of observations (e.g., in clinical chemistry or histopathology). In most cases, such tests were considered to have been conducted adequately because more information would not be expected to alter the conclusions, because existing data were sufficient to evaluate toxicity or calculate an acceptable LD_{50}, or because doses were high enough to give positive results or exceed the limit test prescribed in the guidelines. The Committee on Toxicity Data Elements notes that reference protocols are developed for general application before it is known which of their results will be important. The committee judged the quality of studies after they had been performed and in the light of the results obtained.

Tests that were rated as inadequately conducted often lacked required observations (e.g., test animal description, diet analysis, chemical analysis, clinical chemistry, and histopathology), used too few doses, or lacked sufficiently detailed end points, such as data tabulation and statistical analysis of data. Occasionally, the committee recommended that these studies not be repeated, either because toxicity was sufficiently well established or because more information would be of slight value.

TOXICITY-TESTING NEEDS

For pesticides and inert ingredients of pesticide formulations, the Committee on Toxicity Data Elements considered 18 test types to be necessary according to the standards it adopted. In this category, all studies of acute oral administration in rodents were judged not to require repetition. Some of the 17 remaining test types needed repetition or were not done at all from 20% to 73% of the time. For cosmetic ingredients, this ranged from 67% to 100%; for drugs and excipients in drug formulations, from 25% to 60%; for food additives, from 33% to 80%; and for chemicals in commerce, from 45% to 100%. This information indicates that, for each category of intended use, substantial testing or retesting remains to be performed for all categories of substances if information gaps are to be filled. The major gaps in testing result from failure to do tests required by the committee according to the standards it adopted, rather than from conducting tests improperly (see Tables 12-18 in Chapter 5 of Part 1). If the unknown amount of information that was not available to the committee had been available, the "untested" category would be somewhat smaller than reported here.

In general, chronic studies, inhalation studies, and more complex studies with specific end points (e.g., neurotoxicity, genetic toxicity, and effects on the conceptus) are most frequently needed. These were among the test types considered by the Committee on Toxicity Data Elements to be necessary for conducting a health-hazard assessment according to the standards it adopted. There are some differences in the gaps in toxicity information from one category of substances to another. To some extent, these may reflect the spectrum of individual tests that the committee prescribed as necessary to meet its criteria for adequacy of information in each category.

The three greatest testing needs for health-hazard assessment of pesticides and inert ingredients of pesticide formulations were in teratology, neurobehavioral toxicity, and genetic toxicity (see Table 19 in Chapter 5 of Part 1). For cosmetic ingredients, testing was found to be needed most for subchronic eye toxicity and subchronic neurotoxicity. A large variety of test types were found to be needed for drugs and excipients in drug formulations, for food additives, and for the three production categories of chemicals in commerce.

HEALTH-HAZARD ASSESSMENT

The Committee on Toxicity Data Elements and the Committee on Sampling Strategies made judgments to describe their ability to make health-hazard assessments of substances in each of the seven categories of the select universe as complete, partial, or none. A complete health-hazard assessment was defined as one that provided a full estimate of hazard associated with the safe use of a substance. A partial health-hazard assessment was defined as one that provided a limited characterization of the hazard associated with the safe use of a substance. Therefore, a partial health-hazard assessment had a broad range extending from very limited (e.g., acute-toxicity evaluation by one route of administration) to almost complete (e.g., full acute- and chronic-toxicity evaluation, except for inadequate neurobehavioral-toxicity determination). The estimates of percentages for health-hazard assessment combine data obtained from the sample of 675 substances used to measure the existence of minimal toxicity information and the 100 with minimal toxicity information that were examined for the quality of test protocols. Results of this analysis indicate not only the percentage of substances in each of the seven categories in which sufficient data of adequate quality are available for a health-hazard assessment when judged against the current standards for protocols, but also the percentage that would require additional testing if an assessment were to be performed.

10

The overall status of toxicity information and of the ability to conduct a health-hazard assessment for each use category is presented in Figure 2. In general, proportionately more testing has been undertaken on pesticides and inert ingredients of pesticide formulations and on drugs and excipients in drug formulations than on other substances. In these two categories, 36% and 39% of substances met the requirements for minimal toxicity information, respectively. The Committee on Toxicity Data Elements judged it possible to make at least a partial health-hazard assessment for 94% and 92% of the substances with minimal toxicity information in each of these categories, respectively.

Cosmetic ingredients and food additives have been somewhat less well tested. Minimal toxicity information requirements were met by 26% and 20% of substances in these categories, respectively, and at least a partial health-hazard assessment was judged possible for 62% and 95% of the substances with minimal toxicity information in these categories, respectively.

In contrast, only about 20% of the substances in each of the three categories of chemicals in commerce have minimal toxicity information. In all three categories, at least a partial health-hazard assessment was possible for about half the substances having minimal toxicity information. Virtually all the substances in the three subsample categories of chemicals in commerce with minimal toxicity information would require additional toxicity testing if a complete health-hazard assessment were needed. The frequency and quality of testing of chemicals in commerce were not related to production volume. Chemicals in commerce produced in quantities of 1 million pounds or more in 1977 have not been tested more often or more adequately than those produced in smaller quantities.

INTERPRETATION OF PHYSICOCHEMICAL AND EXPOSURE DATA

The committees attempted to relate the quantity and quality of toxicity testing of each substance in the subsample to breadth of known exposure, expected trends in exposure, physicochemical properties and chemical fate of the substance, strength of evidence of toxicity in humans, and severity of reported chronic human toxicity. In addition, the committees sought information on occupational and environmental exposure and attempted to relate it to the extent of testing needs.

It became evident as the dossiers for the 100 substances in the subsample were examined that characterization of the substances with respect to each of these factors, when possible, was based on scanty information. Most of the available information was on the physicochemical properties of the substances; the least was on exposure. However, no comprehensive method of gathering the needed information could be identified, and, in the end, the principal basis for characterizing exposure was the knowledge and expertise of the committee members. The immediate use of substances in the synthesis of new substances may not result in a reduction of exposure intensity, but will reduce the number of persons exposed.

11

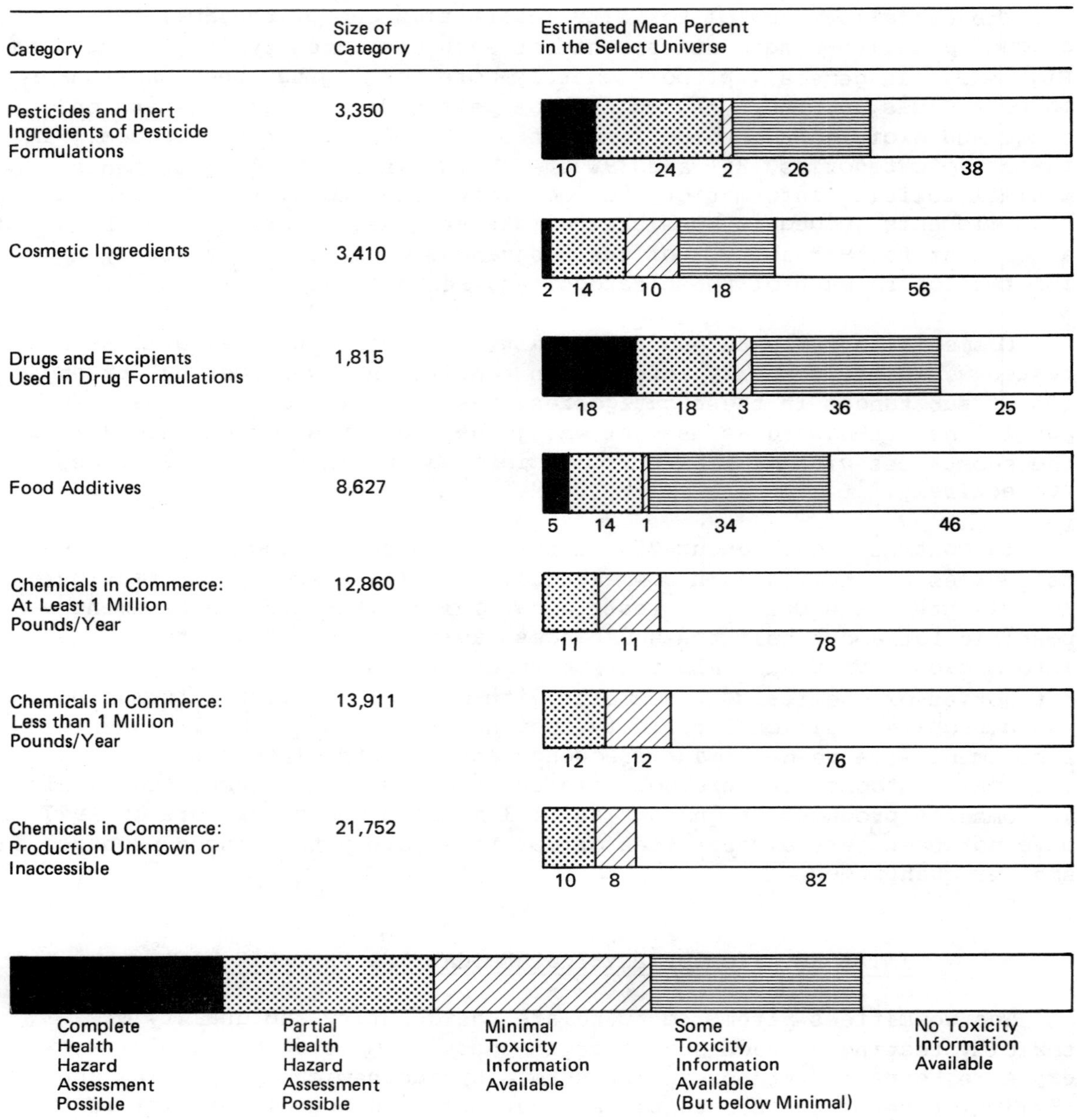

Category	Size of Category	Estimated Mean Percent in the Select Universe
Pesticides and Inert Ingredients of Pesticide Formulations	3,350	10 24 2 26 38
Cosmetic Ingredients	3,410	2 14 10 18 56
Drugs and Excipients Used in Drug Formulations	1,815	18 18 3 36 25
Food Additives	8,627	5 14 1 34 46
Chemicals in Commerce: At Least 1 Million Pounds/Year	12,860	11 11 78
Chemicals in Commerce: Less than 1 Million Pounds/Year	13,911	12 12 76
Chemicals in Commerce: Production Unknown or Inaccessible	21,752	10 8 82

Complete Health Hazard Assessment Possible	Partial Health Hazard Assessment Possible	Minimal Toxicity Information Available	Some Toxicity Information Available (But below Minimal)	No Toxicity Information Available

FIGURE 2 Ability to conduct health-hazard assessment of substances in seven categories of select universe.

It should be emphasized that the study was designed to characterize the status of toxicity-testing needs for substances to which there is known or anticipated exposure, without regard in the selection process to the extent of that exposure. The 100 selected substances contained, in each category, few of the substances known to be produced in the greatest volumes. Hence, this study may not provide an accurate estimate of the status of toxicity information on the principal substances to which humans are exposed.

The following observations emerge from the committees' analysis:

• Of the 100 substances in the subsample, 42 (those with at least minimal toxicity information) were considered to involve widespread exposure. An additional 14 were considered to have limited exposure potential, which would be intensive for specific groups.

• Physiocochemical data on 20 of the 100 substances led to a high concern about potential adverse human health effects. For 32 additional substances, the concern was moderate.

• There was no relation between the amount of testing that had been performed and the degree of concern about a substance based on physicochemical information.

Among the seven categories, it is the chemicals in commerce that have the smallest amount of information relevant to human exposure in the workplace and in the general environment. This is of particular concern, inasmuch as the primary motivation for testing chemicals in commerce is their potential for environmental and occupational exposure.

The committees suggest that a coordinated effort be made to collect information needed to assess potential human exposure in the workplace and in the general environment. Development of analytic methods; systems for monitoring ambient air, water, soil, and foods; personal monitoring systems; and highly sensitive and selective instrumentation for the evaluation of human exposure should be integral parts of this effort.

APPROACHES TO PRIORITY-SETTING

Part 1 of this report shows that, of tens of thousands of commercially important chemicals, only a few have been subjected to extensive toxicity testing and most have scarcely been tested at all. Many other constituents of the human environment, including natural chemicals and various contaminants, are also potential candidates for testing. Although it can be convincingly argued that many chemicals do not need to be tested, because of their low potential for human exposure or for toxic activity, it is clear that thousands or even tens of thousands of chemicals are legitimate candidates for toxicity testing related to a variety of health effects.

Many government and private institutions have responsibilities for toxicity testing, and the National Toxicology Program has a special mission to develop testing methods and to fill in the gaps left by other institutions. However, the resources available to NTP for testing--whether expressed in terms of budget, staff, or facilities-- are limited. Hence it must decide which chemicals to test and which tests to perform. The need for priority-setting is especially acute for lifetime bioassays, which may cost up to a million dollars for a single chemical. Priority should presumably be assigned to chemicals and tests that are in some sense the most important. The Committee on Priority Mechanisms interpreted its charge to include defining "most important" and indicating how to identify the chemicals and tests that satisfy the definition. That is, the committee concluded that definition of the goal of the testing program was essential to designing a priority-setting system and that the goal would largely drive the logic of the design.

Testing priorities have traditionally been assigned on the basis of expert judgment, which is now supplemented with a variety of analytic, data-based techniques, such as scoring systems. The committee believes that this basic pattern should continue, with further improvement in techniques to allow expert judgment to be most effective. The committee recognized that no priority system, scheme, or procedure can be perfect, because the knowledge needed for unerring selection of the most important chemicals and tests is the same as the knowledge resulting from a complete and accurate testing program for all chemicals, which would of course make priority-setting unnecessary. The priority-setting system and the testing program form a continuum whose overall objective is to yield information of maximal value about the overall hazards of chemicals.

In examining traditional approaches, including expert judgment and mechanical priority-setting systems, the committee found some common themes that can be considered conventional wisdom and with which it agrees:

● Long lists of candidate chemicals need to be reduced to short lists through screening, which yields increasing amounts of information on decreasing numbers of chemicals and possible tests.

● The two key elements for screening are estimated human exposure and suspicion of toxic activity. (This priority-setting effort is oriented to human health and not to effects in other species, except insofar as they point toward human effects.)

● Chemicals that have already been tested adequately for a given effect are of low priority for further testing for that effect.

Although documentation on the goals of most systems is somewhat vague, all systems seem to use the goal of <u>reducing the uncertainty about the hazards of the population of chemicals in the human environment as rapidly as possible within the limits of available resources</u>. The key elements embodied in this goal are hazard (determined by both exposure and toxicity) and reduction of uncertainty. Testing is needed most where uncertainty is greatest; there is no need to continue testing for hazards that are already well known. The above objective not only seems to be the common denominator of current procedures for priority-setting, but is obviously a worthy goal, because it allows society the best chance to make decisions about chemicals that will reduce their hazards or at least to accept them with full knowledge of the magnitude of their hazards.

Of course, there are other legitimate goals of the testing program and therefore of priority-setting. On the one hand, general improvement in the understanding of chemical toxicity is worthy in itself; on the other, testing directed at chemicals of great public concern to confirm or deny the concern, and thus reduce anxiety, is worthy. These goals are best addressed by subjective exercise of expert judgment and were not addressed further by the Committee on Priority Mechanisms.

Given a goal for the priority-setting system, the committee needed to decide whether improvements over current procedures for selecting chemicals for testing were possible. It concluded that improvements were possible--at least at the margin--by injecting additional systematic information-gathering and -processing procedures. Many current procedures skirt the issue of the goal of the testing program and therefore are somewhat inconsistent in approaching the goal. In particular, the concept of the value of information is an important contribution to systematic priority-setting. In brief, this concept asserts that the value of any information-gathering activity, such as toxicity testing or searching for information on human exposure to chemicals, lies in the value of the resulting information in guiding decisions. The contribution of this concept is in making explicit that the goal of the testing program should be embodied in the priority-setting system.

In the realm of chemical hazards, the "cost" of not knowing the degree of toxicity of a chemical (or the degree of human exposure to it) lies in misclassifying its hazard--e.g., believing that it is innocuous, when it is actually toxic. Maximizing the value of information about chemical hazards--or, equivalently, minimizing the cost of misclassification of them--is therefore essentially the same as the goal emphasized earlier. Thus, incorporating the value-of-information concept explicitly in the priority-setting system provides an advantage, even though current procedures often use it implicitly, even if erratically.

The committee identified a second category of potentially important improvements. It includes provisions for validating some key estimates produced by the priority-setting system and thereby allowing for a self-improvement cycle to modify the system accordingly. Current systems are difficult to validate, because they rarely yield estimates that are directly verifiable, but simply indicate which chemicals to test for which effects. The committee proposes to redefine the elements of the priority-setting system to allow them to be checked. For example, an ideal system would estimate the percentage of chemicals that would yield positive results if tested; as experience accumulated, it would be possible to modify the system on the basis of the errors in the estimates. We note again that a priority-setting system cannot be free of errors in selecting chemicals for testing or in classifying them by degree of hazard when the test results are in, but any good system should reduce the frequency of errors as information accumulates and improvements become possible.

With these broad kinds of improvement in mind, the Committee on Priority Mechanisms decided to outline an illustrative system that would incorporate the stated goal and general features, building on the experience of previous priority-setting procedures, but trying to make them more systematic, defensible, and robust. The committee recognized that its own resources were inadequate to develop a fully operational priority-setting system with all the desirable features, but it hoped to provide NTP with sufficient guidance and examples to enable it to improve its current selection system while adhering to its institutional operating principles.

Several design principles became evident and were used by the committee in developing an illustrative priority-setting system:

● Any rational system can be conceptually divided into stages, with more information on fewer chemicals and fewer potential tests in each succeeding stage. The committee describes how a four-stage system might be designed. In this system, Stage 1 acts as a coarse screen and depends almost totally on automated information sources; Stage 2 begins the use of expert judgment; Stage 3 relies heavily on traditional expert judgment with only minor changes; and Stage 4 is the testing program itself.

● Both exposure and suspected toxicity considerations are useful in every stage of priority-setting. Information on either would necessarily be relatively crude (that is, there would be relatively little information about degree of hazard) at early stages, but would be correspondingly less expensive to acquire.

● Many indicators of exposure and toxicity are available--e.g., for exposure: production volume, use patterns, and persistence; for toxicity: chemical structure and the results of an acute test.

Whether or not to use a specific indicator at a particular stage of priority-setting depends on its cost, its individual value in characterizing exposure or toxicity, and its combined value with other indicators in characterizing degree of chemical hazards.

- The performance of a system should be evaluated according to its ability to characterize the hazards of groups of chemicals, not only its ability to indicate test-no test decisions.

- To accomplish test objectives, a system must take into account the frequency of occurrence of various properties (e.g., carcinogenicity) and of various indicator values (e.g., a positive result of an acute test) in groups of chemicals.

- An ideal system would be capable not only of dealing with a relatively small number of chemicals nominated to NTP by agencies--as in current practice--but also of dealing with a much larger number of chemicals in the total select universe of concern (53,500, as stated in Part 1 of this report). This capability does not necessarily imply that NTP should be the entity that operates the long-list part of the system.

- Expert judgment is essential for operation of the system beyond the earliest stage, where judgment enters into the design but not into the operation. Simply put, not enough is known about chemical hazards to specify a purely mechanical system, and humans need to integrate diverse data into judgments about the degrees of exposure and suspected toxicity. However, these judgments should be made at the lowest level of aggregation needed, because humans have difficulty in integrating information and concepts that are far outside their normal range of experience.

Beyond these broad principles, possible designs for a priority-setting system are multiple, and the specific choices for design--let alone operation--depend on expert judgments. The Committee on Priority Mechanisms offers in this report a possible design (admittedly sketchy and incapable of immediate implementation) that illustrates the key departures from current practice that seem warranted. The design may look unfamiliar, because of its description in mathematical terms, but it attempts to capture how a rational person or group would set priorities for testing if able to gather, assimilate, and integrate all the relevant pieces of information in a completely informed and objective manner.

The reasoning behind the committee's design is presented in Chapter 2 and Appendix B of Part 2. This reasoning depends in part on a "model" of the system that allowed its operation to be simulated--in highly simplified form--under a variety of assumptions. The reasoning and system simulation are complex and may well challenge the reader; however, they allowed the committee to explore the overall performance of various possible system designs with the goal of reducing uncertainties about chemical hazards--an evaluation that has not been typical in the past.

The result of exercising the model is the simplified system description given in Chapter 3 of Part 2, in which the operation of the illustrative system is described from the viewpoint of an outside observer--what happens, but not why. Thus, Chapter 3 presents the illustrative system as a "black box," whereas Chapter 4 and Appendix B of Part 2 describe the "wiring diagram" for the interior of the box.

The committee believes that a fully developed version of the outlined system not only is a plausible extension of current practice, but also would provide some improvements over existing priority-setting procedures toward the goal defined earlier. Obviously, it might not provide improvements toward other goals, but it should not impede them. Even at the margin, the improvements would probably easily justify the costs of developing, implementing, and operating the system. However, the implementation of these concepts in the illustrative system or one of similar scope would require adjustments in the established patterns of thinking about testing priorities. Specifically, full application of the proposed analytic techniques will require that each information-gathering procedure be described quantitatively with respect to its ability to identify and to characterize potentially toxic chemicals. This requirement is not readily fulfilled in our present state of knowledge. Hence, efforts toward further quantification of the performance characteristics of toxicologic methods would be essential to full implementation of the priority-setting approach proposed herein. For this reason the approach can be pursued initially on a pilot scale, with further implementation depending on the development and availability of the necessary data. The committee believes that it should be possible to institute changes in current procedures gradually without irreversibly committing resources to the novel features of its suggestions.

INTRODUCTION

Human life has always entailed exposure to chemicals. The substances we eat, drink, and breathe are composed of chemicals. The twentieth century has seen substantial growth in the synthesis of new molecules, some of which have proved useful in treating disease, preserving food, and reducing the cost of commodities. The estimates of such substances in the environment range as high as "hundreds of thousands" (NRC, 1975). In recent decades, there has been widespread concern that synthetic chemical substances--increasing in number and concentrations--and natural substances may adversely affect human health. In response to the concern, such agencies as the Environmental Protection Agency, the Consumer Product Safety Commission, the Occupational Safety and Health Administration, the Food and Drug Administration, and the Department of Agriculture began to monitor the use of chemicals.

If one supplements the catalog of man-made substances with the naturally occurring chemicals, such as those which constitute food, the list of substances to which humans are exposed may appear endless. Responding to the Toxic Substances Control Act (TSCA), the Environmental Protection Agency (EPA) has cataloged more than 56,000 substances that are now being manufactured or imported and that enter into various phases of chemical manufacture and formulation in the United States (U.S. Environmental Protection Agency, 1982). Human exposure to these agents is known only in small measure and must be characterized by inference. The TSCA Inventory excludes some classes of agents that are regulated under other statutes. Humans are exposed to food additives, pharmaceutical agents (prescription and over-the-counter), and cosmetic ingredients, but many are regulated under the Food, Drug, and Cosmetic Act, rather than under TSCA.

Recognition of the impact of these and other classes of substances on public health leads society to seek information for determining the magnitude of their potential effects. Such information is based primarily on predictions from results of toxicity studies. The development of a strategy for obtaining appropriate information requires an estimation of the quantity and quality of available toxicity data as criteria for the assessment of human-health hazards, as well as knowledge of the number of substances on which necessary experimental data are not yet available. A characterization of the magnitude of needed testing would inform those who allocate resources for such testing.

However, because resources for developing sound scientific bases for identifying public-health hazards are limited, there is a strong impetus to select the most serious problems for immediate attention. Thus, it is important to establish priorities among chemical and physical agents and to select those thought but not proved to have the greatest potential impact on human health.

The methods currently used by federal agencies for assigning priorities are diverse. Therefore, it is useful to review existing ranking systems and to develop criteria for a priority-setting framework that acknowledge various needs and assist in ordering them.

THE NATIONAL TOXICOLOGY PROGRAM

On November 15, 1978, the Secretary of the Department of Health, Education, and Welfare (DHEW) announced the establishment of the National Toxicology Program (NTP) in DHEW (now the Department of Health and Human Services, DHHS) (U.S. Public Health Service, 1979). The broad goal of NTP is to coordinate DHHS activities in the testing of chemicals of public-health concern and in the development and validation of new and better-integrated test methods. Specific goals of NTP are to extend the toxicologic characterization of chemicals being tested, to increase the rate of chemical testing (within the limits of available resources), and to develop and begin to validate a series of protocols appropriate for regulatory needs. One of its major responsibilities is the development of an information base on the toxic properties of substances that, because of their use or presence in the environment, have a substantial potential for adversely affecting human health. NTP provides access to scientific information about toxic, potentially toxic, and hazardous chemicals. This information is needed by regulatory agencies. It can be used for the prevention of chemically induced disease and for otherwise protecting the health of the American people.

The NTP Executive Committee provides linkage between DHHS research agencies and federal regulatory agencies to ensure that toxicologic research and test development under the aegis of NTP are responsive to the needs of those agencies and to the needs of the public. This unique and important aspect of NTP brings together for the first time on a regular and permanent basis the regulatory agencies and the research agencies that are conducting fundamental biomedical research.

For further information, the reader is referred to the NTP Annual Plans for fiscal years 1979 through 1982.

THE STUDY

In September 1980, the National Research Council (NRC) began a 3-year study whose results were intended to assist NTP in carrying out its mission. Broadly defined, the purposes of the study were to characterize the status of toxicity information on substances to which there is known or anticipated human exposure and to develop and validate uniformly applicable and wide-ranging criteria by which to set priorities for research on substances with potentially adverse public-health impact.

During the course of this study, NRC transmitted to NTP two interim reports entitled Strategies to Determine Needs and Priorities for

Toxicity Testing. The first report, Volume 1: Design, was delivered in October 1981, and the second, Volume 2: Development, in October 1982. This volume, the final report of the entire 3-year effort, encompasses much of the information contained in the two interim reports.

The study, called "Identification of Toxic and Potentially Toxic Chemicals for Consideration by the National Toxicology Program," was undertaken by three committees: the Committee on Statistical Sampling Methods (commonly called the Committee on Sampling Strategies), the Committee on Characterization of Status of Toxicity Data Elements for a Select Universe of Compounds (commonly called the Committee on Toxicity Data Elements), and the Committee on Research of Agents Potentially Hazardous to Human Health (commonly called the Committee on Priority Mechanisms). These committees received guidance and coordination from the Steering Committee on Identification of Toxic and Potentially Toxic Chemicals for Consideration by the National Toxicology Program.

The committees completed five major tasks during the first year: (1) They identified a "select universe" of 65,725 substances chosen because of their potential for human exposure and their ready accessibility through computerized lists. This universe contains pesticides and inert ingredients of pesticide formulations, cosmetic ingredients, drugs and excipients in drug formulations, food additives, and other chemicals in commerce as listed in the TSCA Inventory. (2) They developed and applied a procedure to obtain a representative sample of 675 substances from the select universe and, from this, a subsample of 100 substances. (3) They established criteria by which to judge the quality of individual toxicity studies and of toxicity data bases on individual substances in the subsample, to determine their suitability for assessing hazards to human health. (4) They analyzed previously developed priority-setting approaches. (5) They described performance criteria that might be part of a comprehensive approach to setting priorities for toxicity testing.

During the second year, the Committee on Toxicity Data Elements completed most of its evaluation of the data base on each of the 100 substances in the subsample. It determined in detail the adequacy of testing protocols for all toxicity studies that, according to the standards adopted, it considered necessary for conducting a health-hazard assessment. The committee formulated operating policies pertaining to the usefulness of information not generally available to the public, the role of human toxicity information, and the extent to which information on chemical mixtures should be used. To facilitate the achievement of its objectives, the committee developed procedures for document control (i.e., data identification, acquisition, and organization) and for evaluation of data (including committee decisions and comments from individual committee members). Standardization of format, content, and method of reporting was also established.

The Committee on Sampling Strategies selected a method for estimating the population percentages from the sample, developed procedures for combining results from sample and subsample analyses, and produced statistical summaries of the results of analyses.

The Committee on Priority Mechanisms sought to develop criteria for a priority-setting approach that could function effectively with either long or short lists of substances. To accomplish this, the committee identified methods for assessing toxicity and exposure, so that they could be used in developing priority-setting criteria. It then designed and refined a model for the selection of methods believed to be important in setting priorities, elucidated performance characteristics at various stages of the priority-setting system, indicated how they may be applied, and determined the relative roles of expert judgment and the more mechanized aspect of the priority-setting approach.

In this final report, the Committee on Sampling Strategies and the Committee on Toxicity Data Elements describe in detail the criteria they used to determine the status of testing quality and their collective judgment on the testing needs for substances in the select universe. The Committee on Priority Mechanisms presents a framework that could be used for decision-making to set priorities for research on substances with a potential for adverse public-health impact. The report summarizes the work of the first 2 years and presents the final results of the study. However, readers may wish to refer to Volumes 1 and 2 for greater detail of the design and development of the study.

Although the work of all the committees was related, each group concentrated on the issues that were relevant to its specific charge. The report is therefore in two parts: the first part was the responsibility of and describes the work of the Committee on Sampling Strategies and the Committee on Toxicity Data Elements, and the second part, the work of the Committee on Priority Mechanisms. The committees had the opportunity to read and comment on each other's part of the report, but each part should be considered the independent product of the group that prepared it.

REFERENCES

National Research Council, Committee for the Working Conference on Principles of Protocols for Evaluating Chemicals in the Environment. 1975. Principles for Evaluating Chemicals in the Environment. Washington, D.C.: National Academy of Sciences. 454 pp.

U.S. Environmental Protection Agency. 1982. Toxic Substances Control Act Chemical Substances Inventory. Cumulative Supplement II. Washington, D.C.: U.S. Government Printing Office. 950 pp.

U.S. Public Health Service (DHHS). 1979. National Toxicology Program Annual Plan for FY 1980. Washington, D.C.: U.S. Government Printing Office. 117 pp.

PART 1

TOXICITY-TESTING NEEDS IN THE SELECT UNIVERSE

CONTENTS

Tables

Figures

27

1

INTRODUCTION

A major function of the National Toxicology Program (NTP) is the selection and testing of chemicals for toxicity. NTP considers chemicals for testing on a continuing basis, as candidates are nominated by the federal agencies served by the programs, by state and local governments, and by academic, industrial, and labor groups. Such candidates are of possible interest to NTP because of their potential for human exposure and possible public-health impact.

This portion of the study, the determination of toxicity-testing needs, was undertaken because NTP recognizes that the number of substances, both natural and man-made, in the human environment is large and is increasing with no clear indication of the nature and amount of toxicity information that might be needed on these substances to ascertain potential for adverse effects on human health. It is useful for NTP to know as precisely as practicable the toxicity-testing needs for substances to which humans are potentially exposed. Consequently, NTP asked the National Research Council to address these matters.

A major objective of the study was to estimate the amount and type of toxicity testing needed in the defined select universe of substances. To accomplish this objective, three operating goals had to be met by the Committee on Statistical Sampling Methods (commonly called the Committee on Sampling Strategies) and the Committee on Characterization of Status of Toxicity Data Elements for a Select Universe of Compounds (commonly called the Committee on Toxicity Data Elements):

● To estimate the proportion of compounds in the select universe on which toxicity data exist.

● To estimate the proportion of compounds in the select universe on which toxicity testing has been adequate or that NTP should consider for additional toxicity testing.

● To determine the nature of any additional toxicity testing that may be required for particular classes of compounds.

The size of the select universe precluded retrieving and evaluating existing toxicity data on all its constitutents to determine the extent to which additional data are needed. To approach an understanding of the status of toxicity information on the select universe, a scheme using carefully conceived sampling techniques, as described in Chapter 2, was developed. A probability sample was extracted from the select universe and later analyzed to learn the extent and quality of toxicity data on substances in the sample. The primary objective of creating the sample was to permit the characterization of the status of toxicity information on chemicals in the sample, the characterization of the quantitative distribution of toxicity data in the sample, and the estimation of the proportions of substances in the select universe on which there are various degrees of toxicity data. This knowledge was then used to estimate the types and amounts of toxicity testing that might be required. The Committee on Sampling Strategies, composed primarily of experts in statistics, was responsible for evaluating sampling methods, for selecting the most appropriate sampling approach to be used in this study, for generating a sample, for assembling the data, and for assisting in the interpretation of the results of sample evaluation.

To estimate the percentage of substances that might need further testing, each substance in the final sample was the subject of an exhaustive search for toxicity information, as described in Chapter 3 and in Appendix M. The Committee on Toxicity Data Elements, composed mainly of experts in the toxicologic sciences, was responsible for establishing criteria by which the toxicity information on chemicals in the sample was to be characterized, for applying the criteria to the sample, and for interpreting the results in relation to the select universe.

By blending their expertise, the two committees jointly estimated the toxicity-testing needs of the select universe. The description of toxicity-testing needs includes an estimate of the proportion of the select universe with a degree of information defined later in this report as minimal for the evaluation of possible hazard to human health.

The methods for documentation, protocol evaluation, and testing-need determination used by the committees are presented in this report. These methods were structured and detailed in such a fashion that new information and different judgments can be entered into the process, so that interpretations can be revised.

Part 1 of this report, which contains the approaches and procedures used to ascertain the toxicity-testing needs of a select universe of chemical substances, begins with a detailed description of the contents of the select universe as a sampling frame, from which a sample and subsample were drawn. The sampling process is then presented, with particular attention to the choice of sampling method and the sample size. This is followed by an accounting of procedures for data identification, acquisition, and organization, including a description of the operating policies established by the two committees. The organizational framework for the recording and evaluation of data is documented with a consideration of the committees' adopted principles for the evaluative process. The specific standards and process of data evaluation used by the committees are then described. Finally, after a presentation of the statistical techniques used for data interpretation and analysis, the committees explain in detail the results of their findings, with estimates of the quality and quantity of toxicity testing available for the entire select universe and of unmet needs.

The report of the Committee on Toxicity Data Elements and the Committee on Sampling Strategies does not include an evaluation of the toxicity itself of the substances examined by the two committees, nor of the predictive power (false-positive and false-negative rates) of tests, the nature of toxicity end points (e.g., the proportion of substances in the select universe that are carcinogenic), or the nature of exposure end points (e.g., the proportion of substances that have high exposure potential). To the extent that it includes an examination of individual substances, it does so only for their usefulness as representatives of a larger population, namely the select universe.

The decisions, criteria, and findings of this study were based solely on scientific principles. The study did not include the use of statutory processes of the various federal agencies with regulatory interest in the substances contained in the select universe. The agencies may have regulatory constraints that could alter their use of the information contained in this report; such use would not necessarily be based solely on scientific evaluation.

In the detailed explanations of each phase, the committees present their procedures and assumptions with cautionary statements where appropriate, alerting readers to the possibility of misinterpretation. Because the process used by the committees was complex and integrative, readers are advised to note the details and cautionary statements with equal care to safeguard their understanding of the committees' findings.

2

SAMPLE SELECTION

THE UNIVERSE AND THE SAMPLING FRAME

According to an estimate based on the Chemical Abstracts Service (CAS) Registry, the universe of known chemicals consists of over 5 million entities. The Committee on Sampling Strategies and the Committee on Toxicity Data Elements used a subset of the universe, a select universe, that contains 65,725 listings of substances to which humans may be exposed. The committees considered many options in choosing the most appropriate select universe for accomplishing the task effectively. First, one might use the CAS list and select substances to which extensive human exposure is known or likely. This approach was not practical. Although the CAS list is computerized, it is not based on a potential for human exposure, and manual evaluation of such a large number of substances was beyond the physical resources of the study. Second, one could search for existing lists of substances preselected for human exposure potential and computerized for reasonably easy access. The committees' decision was based on a desire to include to the fullest feasible extent the substances of possible interest to NTP; to avoid the unnecessary inclusion of substances not of interest to NTP; to choose a select universe that, through analysis, most completely reflects true toxicity-testing needs; and to take advantage of chemical-name lists that were computerized, are associated with CAS Registry numbers, and permit retrieval for sampling purposes.

In consideration of these factors, the committees established the select universe by identifying five classes of chemicals: (1) pesticides and inert ingredients of pesticide formulations; (2) cosmetic ingredients; (3) drugs and excipients used in drug formulations, (4) food additives, including direct and indirect additives, colors, flavors, chemicals generally regarded as safe (GRAS), and chemicals listed in the chemical-name category of the Food and Drug Administration (FDA) Bureau of Foods dictionary of chemical names; and chemicals in commerce, as listed in the Toxic Substances Control Act (TSCA) Inventory (U.S. Environmental Protection Agency, 1979, 1980). Chemicals in commerce were separated according to 1977 production reported by manufacturers: (5) production of 1 million pounds (454 metric tons) or more, (6) production of less than 1 million pounds, and (7) production unknown or inaccessible because of manufacturers' claims of confidentiality. The Committee on Toxicity Data Elements and the Committee on Sampling Strategies regarded the contents of the select universe as closely approximating the universe of interest to NTP. The committees recognized that this select universe did not systematically include substances that were environmental decomposition products, manufacturing contaminants, or natural substances (e.g., plant pollens and foods). However, the committees elected not to include a miscellaneous category of such substances, because a suitable list could not be identified. Therefore, the sum of the above, 65,725 entries, was taken as the select universe of substances for purposes of this study. Statistical adjustment for duplications indicated that the select universe contained about 53,500 distinct entities. The samples selected by the Committee on Sampling Strategies were for the use of the committees in fulfilling the study's objectives. They were chosen only as representatives of the select universe, not for any intrinsic interest of their own.

Each of the seven categories can be divided into subcategories based on a variety of criteria. For example, drugs could be divided according to route of administration (e.g., oral, parenteral, or intravenous), strength of pharmacologic activity, intended-use volumes, production volumes, or structure-activity relationships. Some of the subcategories might cross the boundaries of categories of intended use. The Committee on Sampling Strategies recognized that the select universe of chemicals could theoretically have been divided into classes based on criteria other than intended use--e.g., chemical structure, degree of toxicity, extent of exposure, or biologic availability--and that each such class could be further divided into categories that are different from those listed above. For the purposes of this study, however, discussions are based on the seven categories and their possible subcategories. The sources and scope of representative lists of chemicals contained in the seven categories are presented in Table 1.

A major role of the Committee on Sampling Strategies was to develop a sampling procedure that would yield statistically valid estimates of the select universe. The procedure is described later in this chapter. The statistical validity of the estimates was especially critical, because the success of the study required, in part, that the sample be used to estimate the status of other chemicals in the select universe. The seven categories of the select universe embody a large collection of chemicals that formed the sampling frame from which the sample was drawn. An overview of the sampling process is presented in Figure 1.

The sampling procedure began with the selection and preparation of the lists that most accurately presented the seven categories of the select universe (see Table 1). The characteristics of each depended on the purposes of the organization for which it was constructed. The sizes, contents, and formats of the lists varied. The lists of pesticides and inert ingredients in pesticide formulations, cosmetic ingredients, drugs and excipients used in drug formulations, and food additives had characteristics, such as use functions, that were more consistent internally than the characteristics of the TSCA Inventory of chemicals in commerce. They were also substantially smaller. The characteristics of each category in the select universe are described below.

TABLE 1 Characteristics of Chemical Lists from Which Sample Was Drawn

Category and Lists Used in the Select Universe	Source	Scope	Organization	Year	No. Entries	No. Chemicals Sampled	
						Sample	Subsample[a]
Pesticides							
Active ingredients	EPA	Chemicals registered by EPA for use as pesticides; some chemicals in list have pending registrations	Sequential CAS Registry number	1977	2,218	--	--
Registered inert ingredients	EPA	Chemicals registered by EPA for use as inert ingredients in pesticide formulations (fillers, solvents, etc.)	Sequential CAS Registry number	1977	867	--	--
Common to both		--	--	1977	265	--	15
Total					3,350[b]	50	15
Cosmetics							
Chemical dictionary of Cosmetic, Toiletry and Fragrance Association (CFTA)	CFTA	Individual ingredients used by cosmetics industry	Three alphabetical cycles	1981	3,410	100	15
Drugs							
Bureau of Drugs ingredient dictionary	FDA	Prescription drugs, nonprescription drugs, and formulation excipients indicated by FDA Bureau of Drugs as marketed as of March 1982	Alphabetical	1982	1,815	50	15

35

TABLE 1 (continued)

Category and Lists Used in the Select Universe	Source	Scope	Organization	Year	No. Entries	No. Chemicals Sampled Sample	Subsample
Food Additives							
Bureau of Foods ingredient dictionary	FDA	Chemicals regulated or classified by FDA Bureau of Foods as direct food additives, indirect food additives, GRAS substances, colors, and flavors	Alphabetical	1981	8,627	100	15
Chemicals in Commerce							
TSCA:	EPA	Chemicals in commerce in U.S. over 1,000 lb/yr in 1977	Sequential CAS Registry number[c]	1978			
≥1,000,000 lb/yr					12,860	125	10
<1,000,000 lb/yr					13,911	125	10
Production data unknown or inaccessible					21,752	125	20
Total					65,725	675	100

a   Sample of 675 was reduced to subsample of 100 by screening process described in Table 3.

b   Includes substances common to active-ingredients and registered-inert-ingredients lists.

c   CAS Registry numbers were assigned to chemicals nonselectively as received by CAS.

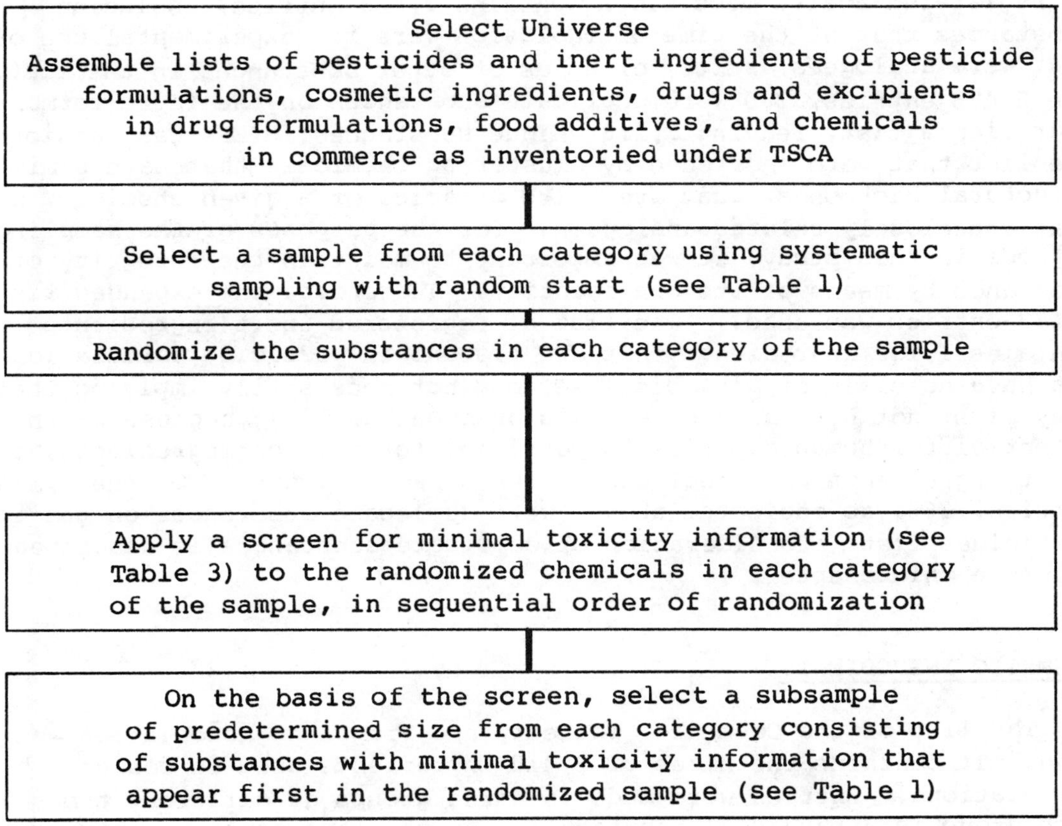

```
┌─────────────────────────────────────────────────────────────┐
│                      Select Universe                          │
│  Assemble lists of pesticides and inert ingredients of       │
│  pesticide formulations, cosmetic ingredients, drugs and     │
│  excipients in drug formulations, food additives, and        │
│  chemicals in commerce as inventoried under TSCA             │
└─────────────────────────────────────────────────────────────┘
                              │
┌─────────────────────────────────────────────────────────────┐
│  Select a sample from each category using systematic          │
│  sampling with random start (see Table 1)                     │
└─────────────────────────────────────────────────────────────┘
┌─────────────────────────────────────────────────────────────┐
│  Randomize the substances in each category of the sample      │
└─────────────────────────────────────────────────────────────┘
                              │
┌─────────────────────────────────────────────────────────────┐
│  Apply a screen for minimal toxicity information (see         │
│  Table 3) to the randomized chemicals in each category        │
│  of the sample, in sequential order of randomization          │
└─────────────────────────────────────────────────────────────┘
                              │
┌─────────────────────────────────────────────────────────────┐
│  On the basis of the screen, select a subsample               │
│  of predetermined size from each category consisting          │
│  of substances with minimal toxicity information that         │
│  appear first in the randomized sample (see Table 1)          │
└─────────────────────────────────────────────────────────────┘
```

FIGURE 1 Process used to draw sample and subsample from select universe.

PESTICIDES AND INERT INGREDIENTS OF PESTICIDE FORMULATIONS

The NIH/EPA Chemical Information System (CIS) lists of registered active pesticides and inert formulation ingredients were used. CIS is a collection of data bases with computer programs to search them. The list of registered active pesticides contains 2,483 entries, including substances that at the time of inclusion were for experimental use or that were analogues, salts, or acids of other substances in the list. Of the 2,483 entries, 265 were duplicate substances on the inert-formulation-ingredients list, leaving 2,218 unique substances. (EPA has versions of the list that were shortened by clustering chemicals that have similar structural backbones, that are salts or acids of a given chemical, or that are closely related analogues. For the purposes of the sampling methods in this study, it was important to maintain the integrity of each substance by means of its own identity. Therefore, the expanded list of 2,483 entries was used.) The list of registered inert ingredients contains 1,132 ingredients that are present in pesticide formulations, but have no claim of pesticidal action (not necessarily implying that they might not have adverse effects on human health); because of the potential for human exposure to pesticide-formulation ingredients, the inert ingredients were included in the select universe. Of the 1,132 entries, 265, as indicated above, were duplicate substances on the active-pesticides list. Consideration of duplicate substances is described later in this chapter.

COSMETIC INGREDIENTS

The list of the Cosmetic, Toiletry and Fragrance Association was used. It contains the names of 3,410 ingredients used in cosmetic formulations. Entries were made in three groups at different times. Each group is arranged alphabetically, so the list contains entries in three alphabetical cycles.

DRUGS AND EXCIPIENTS IN DRUG FORMULATIONS

The FDA Bureau of Drugs provided entries from its chemical-ingredient dictionary as it existed in October 1982. This included all nonproprietary prescription drugs (879) and nonprescription drugs (218) on the U.S. market in October 1982, as well as excipient substances used in prescription-drug formulations (717) and excipient substances used in nonprescription-drug formulations (284) at that time. This total list of 2,098 substances was purged of duplicates--e.g., nonproprietary prescription drugs that are also nonprescription drugs--for a final list of 1,815 substances. It did not contain substances under investigative new drug (IND) status at that time, substances not approved for use in the United States, drugs for veterinary use only, chlorofluorocarbon aerosol propellants used with drugs, or color additives used in drug formulations. Color additives are subject to certification under separate provisions of the Food, Drug, and Cosmetic Act and were largely accounted for in the food-additives list (see below). Because of duplicate removal and specific exclusions, the list used in this study may be smaller than that reported in other sources. Furthermore, other lists of international scope may have different sizes and contents, because different criteria were used in placing substances on them.

FOOD ADDITIVES

The FDA Bureau of Foods chemical dictionary was used. This dictionary contains 19 chemical sorting codes. Six were used to make up the list of food additives from which the sample was drawn (see Table 2). Later, it was found that a code containing 90 animal-drug additives from the food-additives category of the select universe had been excluded. The drugs in this code have veterinary applications for animals consumed by humans. Metabolites of these drugs are contained in the chemical-name code. Because they constituted a small fraction of the number of entries in the food-additives list, the probability of their selection for the sample would have been small, so this omission was not corrected.

Cosmetic ingredients and drugs in the Bureau of Foods dictionary were specifically excluded from the list used to draw the sample, because they were contained in the lists of the cosmetic-ingredient and drug categories. The six components of the dictionary used provided a total of 8,627 entries, from which the sample was drawn. An undetermined number of these entries were altered forms of food additives that may appear in foods, even though their presence has not been confirmed by FDA. These substances, termed "theoreticals" by FDA, are possible products of known chemical pathways. They are not identified by any special designation and could not be removed mechanically from the list before sampling. The list was alphabetical, with an added minor portion of substances whose names began with numerical prefixes; this portion was organized according to ascending value of the numerical prefix. The alphabetical listing precluded separation of ingredients into code categories. Four of the six categories (direct, indirect, color, and flavor additives) implied categorization by use, but the fifth (GRAS) implied a decision by FDA under a "grandfather" clause that all substances in this category were safe for human use. It was an express desire in this study not to presume the degree of toxicity of any substance (such as those in the GRAS category), so that sampling could be based strictly on statistical premises and tenets. Therefore, the alphabetical integrity of the list was maintained, and entries in all six categories were selected by the prescribed random selection.

CHEMICALS IN COMMERCE

The TSCA Inventory of chemicals in commerce gave rise to special problems, because its construction was not restricted by specific use or

TABLE 2 Six Codes of Chemical Classification in FDA Bureau of Foods
Dictionary Used to Form Food-Additives Category from Which Food-Additive
Sample Was Drawn

Code[a]	Fraction of Dictionary[b]
Direct food additives	0.015
Indirect food additives	0.042
Flavors	0.063
Colors	0.003
GRAS[c] substances	0.029
Chemical names[d]	0.174
Total	0.325

[a] Substances in 13 codes of Bureau of Foods chemical dictionary were
excluded from list. 13 codes were animal-drug additives, food
additives, biologics, cosmetic-label ingredients, cosmetic
substances, indirect food additives (temporary file), drugs for human
use, industrial chemicals, pesticide chemicals, and trade names for
food additives, human drugs, pesticides, and veterinary drugs.

[b] As of February 4, 1981, when Bureau of Foods dictionary contained
25,401 preferred terms. Figures are fractions of substances in
corresponding code files.

[c] Generally regarded as safe.

[d] This category contained substances that were included in
food-additive petitions by manufacturers. Included were agents that
(1) were awaiting assignment to more specific category, such as one
of first five of table, or (2) were not assigned to a category,
because they were intermediate products, impurities, or related
compounds of safety interest only.

class, such as drugs, but rather was based on the amount of each chemical produced during 1977, as reported by manufacturers and processors. The Inventory contained (1) substances classified into 10 production ranges; (2) a group on which production data were absent; (3) a group on which production data were inaccessible to the general public, because of manufacturers' claims of confidentiality; (4) a group that was not produced during 1977, the year for which production data were amassed to assemble the Inventory; (5) a group used in processing of other substances (as opposed to their own manufacture), on which production data were not obtained; and (6) a group manufactured by industries that were not required to report production data under the terms of TSCA. Availability of production data in the Inventory was thus not uniform. This problem was exacerbated by other circumstances surrounding the Inventory's construction:

● Production volumes were reported in ranges too wide to permit accurate summation of volumes of all reporting manufacturers. Furthermore, manufacturers were not required to report a substance's production at plant sites where production began after the time of reporting. As a result, the indicated 1977 production of a given substance may have a large error.

● Under the terms of TSCA, an unknown number of chemical manufacturers, such as small businesses, were not required to report that they were producing a given substance. This introduced errors of unknown size in the production data in the Inventory on an unknown number of chemicals of unknown identity.

● Processors and users were not required to report. EPA has estimated that, if this had been required, an additional 750,000 report submissions would have resulted without substantially increasing the number of substances.

● The total production of petroleum products and related substances is not accurately reflected in the Inventory. Although, on the basis of reported production volumes, gasoline is the leading substance and most of the next 10 high-volume substances are also petroleum products, some agents that are major fractions of mixtures are not reported as individual chemicals, but rather as parts of mixtures (e.g., benzene in gasoline).

● Over 85,000 submissions of volume data from manufacturers were not verified by EPA.

● A given substance may be listed more than once with different CAS Registry numbers in the Inventory.

● About 75% of the known production data are on UVCBs (unknown, variable composition, complex, or biologic products), such as petroleum products.

● There are no production or use data on natural substances, such as asbestos, although these substances are listed in the Inventory.

The TSCA Inventory was categorized by the Committee on Sampling Strategies into substances of which 1,000,000 lb or more was produced during 1977, substances of which less than 1,000,000 lb was produced during 1977, and substances on which production data were inaccessible or absent. Because of the large errors in production volumes tabulated in the TSCA Inventory, the sizes of the categories should be thought of as rough approximations.

At the time the sample was drawn, the TSCA Inventory contained 48,523 usable entries distributed among the three categories, as shown in Table 1.

CONSIDERATIONS IN DEVELOPING THE SAMPLING STRATEGY

The development of a sampling plan for a complex study requires balancing of various competing objectives. For example:

● Will the most important inferences apply to the whole population, to independent segments of it, or to comparisons among segments?

● Can the sampling frame be defined in a way that is simultaneously precise and focused on the real objects of inquiry?

● What part of the total effort should be devoted to preparation of the lists from which the sample will be drawn?

● Does the difficulty of data collection vary substantially from one population member to another, and, if so, should the variation be used to reduce total costs (or expand sample size within the constraints of available resources)?

In the present case, these and similar questions were particularly acute, because of the great cost and effort required for each substance to be subjected to full investigation and assessment.

The Committee on Sampling Strategies believes that the sampling plan it adopted satisfies the competing demands and constraints efficiently. However, two issues required additional discussion:

● The partition of the sample among the various lists and sublists.

● The handling of interlist duplication.

The committee determined that the sample should provide estimates for each of the seven categories. The sample sizes were then assigned to the categories with the aim of yielding estimates with roughly the same precision for each category.

In light of similar considerations, the Committee on Sampling Strategies regarded the 15:15:15:15:40 distribution (pesticides and inert ingredients of pesticide formulations:cosmetic ingredients:drugs and excipients in drug formulations:food additives:chemicals in commerce) in the subsample of 100 as illustrated in Table 1 to be a reasonable allocation of effort. The Committee further decided that a 10:10:20 split of the 40 substances from the TSCA Inventory would allow at least minimal inferences regarding specific subcategories, roughly in proportion to the expected need for information on their toxicity. These samples are probably at the lower limit of sample sizes that are usable for the present purposes; making some groups larger at the expense of other groups would have eliminated the latter from separate consideration, although they would still have contributed to inferences regarding the whole select universe.

Some substances appear on more than one of the lists used to define the select universe. Such duplicates could have been identified in the whole universe or only for substances in at least one of the subsamples; and once identified, they could have been left in place or removed. It would have been possible to remove duplication among lists, for example, by establishing a priority order among the five lists and assigning each duplicated substance to the highest priority list of which it was a member. However, this would not have produced unbiased estimates for separate lists without additional steps to flag each substance on each list according to which of the other lists it was on. Careful matching of the entire lists would have involved little statistical advantage, but much effort. The committee also considered a modification of this approach in which only compounds in the sample would be matched against all lists for the purpose of describing the select universe. The advantage of having precise, rather than estimated, numbers of duplicates was judged not to offset the extra effort of matching the entire lists. Therefore, duplicates remain as they appear in the lists, and statistical methods were developed to reflect and adjust for this decision.

The Committee on Sampling Strategies chose an approach to gain an understanding of the probability of the presence of information of adequate quality to enable assessment of the health hazards of a select universe of substances. Selections were made on a statistical basis without specific or deliberate reference to substances to which humans are primarily or extensively exposed.

SAMPLING STRATEGY

SAMPLE SIZE

Both the sampling plan and the sample size were substantially affected by the large resource investment needed to study even a relatively small number of substances. Investigation of substances was a two-phase process: screening of the sample (rapid and inexpensive) to determine whether toxicity studies had been performed and reported, and a detailed search (resource-intensive) for information on the subsample. It appeared that no more than 100 substances could be assessed in the subsample, and preliminary studies suggested that a sample size of approximately 700 would produce at least 100 substances in the subsample for the later detailed study.

A sample of 700 was therefore selected to estimate the proportion of substances on whose toxicity there is sufficient published material to warrant further study. In the course of the study, two adjustments were made in the size of the sample, and that resulted in a subsample of 675. In the first adjustment, the 50 substances selected for each of the cosmetic-ingredients and food-additives categories were insufficient to produce the required 15 substances that would pass the minimal-toxicity-information screen for the categories of the subsample. (Minimal toxicity information is described later.) Therefore, 50 more substances were selected for each of the two categories in a random fashion identical with that used to select the first 50. In the second adjustment, the 250 substances selected for the category of chemicals in commerce with unknown or inaccessible production levels far exceeded the number needed to produce the 20 for the subsample. Therefore, this category was reduced by a random procedure to 125 substances. All 675 entities in the sample were screened to identify substances on which there was minimal toxicity information. The 675 substances were distributed among the seven categories of the select universe in a proportion of 50:100:50:100:125:125:125, representing pesticides (active ingredients and registered inert ingredients of pesticide formulations), cosmetic ingredients, drugs and excipients in drug formulations, food additives, and the three production categories of chemicals in commerce (see Appendix A).

The available resources were enough to examine only about 100 substances on which there was at least minimal toxicity information so that their testing quality could be assessed. The Committee on Sampling Strategies faced the task of allocating the 100 substances among the seven categories of the select universe. Two approaches were considered in dividing the 100 among the seven categories:

● Make each of the seven subsample categories large enough to provide usable information on its corresponding category of origin.

● Allocate category sizes with some appreciation of the relative size and importance of each list to NTP.

The Committee on Sampling Strategies recommended the first approach, and the subsample of 100 with at least minimal toxicity information was drawn according to the proportion of 15:15:15:15:40. Furthermore, on the basis of the division of the chemicals in commerce into three production categories (125 substances produced at 1,000,000 lb/yr or more, 125 produced at less than 1,000,000 lb/yr, and 125 with unknown or inaccessible production), the 40 chemicals in commerce in the subsample were selected in a proportion of 10:10:20, respectively. The sizes of the categories in the select universe, the sample, and the subsample are presented in Table 1. The contents of the sample and the subsample are presented in Appendix A.

SAMPLING PLAN

A stratified sampling plan was used to control the composition of the sample with respect to the seven categories in the select universe. This was critical in attaining the required sample sizes for all categories. For example, although the list representing chemicals in commerce, the TSCA Inventory, is larger than the other lists, it has a disproportionately small number of substances with toxicity information, compared with the other categories. Regulations mandate toxicity information associated with the use of substances in the other categories before registration and marketing. Variations in the toxicity data bases on substances in the seven categories resulting from differences in degree of exposure to them were considered by the Committee on Sampling Strategies to be a second reason for choosing a stratified sampling procedure. By stratifying according to these seven categories, it was possible to specify an adequate number of substances from all lists.

The sample was drawn by using the same procedure seven times, once for each category. A random start was used to select the first sampled substance from a list. The rest of the substances in the category being sampled were drawn from the remainder of the category's list at equal intervals calculated to provide precisely the sample size needed. Thus, the sample was drawn from all parts of each list. This procedure, called "systematic sampling with a random start," is standard in such applications. The interval varied according to the sizes of the lists, because unequal sampling rates were necessary to achieve the required sample size. Each substance was selected according to its position on the list from which it was sampled. Because probability sampling was applied to each list, the sample was still statistically valid.

Although the sample was an unbiased probability sample, it did not conform to the kind of sample design that lends itself to simple existing methods of variance estimation (for example, the committee considered whether periodicities in the lists might invalidate the sample). A systematic sample was selected, rather than a simple random sample, because it is easier and cheaper to select, and it will almost always have lower sampling errors. However, the Committee on Sampling Strategies paid special attention to this aspect of the sampling method and concluded that the effect of the systematic sampling was negligible and that, for variance purposes, the sample was obtained by a simple random process within seven strata.

SCREENING THE SAMPLE FOR MINIMAL TOXICITY INFORMATION

After the selection of substances for the sample, each entry in each category was assigned a random number. These numbers, each with its assigned substance, were numerically ordered within each category, to provide seven randomly ordered samples totaling 675. The sampling plan called for a screening of the substances in the sample for <u>minimal toxicity information</u> to obtain the subsample of 100 substances. This served two functions:

● In the assessment of the adequacy of toxicity-testing protocols by the Committee on Toxicity Data Elements, screening precluded the appearance in the subsample of a substance on which there was either no toxicity information or too little for use by that committee.

● In the ensuing exhaustive search of the literature for all available toxicity information, screening obviated exhaustive searching for literature when it was nearly certain that there was none.

The criteria for minimal toxicity information presented in Table 3 were delineated by the Committee on Toxicity Data Elements. That committee recommended that a combination of toxicity assays as described in the table—acute, chronic, subchronic, genetic, and reproductive or developmental, each including human and animal studies—be considered to constitute minimal toxicity information.

During this critical screening process, two conflicting criteria needed to be satisfied:

● A sufficient body of toxicity-testing data on each substance in the subsample of 100 was desired, so that the committee members would be able to perform an analysis.

● The number of substances that could be screened as candidates for the subsample was constrained by available resources.

Committee members used their collective experience to establish requirements for the amount and type of toxicity testing that would provide minimal toxicity information for each category of substances. These requirements, presented in Table 3, were varied to meet what should be existing toxicity information for categories with differing intended uses.

By the terms outlined for each category in Table 3, the substances were listed in random order and, in that order, subjected to the screen for minimal toxicity information. Each category of the subsample contained the required number of substances that, according to the random order, were the first to pass the screen. The number of substances that were screened to find the number needed for the subsample varied among the seven categories (see Appendix A). Substances that the literature search revealed to have less information than required by the standard for minimal toxicity information were not allowed into the subsample. Frequencies by category for the select universe, the sample, and the subsample are presented in Table 4. Note that within each

TABLE 3 Required Studies (*) in Screen of Sample of 675 Substances for Minimal Toxicity Information for Subsample of 100

Study Type[a]	Pesticides and Inert Ingredients of Pesticide Formulations[b]	Cosmetic Ingredients[c]	Drugs and Excipients in Drug Formulations[d]	Food Additives[e]	Chemicals in Commerce[f]
1. Acute toxicity (by any route)--single administration within 24 h	*	*	*	*	
2. Subchronic toxicity (oral, dermal)--28-d, 90-d, including guinea pig sensitization	*	*	*	*	
3. Reproductive and/or developmental toxicity			*	*	
4. Chronic toxicity				*	
5. Genetic toxicity					

a Human case studies and experiments with humans were included in these five study types.
b If information was present on the two required study types and any one of the three remaining study types, the substance became a member of the final sample of 15.
c If information was present on the two required study types, the substance became a member of the final sample of 15.
d If information was present on the three required study types, the substance became a member of the final sample of 15.
e If information was present on any three of the four required study types, the substance became a member of the final sample of 15.
f If information was present on any of the five study types, the substance became a member of the subsample.

TABLE 4 Sample Sizes for Sample and Subsample

Category	No. Entries in Select Universe	No. Substances in Sample			No. Substances in Subsample
		Total[a]	From Indicated Category	From Other Categories	
Pesticides and inert ingredients of pesticide formulations	3,350	106	50	56	15
Cosmetic ingredients	3,410	162	100	62	15
Drugs and excipients in drug formulations	1,815	95	50	45	15
Food additives	8,627	212	100	112	15
Chemicals in commerce:					
>1 million lb/yr	12,860	259	125	134	10
<1 million lb/yr	13,911	130	125	5	10
Production data unknown or inaccessible	21,752	136	125	11	20
Total	65,725	--	675	--	100

a Reflects multiple listings. For example, 100 food additives were selected from the food-additive list, and an additional 112 selected from the other lists were found to be on the food-additive list as well.

48

category the number of substances selected in the sample differs from the total number of substances used for later analyses. The differences are due to the overlap among lists constituting the select universe. For example, 106 pesticides and inert ingredients of pesticide formulations were used in the analysis of the sample. They included 50 substances that were selected for that category from the lists of pesticides and inert ingredients of pesticide formulations and an additional 56 substances that were selected from lists for the other categories but found to appear also in the lists of pesticides and inert ingredients of pesticide formulations. All 56 substances and some of the 50 had higher probabilities of selection than substances that appeared only on the list of pesticides and inert ingredients of pesticide formulations. The weights used in constructing the estimates that are based on the whole sample took these various probabilities into account to produce unbiased results. The statistical procedure used to prepare estimates is described in Chapter 4.

Three chemical indexes were used in the screen for minimal toxicity information: CIS/Structure and Nomenclature Search System (SANSS), the National Library of Medicine's Chemline, and the Chemical Abstracts Service's Chemname. These indexes collectively contain over 5 million unique chemical substances by CAS Registry numbers, with synonyms and trade names for each chemical. The Registry of Toxic Effects of Chemical Substances (RTECS) and the Toxicology Data Base (TDB) offer toxicity data extracted from published research findings. TOXLINE houses 11 subfiles, including those generated for chemical-biologic activities, air pollution and industrial hygiene, toxicity bibliography abstracts on health effects of environmental pollutants, pharmaceutical abstracts, pesticide abstracts, Environmental Mutagen Information Center and Environmental Teratology Information Center files, and the toxicology section of Chemical Abstracts. These contain literature from 1965 to the present. For literature published from 1950 to 1965, the NRC Toxicology Information Center (TIC) card catalog was searched manually.

The limited-search strategy used in screening had three steps:

● CIS/SANSS was searched to find alternative names of each substance in question and to point to other data bases that might contain information on that substance.

● If CIS did not provide this information, the National Library of Medicine's Chemline and the Chemical Abstracts Service's Chemname were searched.

● Once the location of available information was ascertained, the following sequence was implemented in an attempt to acquire minimal toxicity information on a substance:

-- RTECS and TDB were searched for basic toxicity information (skin irritation, eye irritation, LD_{50}, LC_{50}, TD_{LO}, etc.).

-- If the minimal information requirement was not met by searching RTECS and TDB, the National Library of Medicine's TOXLINE was searched for information on acute, chronic, subchronic, genetic, and reproductive or developmental toxicity.

-- If the requirement still was not met, the chemical name was searched in the card catalog of TIC.

-- If the requirement still was not met after all preceding parts of the search strategy, the books and other resources of TIC were used--at least two appropriate toxicology reference books in each case.

In this manner, 100 substances constituting a subsample with minimal toxicity information as defined by the Committee on Toxicity Data Elements were selected from a larger, randomly ordered, stratified sample of 675, which was itself a product of a large select universe of 65,725 listings. After selection of the 100 substances for the subsample, the screen for minimal toxicity information was applied to the remainder of the 675 substances in the sample. Thus, all statistical analyses of the sample included the data on all 675 substances. Under the direction of the Committee on Toxicity Data Elements, the subsample of 100 substances was then subjected to a comprehensive literature search for the assessment of the quality of toxicity testing by that committee.

3

OPERATING POLICIES FOR IDENTIFICATION, ACQUISITION, AND ORGANIZATION OF DATA

The available information on each of the 100 substances in the subsample was collected and organized to facilitate assessment of the quality and quantity of the toxicity data base. The algorithm for the identification, acquisition, and organization of the data is summarized schematically in Figure 2. This algorithm is the result of evolution and refinement through a series of procedures, whose rationale and results are described in Appendix M. The search pattern included several computerized data bases and the resources of the NRC Toxicology Information Center. Information was also sought from government and industry files. During this search, the NRC personnel in charge of the initial screening of data consistently tended to retain more information than committee members eventually found to be relevant.

The Committee on Toxicity Data Elements formulated ground rules to:

● Ensure that documents selected were pertinent to the relevant data base.

● Circumscribe precisely the limits of the data base to be used (e.g., what was and what was not to be included).

● Encourage early resolution of difficult problems encountered by the committee (e.g., the use of articles in foreign languages).

● Avoid unnecessary repetition of steps in data identification, acquisition, and evaluation.

Operating policies were established regarding the use of restricted-access or confidential information; toxicity information on humans; data on physical and chemical characteristics, manufacturing processes, production volumes, intended uses, and human exposures; chemical review articles; articles in foreign languages; and information on compounds structurally similar to those in the subsample. These policies are discussed below.

RESTRICTED-ACCESS OR CONFIDENTIAL INFORMATION

Restricted-access or confidential information, which includes proprietary or confidential business information supplied in the process of chemical registration, is contained in the files of the Food and Drug Administration (FDA) and the Environmental Protection Agency (EPA). Because the committees wanted the search for information to be as complete

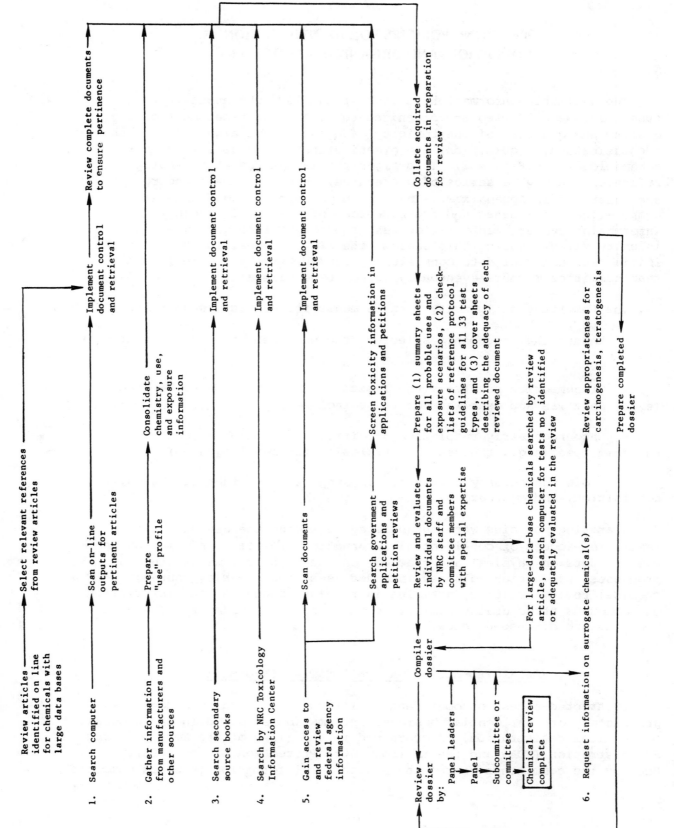

FIGURE 2 Steps in process of evaluating adequacy of toxicity testing for substances.

as possible, they requested from these agencies the maximal permissible access to restricted information related to the chemical and physical characteristics, manufacturing processes, production volumes, intended uses, environmental fate, human-exposure potentials, and animal and human toxicity of substances in the subsample. Confidential or restricted-access data were made available to the committees under two conditions, and two operating policies were adopted to suit these circumstances:

- Some confidential or restricted-access information was made accessible only to specific committee members or staff of NRC. In such cases, the information was evaluated by those persons according to the prescribed procedure for all data evaluation, but only the results of the evaluation were presented to and used by the Committee on Toxicity Data Elements and its panels. The evaluated documents remained confidential and unseen by other members of the committee. Thus, some confidential or restricted-access information was incorporated into the evaluation process.

- The committee was aware that confidential or restricted-access information was held in files that were not available for use in this study. Under these circumstances, the data could not be considered in the evaluation of a substance's data base. The committee's policy on this matter reflected its general view that <u>data not available for its confidential review were presumably not available for legitimate review by other interested parties; hence, in an operational sense, such data no longer exist</u>, although staff members of the government agencies holding such information might have been able to use the data for statutory purposes.

INFORMATION ON TOXICITY IN HUMANS

To maintain consistency, all 100 substances in the subsample were evaluated on the basis of animal toxicity data and human sensitization and skin-penetration studies, when available. Other toxicity studies in humans were not judged for the adequacy of their testing protocols, because reference protocol guidelines do not exist for them. Information contained in reports of these studies was, however, considered in the final appraisal of a substance's toxicity-testing needs. The side effects of drugs in humans were determined from standard source books (see Appendix M); this procedure may have excluded some preclinical and clinical data contained in inaccessible FDA files.

CHEMICAL AND PHYSICAL CHARACTERISTICS, MANUFACTURING PROCESSES, PRODUCTION VOLUMES, INTENDED USES, AND EXPOSURE OF HUMANS

Information on these subjects included all the data elements listed in the "Physicochemical, Use, and Exposure Information" section of Appendix L. Some elements were regarded by the committee as more likely to exist or be readily obtainable than others and were consequently pursued more aggressively. These were chemical identification (structure, nomenclature, and defined purity), intended use, production volume, chemical and physical data (e.g., state, density, vapor pressure, water and organic solubilities, octanol:water coefficient, and pH), and exposure potential in the various intended-use, environmental, and occupational settings. This last item included the probable routes of exposure and the manufacturing system used (open or closed and with or without an exhaust control mechanism and/or local ventilation). The various sources and procedures used to obtain this information are described in Appendix M.

CHEMICAL REVIEW ARTICLES

In every instance, the Committee on Toxicity Data Elements used primary-source literature in its evaluations. For substances with large data bases (more than 250 computer citations), comprehensive review articles were used as an aid to finding the primary-source literature. Primary-source references cited in the reviews were obtained; once those on the required test types (described later) for a given substance had been identified, the data search was continued only for the required test types that were still missing. The committee recognizes that unrequired test types may not have been found for substances with large data bases, but any such omissions would not affect the committee's assessment of toxicity-testing needs.

ARTICLES IN FOREIGN LANGUAGES

English summaries or translated abstracts of articles in foreign languages were examined to estimate their usefulness for the committee's task. When it was appropriate, complete articles were translated. Reasons for judgments concerning each abstract's usefulness (or lack of it) were documented.

SUBSTANCES STRUCTURALLY SIMILAR TO THOSE IN THE SAMPLE

The use of surrogate chemicals (those with similar chemical structures) was considered case by case. The policy established by the committee required that surrogate information not be used until all data on the substance in question had been completely evaluated. At that point, specific information on surrogates could be used to fill clearly defined voids where the data base was considered not to be adequate for conducting a health-hazard assessment. Information on surrogate substances was used on only two occasions.

4

DATA EVALUATION

THE DOSSIER CONCEPT

The available information on each of the 100 substances was organized in a working document or dossier with a standardized format, content, and method of reporting. These dossiers were the focal point for the committees' operating policies, all document control efforts, and all evaluations of data adequacy. Each dossier was the unit of record for all committee decisions and actions. Each dossier contained a synopsis of the substance's physicochemical properties, manufacturing processes, production and consumption volumes, chemical fate, intended and other uses, and exposure potential; a summary of the toxicity data base; and a statement of adequacy of the complete data base. As the evaluation of the toxicity information progressed, additional documentation was added to each dossier until, in addition to the synopsis, it contained:

● A summary of adequacy ratings for tests that were required for a substance's intended uses according to the standards adopted by the Committee on Toxicity Data Elements, as well as the tests required for occupational and environmental exposure, indicating which tests had been performed, the documents in which they were reported, and judgments of adequacy of the test protocols.

● A summary of the amount and quality of all information in the dossier for assessment of the substance's potential hazard to human health.

● Each complete document (or identifying first pages or English summaries thereof) dealing with the substance's human toxicity and exposure and evaluations of the toxicologic information contained in confidential files and, for each toxicity study:

    -- An annotated comparison of the study's protocol with the appropriate reference protocol guidelines (described later in this chapter).

    -- A cover sheet preceding each study and its protocol comparison identifying the type of testing reported in the paper, the committee's judgments of adequacy, and reasons for judgments.

● A data sheet detailing chemical and physical properties, chemical reactivity in nonbiologic systems, bioavailability, analytic methods available for detection, and known uses and exposure.

● A list of synonyms for the compound found in the Chemline, CIS, Chemname, RTECS, and TDB automated data bases and/or the Merck Index, Hawley's Condensed Chemical Dictionary, and the Cosmetic Ingredient Dictionary.

• The names of manufacturers listed in the TSCA Inventory.

The major components of a dossier and their contents are presented in Appendix L.

GENERAL PRINCIPLES FOR EVALUATION OF TOXICITY-TESTING PROTOCOLS

An ideal data base on the toxicity of a chemical would contain enough information to permit the assessment of hazards and safety associated with anticipated use and other exposure. Toxicity information obtained from the experience of exposed humans usually is not available, and it is common practice to use information obtained from tests on laboratory animals. Deficiencies in a toxicity data base do not always invalidate the use of the information to predict at least some human health effects, but may reduce the certainty of a health-hazard estimate for that substance.

The Committee on Toxicity Data Elements used three steps to develop a suitable approach to the determination of toxicity-testing needs. First, the committee reached an agreement on a strategy for judging the adequacy of toxicity data. Second, it established guidelines for assessing the quality of individual toxicity studies. Third, it created a decision-making system to review and evaluate the total data base on the toxicity of a substance. These three steps were used to determine the extent of needed additional toxicity testing for the subsample of 100 substances. Results were used to estimate testing needs for the select universe. Answers to three fundamental questions describe the adequacy of the toxicity data base on a substance:

• What toxicity tests are needed for the substance?

• What tests have been performed and how well have they been done?

• Does the quality of the information permit assessment of the human health hazard?

Although these three questions are fundamental to the overall procedure for evaluating the adequacy of a data base, several additional, more detailed or specific questions may be asked as each substance is examined:

• Is there at least a minimal amount of toxicity information on the substance?

• Is there exposure information on the substance?

• Have all the tests identified as necessary been conducted?

● Has each required toxicity test been conducted in a manner conforming to reference protocols or, if not, did its quality satisfy basic criteria of scientific methods?

● If so, are the nature and quality of test data adequate for the assessment of health hazard?

● What documentation supports the conclusion that available data are of sufficient quality for a health-hazard assessment or that more tests are required?

The committee developed a procedure for determining the adequacy of available toxicity information on a substance (see Figure 3). First, a substance was chosen from the select universe on the basis of the availability of minimal toxicity data, as described earlier. The next step was a search for pertinent information, as listed in Table 5, followed by a determination of intended uses. Next, on the basis of the category to which the substance belonged and the exposure settings, specific tests required to define the toxicity of the substance for each exposure setting were identified (see Appendixes B through G). After establishing which tests were required, the committee examined the available information to identify both the availability and the quality of the required tests. To estimate quality, the report of each test was compared with a set of reference protocol guidelines. Finally, the information was judged to be sufficient to assess the health hazard, in which case further testing would not be needed, or insufficient, in which case further testing would be needed.

The committee not only used data from laboratory studies for hazard assessment, but also examined any epidemiologic studies and information on the extent of exposure to a substance. The committee felt not only that the results of animal experiments may provide guidance for planning epidemiologic investigation, but, more importantly, that animal data can be most valuable when epidemiologic evidence is weak, nonspecific, or relatively insensitive. Conversely, good epidemiologic data minimize the need for animal data.

CONSIDERATION OF EXPOSURE

Three exposure situations largely determine the type of potential hazard and hence the spectrum of data appropriate to evaluate a hazard: exposure via intended use, occupational exposure, and ambient environmental exposure. For example, food additives are meant to be ingested, cosmetics are applied to the skin, and drugs are administered in several forms by several appropriate routes. Humans can also be exposed to food additives, cosmetics, and drugs unintentionally during their manufacture and purification; during packaging, transportation, and storage before their intended use; and during disposal of residues and wastes. There are few intentional exposures of people to most pesticides

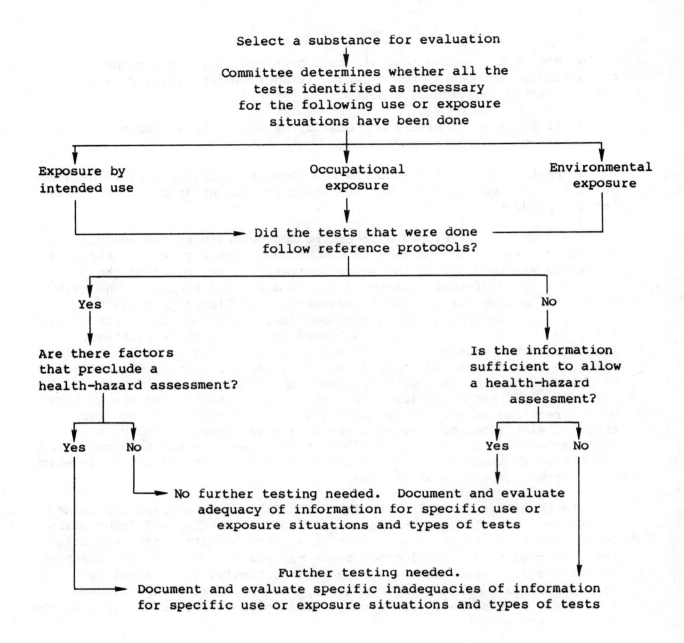

FIGURE 3 Outline of procedure for decision-making in evaluating adequacy of toxicity information on specific substance

TABLE 5 Information Sought in Exhaustive Literature Search for Each Substance
in Subsample of 100

Information Category	Information
Chemistry	Synonyms, trade names, structural formula, molecular formula, CAS Registry number, purity, identification and quantity of contaminants, melting and boiling points, specific gravity, vapor pressure, particle size, water solubility, solubility in organic solvent, complexity of the chemical species, partition coefficient, pH, dissociation constant, shelf-life, stability, potential for undergoing oxidation and reduction, potential for undergoing hydrolysis under various pH conditions, photolytic reactivity, adsorptivity, desorptivity
Process	Synthetic pathways (chemical origin, starting materials, stage of appearance in pathways, final product in pathways)
Production	Companies that produce substance, sites of production, quantity volume (per site total); percent imported, volume trend
Use	Percent produced for commercial use and for consumer uses, percent degraded, number and kinds of uses, unintentional release (during storage, transport, disposal, packaging, manufacture, industrial use)
Chemical fate	Demographic and geographic distribution, environmental pathway, environmental stability, turnover (half-life), degradation, persistence, partition (in soil, water, air), bioaccumulation, environmental transport, environmental bioavailability
Human exposure	Routes, form, mode (occupational, consumer, etc.), number exposed, frequency of exposure, extent of contact (each episode, total), dose and duration of dose (each episode, total), rate of absorption
Toxicity	Summary of all available toxicity information (see Appendixes B through G)

and many other chemicals in commerce, but exposures do occur during production, distribution, use, and disposal. The term "environmental exposure" is used to include all potential human exposures other than those related to the workplace or inherent in the intended use.

The tests that the committee selected to support health-hazard assessments for substances in various classes of use are listed in Appendixes B through G. Batteries of required tests from among the 33 test types listed are identified for direct and indirect food additives (including colors), drugs and excipients in drug formulations (oral, parenteral, dermal, inhalation, ophthalmic, vaginal-rectal, over-the-counter, and veterinary), pesticides and inert ingredients of pesticide formulations, cosmetic ingredients, and other chemicals in commerce. To the extent feasible, the committee selected tests with routes of exposure similar to routes of exposure of humans under various circumstances.

The Committee on Toxicity Data Elements recognizes that duration of exposure, as well as route, is intrinsically important in the manifestation and intensity of toxicity in test species and in the prediction of hazards to humans. It therefore incorporated duration of exposure--acute, subchronic, and chronic--into its selection of toxicity tests for predicting hazard. For example, if a substance is believed to be present consistently in common foods and lifetime exposure of humans is highly likely, data from chronic-feeding studies are appropriate for the substance. Similarly, if a substance is likely to be in the environment of women of child-bearing age, laboratory studies that investigate possible reproductive/developmental injury are appropriate for assessing hazards to humans.

During the construction of the dossiers, the quality of a given toxicity-testing protocol was evaluated without regard for the different potential uses and different exposure settings of the substance. These two factors were taken into consideration during later judgments as to toxicity-testing needs for each substance. In the test summaries, different measures of quality might have been used for different exposure settings (intended-use, occupational, and environmental) because the adequacy of a protocol might vary with the setting (e.g., a protocol might be considered adequate for low-level environmental exposure, but inadequate for high-level occupational exposure).

PURITY OF SELECTED SUBSTANCES

Chemical purity is a nonquantifiable variable that must be considered in each evaluation, and some impurities might have toxicity very different from that of the selected substance. There are three reasons for such variability: (1) the names of some substances were not clearly stated in the lists or by investigators studying them; (2) impurities might vary in composition or concentration with different methods of production or from lot to lot; and (3) some of the substances selected (e.g., vegetable oils) may contain other compounds (e.g., pesticide residues). Although this variability impedes attempts to attain consistency in judgments of adequacy, it would affect any other judgments of toxicity equally and might be useful to the extent that it reflects exposure of humans to similarly contaminated or undefined substances.

The committee also recognized that exposure is often to mixtures of substances, rather than to single chemical entities. Mixtures have the potential for synergistic interactions that potentiate or antagonize the toxic effects of individual components. Whether special studies of toxic interactions are necessary for adequate evaluation of health hazards to humans is a matter of scientific judgment.

In this report, the terms "chemical" and "substance" refer to any item that appears on any of the lists that constitute the select universe, although many of these items are not single chemical compounds. Undefined substances drawn from the select universe presented problems early in the review of the subsample of 100. Some of them were chemically so undefined (e.g., "solvent dewaxed, light paraffinic petroleum distillates") or were so variable (e.g., "zeolites containing calcium, iron, magnesium, or vanadium") that they could not be evaluated according to the established procedure. The statistical analyses and estimates reflect this procedure and inferences from the subsample apply strictly only to better-defined substances. This limitation does not apply to inferences from the sample.

GUIDELINES FOR ASSESSING THE QUALITY OF INDIVIDUAL STUDIES

BASIC CRITERIA FOR SCIENTIFIC METHODS

The Committee on Toxicity Data Elements believes that it is not appropriate to judge the adequacy of past and future studies solely by matching them against protocols that are considered acceptable today. The committee suggests that a study be considered adequate for use in a health-hazard assessment if it meets the following basic criteria:

● All elements of exposure are clearly described, including characteristics of the substance's purity and stability, and dose, route, and duration of administration.

● Results in test subjects are predictive of human responses and test subjects are sensitive to the effects of the substance. In toxicity tests of a substance involving several species, data obtained with the most sensitive species are often used for making health-hazard estimates. This is often a conservative approach. When metabolic activation is necessary to produce toxicity and there is evidence that the metabolic pathway in the most sensitive species is different from that in man or the target species, results in a species with metabolic pathways similar to those of man should be given particular consideration.

● Controls are comparable with test subjects in all respects except the treatment variable. Depending on the study, appropriate controls may be positive, negative, or historical. Historical controls, however, rarely meet this criterion.

● End points answer the specific question addressed in the study and observed effects are sufficient in number or degree to establish a dose-response relationship that can be used in estimating the hazard to the target species.

● Due consideration in both the design and the interpretation of studies must be given for appropriate statistical analysis of the data.

Although these criteria do not capture all potentially critical aspects of scientific judgment, the available data on a given substance may be considered of adequate quality if tests have been performed and reported according to these basic scientific principles. Several additional factors, although not often critical in deciding whether a given test is adequate, are highly desirable and should be taken into account:

● Subjective elements in scoring should be minimized. Quantitative grading of an effect should be used whenever possible. Sometimes, this is not feasible, as when pathologists attempt to judge the nature and extent of a malignant neoplasm. Such evaluations depend on the experience and training of the pathologists.

● Peer review of scientific papers and of reports is desirable and increases confidence in the adequacy of the work.

● Reported results have increased credibility if they are supported by findings in other investigations.

● Similarity of results to those of tests conducted on structurally related compounds increases credibility.

● Evidence of adherence to good laboratory practices improves confidence in the results.

SELECTION OF REFERENCE PROTOCOLS

The quality of individual toxicity tests may be assessed by answering the question: Does the quality of the information permit a health-hazard assessment that is acceptable? In recognition of the need for accepted and reproducible standards, the committee chose as its first step in the qualitative evaluation of toxicity data on a given substance a comparison of the study with a reference protocol. Because a requirement for inclusion of the substances in the subsample was the existence of minimal toxicity information, there were no selected substances without some information for the assessment of the quality of testing protocols. However, for each substance in the subsample, some toxicity information was missing or some data were derived from studies that did not meet the reference protocol guidelines. A comparison of available tests with reference protocols, combined with the judgment of the committee relative to the basic criteria of scientifc methods, enabled the categorization of substances with respect to the quality of toxicity-testing protocols.

In selecting reference protocols for judging the quality of individual studies, the committee used various resource documents on short-term and long-term toxicity testing, with emphasis on those constructed through national and international collaborative efforts. The committee identified the reference protocols of the Organisation for Economic Co-operation and Development (1979, 1981), the Interagency Regulatory Liaison Group (1981a, 1981b, 1981c, 1981d, 1981e), and the National Research Council (1975, 1977a, 1977b, 1980) as the most appropriate in this regard (see Appendix H). It should be understood that it was not the committee's intent to endorse any particular test protocol. Rather, on a pragmatic basis, particular tests were selected as appropriate for judging the adequacy of testing of chemicals. Although over-rigid protocols are impractical, the reference protocols provide descriptions of standard test methods with sufficient detail to establish a basis for sound study design while permitting flexibility where scientific judgment was advantageous. The committee used the most current documents, sometimes with changes or additions based on its own judgment, as presented in Appendixes I through K. The committee believes that these modifications and additions will be useful for future development of a data base for health-hazard assessment. A published document describing each modified test system is cited in Appendix H.

Not every toxicologist might agree on every detail in the guidelines, but only reference protocols widely reviewed and generally accepted were used in this study. The list is not intended to reflect the attitudes or practices of regulatory agencies.

Because some toxicity reports did not contain terminology directly compatible with the specifications of the reference protocol guidelines, it was often necessary to make judgments on whether the study adequately followed the guidelines. In general, these judgments were relatively easy to make and engendered little or no controversy within the committee.

Reference Protocol Guidelines for Neurobehavioral-Toxicity Tests

The committee recognized that the neurotoxicity-testing protocols developed by the OECD (Organisation for Economic Co-operation and Development, 1981) are appropriate only for evaluating the neurotoxicity of organophosphorus compounds. These protocols cannot be used to evaluate mammalian neurotoxicity for other substances, nor are they appropriate for studying functional behavioral changes produced by substances other than organophosphates for which no specific neural lesion has been identified. The OECD expert group on neurotoxicity also recognized this matter and, at its meeting in April 1982, took two actions: it changed the titles and scopes of the neurotoxicity tests proposed in the OECD guidelines to reflect their applicability only to organophosphorus compounds, and it recommended the development of guidelines for more general neurotoxicity testing. There was a consensus in the OECD group that neurotoxicity testing should include functional behavioral assessments outside the laboratory holding facility and neuropathologic examination of various neural tissues after in situ perfusion. The Committee on Toxicity Data Elements agrees.

For delayed-neurotoxicity tests of organophosphorus compounds, the committee used previously established reference protocols. For other classes of compounds, a detailed protocol for neurobehavioral-toxicity testing has not been completed and approved by OECD. Therefore, the committee adopted for its own interim use an alternative set of protocol guidelines that have attained some degree of general acceptance in the scientific community (Appendix I).

Reference Protocol Guidelines for Genetic-Toxicity Tests

After the start of this study, the OECD drafted guidelines for 10 genetic-toxicity tests. These were later adopted by the committee. The 10 tests were the Ames Salmonella/liver microsome reverse-mutation assay, Escherichia coli reverse-mutation assay, rodent micronucleus assay, in vitro chromosomal-aberration assay in mammalian cells, sex-linked recessive-lethal assay in Drosophila melanogaster, forward gene-mutation assay in mouse lymphoma L5178Y ($TK^{+/-}$) cells, forward gene-mutation assay in Chinese hamster ovary (HGPRT) cells, forward gene-mutation assay in Chinese hamster V79 (HGPRT) cells, in vivo chromosomal-aberration analysis in rodent bone marrow, and rodent dominant-lethal assay. The committee also adopted a policy of judging the testing protocol of each genetic-toxicity study for its adequacy and then of judging the overall adequacy of all genetic-toxicity protocols for a given substance according to the requirements described in Appendix J.

PROCEDURES FOR EVALUATION OF THE DATA BASE

INITIAL CONSIDERATIONS

Existing information was evaluated against two sets of criteria to judge its quality and completeness. The first set was a series of reference protocol guidelines that have received widespread review and general acceptance. This array of protocols was selected not as the most reliable and efficient group of tests, but rather, by convention, as the best available for chemical-safety assessments. The second set of criteria was based on the accumulated experience and expertise of committee members, whose combined judgment was used to determine the adequacy of an individual study if it did not meet the reference protocol guidelines.

The second set of criteria was established by the committee in the expectation that the data bases of only a few substances would meet all the requirements of the reference protocol guidelines, partly because much toxicity information was generated before the guidelines were developed. The committee expected that sufficient data might often be available for evaluation, even though some toxicity information would be missing and some data would be derived from experimental designs other than those prescribed in the reference protocols. Therefore, the committee intended that its determination of the adequacy of toxicity-testing data for conducting a health-hazard assessment would be based sometimes on information derived from experiments that followed the reference protocols and sometimes on other information that met the committee's own subjective criteria for evaluating scientific methods. Using this combination, the committee assessed the adequacy of the toxicity-testing protocols for all chemicals in the sample.

The committee felt that the evaluation of toxicity data bases to predict hazard to human health must be approached with caution and flexibility. In general, data from properly conducted animal studies are often predictive of the degree of hazard to humans; however, for individual substances, such laboratory investigations may be misleading with regard to target organ, potency, or type of effect. Thus, expert judgment to ensure the proper use of all available data is an essential part of each analysis. For example, the metabolism of a toxicant may differ between test species and humans in ways that produce false-negative or false-positive results with regard to possible human hazard. The appropriate test battery may be incompletely performed, but there may be other data, such as extensive information on the mechanisms of action in several species, to obviate a need for additional tests. And data from human studies, both epidemiologic and clinical, may be essential in deciding whether to conduct a test on a substance merely for the purpose of completing the recommended battery of tests for that substance. For example, there may already have been human studies and exposure of sufficient breadth and sensitivity to reduce the need for toxicity studies in laboratory animals, or clinical studies may have detected skin sensitization or toxicity so that similar investigations in laboratory models would be unnecessary. To the extent feasible, therefore, the committee analyzed data available from human experience (including case studies and retrospective and prospective epidemiologic studies) to delineate the need for further testing.

The nature of the substance examined might also affect the type and amount of toxicity tests required to assess human health hazard. If natural products were examined in a rote, rigid fashion, they might appear to be inadequately tested; however, a long history of widespread use without reported toxicity might suggest that no additional testing is needed, even though most recommended tests had not been conducted or had not been conducted according to the reference protocol guidelines. Alternatively, it is not always appropriate to assume that toxicity data are adequate and of satisfactory quality just because a substance is a natural product or has a long history of apparently safe use. Furthermore, adequate toxicity testing of a substance in the intended-use setting is not always a sufficient basis for concluding that there are adequate data on occupational exposure (such as industrial exposure during its manufacture) or environmental exposure derived from its liberation during use, disposal, or destruction. An example of such a substance might be a drug intended for one-time or very limited use, but on which additional information might be needed to evaluate its potential toxicity for the workers who are chronically exposed while they produce or package it. The committee's judgments on the quality of a substance's toxicity-testing protocols involved complex decisions. Substances were considered case by case on their own merits.

Adequacy Ratings

Evaluation of the quality of the toxicity-testing protocols that have been used for each of the 100 substances in the subsample required that the information be obtained, assembled in a documentable form, reviewed, and judged for adequacy of the data base. The Committee on Toxicity Data Elements always used studies of highest quality, even when other studies of the same test type were done. However, the committee recognized that very few studies would be performed according to current guidelines and developed a ranking system that assessed the quality of the toxicity-testing protocols of each study:

• G, for a study that was performed according to current reference protocol guidelines.

• A, for a study that was not performed according to reference protocol guidelines, but was nevertheless adequate for conducting a health-hazard assessment.

• IN, for a study protocol that was inadequate for conducting a health-hazard assessment, but was judged not to need repetition. This rating was assigned in either of two situations:

(1) Where the available information was deemed to be sufficient to allow an assessment of health hazard of the substance, tests not done or done inadequately were considered by the committee to be no longer required or not to require repetition.

(2) When the observations of one test type not done or done inadequately were encompassed in the observations of another test type done adequately, the former test type was considered by the committee to be no longer required or not to require repetition.

● IR, for a study protocol that was inadequate for conducting a health-hazard assessment and judged to need repetition.

● C, for a test of indeterminate quality whose adequacy could not be assessed. This rating was most frequently given to abstracts, review articles, or other reports in which protocols were not fully described.

Measures of Adequacy for Tests Not Meeting the Reference Protocol Guidelines

As discussed above, toxicity tests that did not precisely follow reference protocol guidelines may still have been judged adequate if the committee determined that the deviations were not important. For example, if the number of animals used in a study was short of that specified by the appropriate reference protocol guidelines, but the results were so definitive that the addition of more test animals would almost certainly not have affected the conclusion, the study was considered to be adequate for conducting a health-hazard assessment. This example illustrates that adequacy is necessarily judged in the context of results; the fact that many tests were judged to be adequate despite deviations from reference protocols does not mean that the reference protocols are unnecessarily rigid. Sound protocols are most important precisely when results are less than clear-cut--a matter that cannot be known at the time the protocol is selected.

Other judgments involved evaluations of chemical and physical properties, study design, study execution, selection of dose or exposure, statistical analysis, and reporting completeness. Studies acceptable in those respects, but deficient in one or a few guideline specifications, were likely to be judged adequate. Studies that deviated from several or many guidelines were judged to be inadequate. Precise reasons for these expert judgments could usually be given only case by case, and they varied with test type (e.g., subchronic toxicity, reproductive/developmental toxicity, skin and eye irritation, and carcinogenicity). However, some of the more common deficiencies for all types of toxicity testing were in numbers of animals used, extent of histopathologic examination, extent of clinical-chemistry laboratory measurements, mathematical treatment of data, scoring systems for irritancy, and survival of animals in long-term studies.

Evaluation of the toxicokinetic properties of chemicals presented special problems. A completely adequate toxicokinetic study would include not only data on absorption, distribution, and excretion, but also identification of metabolites of the substance and their distribution and excretion. Seldom, if ever, were all these data presented in a single publication; such information was more typically published in a series of communications (by the same or different authors). In most cases, the one or two toxicokinetic studies reviewed by the committee may have been adequately performed, but they did not in themselves constitute an adequate--i.e., complete--toxicokinetic study. Thus, an assessment of adequacy for data on this item referred to the whole of available information, not to any single report.

Assessment of studies on reproductive/developmental toxicity requires the interpretation of results and effects on the conceptus. These are described in Appendix K. Frequently, only one study related to developmental effects of a substance was found. Such a study was considered to be adequately conducted if it was a "limit" test--showing a lack of developmental toxicity at a dose of at least 1,000 mg/kg of body weight--so that no further testing of developmental effects may have been necessary. However, in the absence of a "limit" test, the study was considered to be inadequate for determining potential hazard to the conceptus if the highest dose reported did not produce an adverse effect on the dam or if the lowest could not be reasonably interpreted as a no-observed-effect level (NOEL) for the conceptus. Once the minimal dose or concentration of a test substance needed to produce overt toxicity in the adult was established and the NOEL in the conceptus, later studies might be judged adequate for identifying the most sensitive species to be used for interspecies extrapolation and health-hazard estimation.

Superseding all such individual variations was the overall assessment by the experts on the committee, taking into account not only the types of variation and their magnitudes, but also the potential for interactions among the variables.

DOSSIER REVIEW PROCESS

During the review process, each document in each dossier was evaluated according to the criteria developed above. The evaluations were recorded in the form shown in Table 6 with the reasons for assigning a specific study to a particular rating category. This permitted the acceptance of convincing experimental protocols that did not meet every detail of the guidelines and provided an opportunity for identifying serious flaws that diminished the credibility of the results.

Dossier analysis required that a very large amount of information be collected, documented, reviewed, and judged for adequacy within a short

TABLE 6 Committee Format for Recording Judgments about Protocol of Toxicity Study

Test Adequate	Test Adequate	Test Inadequate	Test Adequacy Cannot Be Judged	Nature of Document Judged
Meets reference protocol guidelines (G)	Does not meet reference protocol guidelines, but nevertheless adequate; state why classified as adequate (A)	State why classified as inadequate; if repetition not required, state why (IR = repetition required; IN = repetition not required)	State why adequacy cannot be judged	Abstract, published paper, review article, etc.

69

time. Once the data were obtained, the method and depth of their evaluation by the committee and the consistency of judgment had to be determined. The judgments made by the committee were coded and recorded in preparation for subsample analysis and extrapolation to the select universe.

The committee recognized that the quality of its comprehensive literature search and its detailed evaluations of the data bases would be the most important determinants in estimating testing needs in the select universe. Recognizing that this was a large and important task, the committee established five working panels, each with a designated leader and two or three other committee members. Assignments were not considered effectively random, because several substances were selected by panel leaders who were familiar with the substances' toxicity data bases. Remaining substances in the subsample were assigned in rotation among the five panels (20 each), including substances from each of the seven categories. Each group then had responsibility for reviewing the data bases. At a series of planning meetings, the panel leaders collectively established standard procedures for data review and evaluation and developed practices to ensure consistency in decision-making.

The review process was time-consuming. The committee recognized early in the second year that it could not carry out the entire review itself on a volunteer basis. Therefore, to expedite the process, the initial phases of the review were carried out by NRC staff and consultants. It is estimated that these initial phases required about 1.5 scientific person-years of effort.

The procedure for the review of the data base on a substance consisted of the following steps:

● Each document was individually reviewed and compared with the appropriate reference protocol.

● A summary sheet was prepared for each document, outlining the pertinent details of the protocol, assigning a preliminary ranking for the quality of the study and reasons for this judgment, stating the nature of the document (abstract, review, etc.), and stating which of the prescribed 33 test types was (were) reported in the document.

● In many cases, the quality of individual study protocols was determined by individual panel members who had applicable expertise. All reviewers were required to document their findings and to provide reasons for them based on the criteria established by the committee. Such documentation was especially important when a reviewer had intimate knowledge of a substance.

The dossier prepared by NRC staff and consultants, including these judgments on individual report, was reviewed by the appropriate panel leader, who then presented it to the panel members for review and modifications deemed necessary. Twelve panel-approved dossiers were discussed by the entire committee to ensure that there was concurrence in the approaches used. The other 88, after review by a panel's leader and members, were reviewed by a subcommittee of at least five designated committee members. Important issues concerning any dossier were placed before the entire committee. Otherwise, judgments of the subcommittee concerning review of dossiers from the panels were regarded as final. The above process ensured that each decision with regard to the quality of every study was reviewed at three levels: by a panel chairman, the panel's other members, and either the entire committee or its designated review subcommittee. The relevant data base was present or easily accessible at each step of this multistage review process; that allowed the panel leader, the members of each panel, the subcommittee, or the committee to conduct an independent review of the original material when any person deemed it necessary.

The committee recognized the need to maintain uniformity and to ensure quality in the review of documents. Standardized procedures for documenting decisions regarding data adequacy provided quality control for decision-making. Variations in the consistency of decisions were reduced first by judging a study's adequacy against the uniformly applied set of reference protocol guidelines. These standards were used for studies of the same test type across all substances and by all persons making the judgments. In effect, all reviewers were making measurements with the same yardstick. Deviations from the guidelines were then noted according to the scheme presented in Table 6, so that one person's reasons for judgment on a chemical could be examined by others making similar judgments on other chemicals. To ensure consistency, the five panel leaders often compared their reasons for judgments on the quality of protocols.

Because committee members often had to exercise scientific judgment when information was inconclusive, it was necessary to provide mechanisms to document their judgment and to ensure that they remained consistent and that testing protocols and other information were always judged as uniformly as possible. The system of multilevel review described above was designed to reduce the errors and differences involved in the committee's use of scientific judgment.

The process led to decisions of whether further testing of a given substance was needed. Before such decisions were reached, the committee considered the types of exposures to substances likely to be encountered, their chemical and physical characteristics, their manufacturing processes, their production volumes, their uses, their chemical fates, their toxicity in animals, and their potential or known toxicity in humans. The committee's detailed evaluation of the data base for each substance in the subsample included determination of the adequacy of each required toxicity test specified in Appendixes B through G. The completed dossiers collectively were used as the committee's record to characterize the subsample.

The committee analyzed the decisions about the quality, quantity, and extent of the subsample's toxicity data base to assess the toxicity-testing needs related to the larger select universe. This extrapolation was a joint effort of the Committee on Toxicity Data Elements and the Committee on Sampling Strategies. The tabulations and interpretations of the evaluated data bases were used as a bridge for applying statistical inferences derived from the subsample to the select universe from which it was drawn.

LIMITATIONS OF THE DATA GATHERING PROCESS

The approach developed to collect data on each substance included searches of the open literature through automated, on-line data files, such secondary sources as reference manuals and textbooks, government technical reports, files of the regulatory agencies (where available), and files provided by some chemical manufacturers and trade associations. The data obtained from the searches of the primary and secondary open literature accounted for the bulk of the information in the dossiers. Search strategies were carefully developed to ensure the most efficient screening of the selected data bases. However, some of the data bases failed to include the most recent research.

The degree of accessibility of government agency files to the committee varied. Some information was obtained from the regulatory agencies, and several research reports from military sources yielded useful information, especially on exposure of humans. At times, confidential data were made available to selected NRC staff members or to specific committee members. However, nonconfidential health and safety data embedded in commercial confidential files possessed by the FDA Bureau of Drugs were unavailable (see Appendix M for further detail).

Responses from manufacturers and trade associations were also mixed. A few manufacturers were extremely cooperative in providing information that supplemented the open literature; however, the total amount of information from this source was relatively small. The committee believes that some relevant but unreviewed toxicity information, especially of a confidential nature, exists in the files of manufacturers.

Evaluations of the 100 substances in the subsample therefore were based largely, but not exclusively, on published data or other publicly available information, which may be somewhat short of the amount and diversity of data contained in the confidential files. The absence of specific information from the dossiers reflected both the inaccessibility of some data bases and the lack of relevant testing. Data not available for the committee's confidential review are presumably not available for legitimate review by other interested parties; hence, in an operational sense, they do not exist.

Most of the exposure estimates were based on intended uses and knowledge about products that contain the materials of interest. Some of the occupational- and environmental-exposure estimates were based on production volumes, environmental fate, and disposal data, but few data of these types were available.

Other kinds of information that were rarely encountered concerned production trends, production processes, and percentage of total production allocated to each intended end use. More information of this type would have contributed greatly to estimates of exposure. Again, it was assumed that much of this information exists, but access to it was limited or restricted. The committee found little or no epidemiologic information or information on environmental fate (e.g., biodegradation and bioaccumulation) for most compounds in the sample.

The data base was limited by the paucity of information on toxic effects in humans. Because observational studies on humans are expensive and involve special difficulties, they cannot be undertaken routinely. Even if extensive resources became available, it would be impossible to acquire conclusive data on many possible outcomes under all different conditions of exposure. Epidemiologic studies involve factors that are different from animal toxicologic protocols. Investigators must know not only the chemical and physical properties of substances and the quantitative and qualitative toxicity data from studies in animals, but also the extent of human exposure, its intensity, and other qualities of exposure that are needed to conduct an adequate study. It is necessary to define pathologic end points or effects, define a control population, conduct followup or retrospective studies, ensure that there is a suitable exposed population with enough exposed subjects to provide reasonable statistical power, and develop mechanisms to minimize or quantify sources of confounding or bias. Sometimes, epidemiologic studies in different settings have each contributed information to increase the credibility of a cause-and-effect relationship, but are flawed because exposure to the substances under study and exposure to some other possible cause of the same end point have occurred simultaneously. For these reasons, the committee did not assess the adequacy of most observational studies of humans, but it did consider information from case reports citing adverse effects in humans and, for some classes of substances, data from human sensitivity tests and available epidemiologic information.

Extensive data might exist on human exposure to some substances (e.g., drugs) with intended uses limited to a few exposures in a lifetime. Limited toxicity testing might be adequate for such end uses, but insufficient for developing safety standards to protect industrial workers producing or using the materials or medical personnel who might handle them frequently in the course of their professional activities. Very few data on potential occupational or environmental exposure were accessible.

For a substance to be reviewed, it had to be well defined, readily characterizable, and identifiable. Thus, some large classes of substances (such as plant products, minerals or ores, and unidentified mixtures) were excluded from consideration in the subsample. They were, however, included in the sample.

Three deficiencies of the TSCA Inventory as a source of chemicals in commerce were most apparent during sample selection and evaluation:

● Poorly defined chemical mixtures in the sample were not sufficiently uniform in composition or could not be sufficiently characterized to determine the extent of toxicity testing performed, much less its adequacy.

● Some substances, according to manufacturers, were no longer in production in 1977 and therefore were no longer "chemicals in commerce."

● Many companies listed in the Inventory as manufacturers of chemicals in 1977 claimed never to have made those chemicals, although in some cases they had made related substances.

Therefore, there was little information on many chemicals in commerce, and it was often impossible to determine whether a substance had minimal toxicity information.

Substances are often selected for toxicity tests because there is a particular interest in them (e.g., because some toxic effect has been observed). Thus, selection of substances that have already had some testing must not be considered a random sampling of all substances in the select universe.

INTERPRETATION OF DATA ON TESTING QUALITY

Although the presence of toxicity information on each of the 100 substances in the subsample and the reviews of that information may have some intrinsic value, the reviews were not intended to provide specific information concerning the need for additional testing of these specific compounds. They served only as a basis for inference about the select universe and the seven categories of substances in it. The substances in the select universe are themselves a nonrandom sample of all substances in the entire universe of chemicals known or used at a specific time (mid-1981). Because of the manner in which toxicity testing can be conducted and has been in the past, some groups of substances were not included as specific classes among the substances in the select universe. Examples are some natural products of largely undefined nature or chemical structure, some mixtures of chemicals, and some industrially used chemicals of variable composition.

Furthermore, the review reported herein was not undertaken as an "audit" of the adequacy of past regulatory policies or procedures, and the data developed in this study are not likely to be immediately useful for regulatory purposes. For example, some regulatory decisions might have been based on information developed from long-standing exposure data not available to the committee during its evaluation of the toxicity data base. It would be inappropriate to judge past decisions against current standards of toxicity evaluation. The committee also recognizes that regulatory standards must be set in accordance with law and federal agency policies. These standards are based on more than toxicity data, and regulations, once set, cannot be lightly or easily changed on the basis of a modest increase in information about effects. In some cases, regulatory decisions are based on proprietary information that is available only to the concerned agencies.

CHARACTERIZING THE SAMPLE AND OPTIONS FOR DRAWING INFERENCES TO THE SELECT UNIVERSE

In the second year of operation, the Committee on Toxicity Data Elements worked with the Committee on Sampling Strategies to identify encoded data in the dossiers that were critical to analysis of the sample and to extrapolation of the analysis to the select universe. Some of the basic critical data in the dossiers described the toxicity tests conducted and their quality. Descriptions of this nature--combined with information on intended use, physical and chemical properties, and potential exposures--were tabulated by the two committees in the third year of the study. The tables serve as springboards to more detailed analyses of the categories in the sample. Algorithms for these analyses were developed by the Committee on Sampling Strategies.

The select universe was sampled in two phases. First, independent systematic random samples were drawn from each of the seven categories. The components of each sample were arranged in random order and then examined one by one for the existence of minimal toxicity information (as defined for each of the five intended-use classes, described earlier) until a specified number of substances with at least minimal toxicity information were identified. The substances were not reviewed further if the literature search uncovered less than the prescribed minimal toxicity information or if the substances were so ill-defined as to preclude evaluation. The latter group included substances in the select universe whose names referred to sets of substances (e.g., "alkyl derivatives of dimethylbenzylammonium chloride") with possibly different toxic properties, on which it would be impossible to characterize the quality of toxicity information. The substances set aside for this reason remained in the sample and are an important part of the data base for evaluating the extent to which substances meet the minimal-toxicity-data criteria defined by the Committee on Toxicity Data Elements.

CONSTRUCTION OF TABLES FOR ANALYSIS

By examining data in both the sample and the subsample, the Committee on Sampling Strategies could obtain an estimate of the amount and quality of information on toxicity testing of substances in the select universe. The process of examining these substances generated other information of interest in the examination of toxicity testing. Some of these data, such as the frequency with which a given toxicity-test type could be found for each of the 100 substances in the subsample, were available from machine-readable files. Other information, such as the quality of reporting of toxicity tests or the ways in which the Committee on Toxicity Data Elements determined the adequacy of a given toxicity test, was not suitable for statistical analysis, but is presented in a qualitative form in the conclusions and recommendations.

STATISTICAL ANALYSIS OF DATA

The dossiers on substances in the subsample provided sets of tabulations, measures, and various descriptive items of information that the Committee on Sampling Strategies used to estimate the testing adequacy for substances in each of the seven intended-use categories of the select universe or in other well-defined sets of substances in that universe.

ESTIMATES BASED ON THE SAMPLE ALONE

To estimate the properties of substances in a well-defined category of the select universe and the variances of those estimates for the sample, the Committee on Sampling Strategies adopted procedures that allowed for the use of all available information on substances in any well-defined subset of the select universe. All substances in the select universe could be placed in different combinations of the original seven categories. For any substance, j, it is straightforward to determine the probability, p_j, of selection into the sample. Because samples were drawn independently from each of the seven categories, i, the probability $1 - p_j$, of not being selected is precisely the product of the probabilities of not being selected from any category, so that if p_{ij} is the probability that substance j is selected from category i, then

$$p_j = 1 - \prod_{i=1}^{7} (1 - p_{ij})$$

(1)

76

Note that, if substance j is not a member of category i, then $p_{ij} = 0$, so the value of p_j is unaffected by that category.

The subcategory to which any substance belongs is defined by the set of categories of which it is a member. There are 64 possible combinations of categories (2 x 2 x 2 x 2 x 4), i.e., 63 possible subcategories with the exclusion of the one classification of chemicals that are not in the select universe because they do not fall into any of the categories. Some of the 63 possible subcategories may, of course, contain no substances from the select universe, and other subcategories may include substances from the select universe but none from the sample.

For any specific analysis of the sample, the subcategories of interest are first determined. For example, an analysis of all substances on the list of drugs and excipients in drug formulations could include as many as 32 subcategories defined by being on that list but on or off the lists of pesticides and inert ingredients of pesticide formulations, food additives, cosmetic ingredients, or chemicals in commerce. Similarly, an analysis of the entire select universe from which the sample was drawn would include up to 63 subcategories.

In this discussion,
$\quad$ h = subcategory (h = 1,2, . . ., 63)

and N_h = number of substances in subcategory h.

Let H be the collection of subcategory members, h, that are of interest for a particular analysis. Let $\bar{x}_h$ be the mean (here, the proportion) of some property of the sample substances in subcategory h. Then an unbiased estimate of that proportion over the whole set of subcategories, H, is

$$\bar{x} = \sum_H \frac{N_h}{N} \bar{x}_h, \tag{2}$$

where $N = \sum N_h$, and the variance of this estimate is

$$\sigma_{\bar{x}}^2 = \sum_H \left[\frac{N_h}{N}\right]^2 \sigma_{\bar{x}_h}^2 . \tag{3}$$

Unfortunately, this ideal formula is not usable in practice, because limited resources precluded exhaustive searches for duplication of substances among categories, so the values for N_h are not precisely known. Information on category-to-category duplication is, however, available for all items in the sample and, for all compounds, j, that were actually found in subcategory h, can be used to estimate N_h as

$$\hat{N}_h = \sum_{j \epsilon h} \frac{1}{p_j} . \tag{4}$$

Replacement of N_h/N by the estimate $\hat{N}_h/\Sigma\hat{N}_h$ introduces an additional source of variation in x, in that, where Σ is the mathematical expectation value,

$$\text{Var}\ (\bar{x}) = E\ \text{Var}\ (\bar{x}|\{\hat{N}_h\}) + \text{Var}\ E\ (\bar{x}|\{\hat{N}_h\}). \tag{5}$$

However, the committee believes that the second term in Equation 5 is likely to be small. Because, in all cases presented in the tables of this report, $\bar{x}_h$ (and hence $\bar{\mu}$) is the estimated proportion of substances with a specific property, when n_h is the sample size in subcategory h and $p_h = \bar{x}_h$ is the proportion observed in subcategory h,

$$\hat{\sigma}_{\bar{x}_h}^2 = \hat{p}_h(1 - \hat{p}_h)/n_h . \tag{6}$$

Equation 6 assumes that the selection was essentially equivalent to a simple random sample, as discussed previously. Thus, the computing formulas that the Committee on Sampling Strategies used for estimates based on the sample are

$$\hat{\mu} = \sum_H \left[\frac{\hat{N}_h}{\Sigma\ \hat{N}_h}\right]\bar{x}_h \tag{7}$$

and

$$\hat{\sigma}_{\hat{\mu}}^2 = \sum_H \left[\frac{\hat{N}_h}{\underset{H}{\Sigma}\ \hat{N}_h}\right]^2 \hat{\sigma}_{\bar{x}_h}^2 . \tag{8}$$

The sample sizes for any category are sufficiently large, with the estimated proportions not too near 0 or 1, that the sample distribution for a category is approximately normal. Therefore, 90% confidence intervals are presented--that is, intervals with at least a 90% probability of including the true value of the proportion, estimated as $\hat{\mu} - 1.65\ \hat{\sigma}_{\hat{\mu}}$ and $\hat{\mu} + 1.65\hat{\sigma}_{\hat{\mu}}$.

ESTIMATES BASED ON THE SUBSAMPLE ALONE

Most of the tables presented in this report give estimates of the proportions of substances in a category that have specified characteristics. Such estimates are based on the sample of substances selected from the corresponding segment of the select universe. Some substances belonging to a category do occur in the subsample selected from other lists and satisfy the minimal-toxicity-information screening criterion for the specified category. It would have been possible to use these additional sample substances in making the estimates; that would probably have resulted in somewhat smaller sampling errors. However, because the screening procedure was not identical for each intended-use category, it would have introduced biases into the results. The Committee on Sampling Strategies therefore decided to base estimates for each category solely on the sample selected for that category. This limitation of analysis further implies that <u>it is not valid to use the results given here to derive estimates across categories in the final sample.</u> Such combined results can be calculated, but would require the preparation of special tabulations.

The probability of selection within a category is constant, so an unbiased estimate of the proportion of screened substances in the ith category that could be evaluated and that have a given characteristic is $\hat{\mu}_i = x_i/n_i$, where n_i is the sample size and x_i denotes the number in the sample that have the characteristic. If P_i denotes the true proportion, the statistic $\hat{\mu}_i$ has the Bernoulli distribution with parameters P_i, n_i. It is then possible to calculate two numbers, L_i and U_i, as functions of $\hat{\mu}$ and n_i, so that the probability that $L_i \leq P_i \leq U_i$ is at least 90%. These are the confidence limits shown in the tables.

ESTIMATES BASED ON BOTH THE SAMPLE AND THE SUBSAMPLE

Some estimates in this report make use of information provided by both the sample and the subsample. Such estimates are for the proportions of some category in the select universe that both satisfied the screening criteria and have one or more additional specified characteristics. The point estimate of such a proportion is the product of two factors that are statistically independent as a consequence of the sampling procedure. The first factor is the estimated proportion that satisfies the screening criteria for the category and is based on the sample alone. As stated above, its distribution is approximately normal, and its variance has been estimated from the sample results. The second factor is the estimated proportion, among substances that satisfy the screen, that also have the specified characteristic; it has a Bernoulli distribution, and lower and upper confidence limits have been calculated.

Confidence limits for the product of the two factors were approximated as follows. Let x denote the first of the two factors referred to in the preceding paragraph, and y the second. Where X and Y denote the mathematical expectations of x and y, respectively, note that the variance of the product of independent variates x and y is given by

$$\sigma_{xy}^2 = \sigma_x^2 \sigma_y^2 + X^2 \sigma_y^2 + Y^2 \sigma_x^2 \quad .$$

(9)

Because the distribution of y is asymmetric in general, estimates of the lower and upper confidence limits are computed separately. To calculate the lower confidence limit, σ_{xy} in Equation 9 is calculated by replacing σ_y with $(y - L_y)/1.65$, where L_y denotes the lower limit for Y, and X and Y in Equation 9 are replaced with their estimates x and y. The lower limit for the product is then calculated as xy - 1.65 σ_{xy}. For the upper limit, σ_y in Equation 9 is replaced with $(U_y - y)/1.65$, where U_y denotes the upper limit for Y. The upper limit for the product is then calculated with this new estimate of σ_{xy} as xy + 1.65 $_{xy}$.

The very substantial costs and demands required to amass and analyze data on 100 substances in a short time permitted a subsample of only 100 substances for the seven categories of the select universe. As a result, confidence intervals are large. Any analysis of the data should be based on an awareness of the limited statistical precision of results from the subsample.

MACHINE-READABLE FILES

To facilitate data analysis, information on the substances in the sample and subsample was assembled in machine-readable files. The presence or absence of the five types of toxicity tests in the minimal-toxicity-information screen (acute, subchronic, chronic, reproductive/developmental, and mutagenicity) was tallied. In addition, the entire list of substances in the sample was scanned to determine which of the seven intended-use categories contained each substance. These 12 items and a numeric identifier for each item in the sample were entered into the computer for analysis.

Dossiers compiled on substances in the subsample contained substantially more items of information than were available on the sample. These items were tallied to provide measures of the amount and adequacy of data available on substances in the subsample.

The seven intended-use categories were expanded into partially overlapping subcategories, as listed in Appendixes B through G. A more complete roster of test types was available, and the test protocols for each test type deemed necessary for a substance's designated subcategories of intended use were evaluated for adequacy. Chemical and physical properties of each substance were sought, as well as its manufacturing process or processes, production volume, potential for exposure, and environmental fate. Overall judgments, such as the ability to assess the potential hazard to human health, were determined. Although much of this information was already in tabular form, some items in the dossier were descriptive. This type of information, primarily nontoxicologic, was intended to assist in assessment of potential for hazard. Although the presence or absence of this information was recorded for later numeric analysis, no judgment of the quality of the nontoxicologic data was made.

5

RESULTS

A select universe of 65,725 listings of chemical substances was compiled from lists of pesticides and registered inert ingredients of pesticide formulations, cosmetic ingredients, drugs and excipients in drug formulations, food additives, and other chemicals in commerce as listed in the Inventory of the Toxic Substances Control Act. A sample of 675 substances chosen by a stratified random process was selected from the 65,725 entries. A random subsample of 100 substances with at least prescribed minimal toxicity information (see Chapter 2) was then selected from the random sample. The sample and subsample contained representatives of seven categories of substances: (1) pesticides and inert ingredients of pesticide formulations, (2) cosmetic ingredients, (3) drugs and excipients in drug formulations, (4) food additives, and chemicals in commerce, which were divided into substances with (5) 1977 production of at least 1 million pounds, (6) 1977 production of less than 1 million pounds, and (7) 1977 production unknown or inaccessible owing to manufacturers' claims of confidentiality.

The findings presented in this chapter are based on analyses of the sample of 675 substances randomly selected from the select universe, including data on the subsample of 100 (selected from the random sample of 675) by application of criteria regarding the presence of minimal toxicity information. Some specific analyses were based solely on the sample or the subsample. Others were based on combined information on both the sample and the subsample. In each instance, the origin of the information is stated to aid the reader in understanding the nature and implications of the data.

The Committee on Sampling Strategies and the Committee on Toxicity Data Elements recognize that their estimates of testing need may be less than the reference protocol guidelines call for, even though some tests may have been done not at all or not in compliance with the guidelines. Although additional testing is desirable, some circumstances make it unnecessary to perform some or all tests because other available information--including that from prior testing of a substance, its history of intended use, and other anticipated exposures to it (such as occupational and environmental)--may permit some judgment of health hazard in the absence of full data.

The reference protocol guidelines can be expected to change with time, as public and scientific perceptions of hazards change, as the technologic ability to detect hazard improves, and as willingness to accept particular hazards changes.

Substances on the list of one category were often on lists of other categories. The overlap among the seven categories was determined for substances in the sample. This information provided the basis for estimating that the select universe contained about 53,500 distinct substances, the lower and upper 90% confidence limits being 49,800 and 57,900, respectively. However, all reported findings are for specific categories of the select universe, rather than for the select universe as a whole. When categories or parts of them were combined, appropriate weighting factors were assigned to account for duplication among lists.

Inert ingredients of pesticide formulations and excipients of drug formulations were included in the select universe, because the Committee on Toxicity Data Elements desired to include to the fullest extent in the select universe the substances that were of possible interest to NTP because of their potential for human exposure. Of the 50 pesticides and inert ingredients of pesticide formulations in the sample, 37 were on the list of active pesticides, 11 were on the list of inert ingredients, and 2 were on both lists; of the 15 in the subsample, 12 were on the list of active pesticides, 2 were on the list of inert ingredients, and 1 was on both. Of the 50 drugs and excipients in drug formulations, 22 were on the list of active drugs, 18 were on the list of excipients, and 10 were on both; of the 32 active drugs, 23 were listed as prescription drugs, 3 as nonprescription drugs, and 3 as both; of the 28 excipients, 13 were found in prescription formulations, 1 was found in nonprescription formulations, and 7 were found in both. Of the 15 in the subsample, 7 were listed as active drugs, 4 as excipients, and 4 as both; of the 11 active drugs, 8 were listed as prescription drugs, 1 as a nonprescription drug, and 1 as both; of the 8 excipients, 2 were found in prescription formulations, none was found only in nonprescription formulations, and 3 were found in both.

These numbers are presented to give the ensuing descriptive analyses a sense of proportion regarding active pesticides, drugs, and nonactive components of their formulations. The committees did not conduct separate analyses to distinguish the findings for active and nonactive substances, although the data may permit others to consider such separate analyses for their own purposes.

QUANTITY AND NATURE OF TESTING

It was recognized from the beginning that the quantity and nature of testing needs were such that they could never be fulfilled adequately only by the use of specific testing regimens. Although tests of substances will always be needed, a better understanding of the "how" and "why" of toxic injury itself at the subcellular, cellular, organ, and whole-animal levels will be necessary to fulfill future needs in the most efficient and economical manner. The Committee on Toxicity Data Elements used a battery of toxicity tests as the basic "measuring stick" for quantitation of testing needs. At the same time, it rejected the concept that every substance in the select universe required the adequate performance of a complete battery of toxicity tests to make possible a human health-hazard assessment, even if that were practical. Thus, other criteria, including data from human exposures, were also used for judgments about testing adequacy. The Committee on Toxicity Data Elements recognizes that meeting the testing needs will require the establishment of priorities for the tests and the substances needing them.

SAMPLE OF 675 SUBSTANCES

Minimal toxicity information was defined as specific combinations of five basic types of tests prescribed by the Committee on Toxicity Data Elements: acute, subchronic, chronic, reproductive/developmental, and mutagenicity. The tests most frequently encountered for each category of substances in the sample of 675 during the search for prescribed minimal toxicity information are presented in Table 7. The data in these tables include the best estimates of the Committee on Sampling Strategies and the Committee on Toxicity Data Elements with upper and lower 90% confidence limits. In some cases, confidence limits are wide because of the unavoidable restrictions on sample sizes in specific segments of the select universe.

As indicated in Table 7, 25-82% of the select universe was estimated to have no toxicity information, on the basis of what the committees were able to discover from the published and unpublished literature available to them. In each of the seven categories of the sample of 675 substances, testing for acute, subchronic, and mutagenic effects was present more frequently than testing for chronic or reproductive/developmental effects. For the select universe of drugs and excipients in drug formulations, about 75% were estimated to have information on acute toxicity and about 62% were estimated to have information on subchronic testing. For pesticides and inert ingredients of pesticide formulations, these values were about 59% and 51%, respectively. Testing was absent most frequently for chemicals in commerce, particularly for chronic and reproductive/developmental effects. All frequencies, even though some were higher than others, are based on a limited number of substances that had minimal toxicity information.

The degree of testing, as determined by the search for prescribed minimal toxicity information and as presented in Table 7, was based on a search strategy that was designed to qualify substances for the subsample of 100 by rapidly identifying information where it was most likely to be found. The rationale for this approach and the search strategy used are presented in Chapter 3. The search was conducted for each of the 675 substances in the sample and may be assumed to be efficient in identifying all or most of the existing information. However, unlike the search strategy for the subsample of 100, it was not exhaustive. As a result, the estimates presented in Table 7 may be slight understatements of the amount of minimal testing that has been conducted.

83

TABLE 7 Estimated Percentages of Substances in Seven Categories of Select Universe That Have Five Basic Tests Used in Screen for Prescribed Minimal Toxicity Information[a]

Category	Sample Size[b]	Percent with Prescribed Test					Percent with Minimal Toxicity Information[d]	Percent With No Toxicity Information[e]
		Acute[c]	Subchronic[c]	Chronic[c]	Reproductive or Developmental Biology	Mutagenicity		
Pesticides and inert ingredients of pesticide formulations	106	49-59-70	41-51-62	15-23-32	25-34-44	20-28-38	28-36-46	28-38-49
Cosmetic ingredients	162	32-39-47	23-29-36	11-16-21	16-22-28	17-23-30	21-26-32	49-56-64
Drugs and excipients in drug formulations	95	67-75-85	53-62-73	30-39-50	35-45-56	23-32-42	30-39-49	16-25-34
Food additives	212	40-42-55	27-34-41	9-13-19	14-20-26	18-23-30	16-20-25	39-46-54
Chemicals in commerce with production:								
≥1 million lb/yr	259	15-20-25	7-10-14	3-4-7	3-6-9	6-9-13	18-22-26	73-78-84
<1 million lb/yr	136	15-22-29	5-8-13	2-3-6	2-4-7	5-10-15	18-24-30	69-76-83
unknown or inaccessible	136	9-15-21	3-7-11	0-3-6	3-7-12	4-8-13	12-18-24	76-82-89

a Estimates expressed as means with upper and lower 90% confidence limits.

b For statistical analysis, substances present in more than one category are added to each such category, regardless of category of their selection. Thus, for analysis here, sizes of all seven categories are greater than sizes used for sample selection.

c Precise definitions for acute, subchronic, and chronic tests as used in this screen are given in Chapter 2.

d Definitions for minimal toxicity information are given in Chapter 2.

e These values are based on search strategy used to identify prescribed minimal toxicity information on substances in sample of 675. This strategy, whose rationale and characteristics are described in Chapter 3, was not exhaustive. Additional information may be in files of industries and government agencies.

SUBSAMPLE OF 100 SUBSTANCES

Table 8 lists the test types used in the detailed analysis of the subsample according to the degree to which they were performed on the substances in each intended-use category that had the prescribed minimal toxicity information. Acute oral rodent studies and acute parenteral studies had been done most frequently. Except for drugs and excipients in drug formulations, the next most commonly conducted test was for genetic toxicity. Toxicokinetic, dermal, and eye irritation studies were done on some chemicals in commerce. Except for the three categories of chemicals in commerce, the test types encountered next most frequently included investigations of effects of subchronic oral administration for 14, 28, or 90 d in rodents and of chronic oral administration. These were followed in frequency by tests of teratology, acute skin and eye irritation, carcinogenicity in rodents, acute dermal effects, and acute inhalation effects. Acute oral administration in nonrodents and subchronic oral administration for 90 d in nonrodents were also performed occasionally. For the three categories of chemicals in commerce, studies investigating carcinogenicity and subchronic toxicity in rodents were also performed. The locations of the 33 test types in Table 8 are based on small random samples, and the percentages of testing for specific types might change if new random samples were examined.

It is evident from the results presented in Tables 7 and 8 that the amount of testing information available on any category of substances is related to the regulatory history of that category. In general, proportionately more testing has been undertaken on drugs and excipients in drug formulations and on pesticides and inert ingredients of pesticide formulations than on other substances. Among all categories, drugs and excipients in drug formulations have the longest history of regulatory interest. Of the substances examined in this category, about 39% were found to meet the requirements adopted by the Committee on Toxicity Data Elements for minimal toxicity information. In contrast, only about 20% of the compounds in the three categories of chemicals in commerce were found to meet the requirements for minimal toxicity information, although these requirements were much less strict than those adopted for drugs and excipients in drug formulations.

The committees recognize that some toxicity tests may not have been known to them. Although toxicity-test information on the subsample of 100 substances was sought from industries and other interested parties via a Federal Register notice and by direct contact with manufacturers and importers of sampled chemicals in commerce, some industrial information probably remained unavailable. Similarly, the committees were not able to examine toxicity, physical, and chemical information on cosmetic ingredients, drugs, excipients in drug formulations, and food additives contained in the files of FDA, except for food additives listed as generally regarded as safe (GRAS).

TABLE 8 Proportion of Substances Tested in Each Category of Subsample According to Each of Defined 33 Test Types

Percent Tested	Category and Subsample Size			
	Pesticides and Inert Ingredients of Pesticide Formulations (15)	Cosmetic Ingredients (15)	Drugs and Excipients in Drug Formulations (15)	Food Additives (15)
100	Acute oral toxicity--rodent	--	--	--
90-99	--	--	--	--
80-89	--	--	--	Genetic toxicity
70-79	Acute parenteral toxicity Genetic toxicity	--	--	
60-69	Acute dermal toxicity Subchronic oral toxicity--rodent: 90-d Toxicokinetics	Acute oral toxicity--rodent Acute parenteral toxicity	Acute parenteral toxicity Acute oral toxicity Teratology--rodent, rabbit	Acute oral toxicity--rodent Acute parenteral toxicity Subchronic oral toxicity--rodent: 14- or 28-d
50-59	Acute inhalation toxicity Acute eye irritation--corrosivity Subchronic oral toxicity--nonrodent: 90-d Chronic toxicity	Genetic toxicity	Toxicokinetics	--
40-49	Acute oral toxicity--nonrodent Subchronic oral toxicity--rodent: 14- or 28-d Acute dermal irritation--corrosivity Teratology--rodent, rabbit	Subchronic oral toxicity--rodent: 14- or 28-d Carcinogenicity--rodent	Subchronic oral toxicity--rodent: 90-d Chronic toxicity Acute oral toxicity--nonrodent	Acute eye irritation--corrosivity Acute dermal irritation--corrosivity Toxicokinetics Carcinogenicity--rodent
30-39	--	Acute dermal toxicity Acute dermal irritation--corrosivity Chronic toxicity Human sensitization	--	Subchronic oral toxicity--rodent: 90-d Teratology--rodent, rabbit

20-29	Skin sensitization--guinea pig Subchronic inhalation toxicity: 90-d Multigeneration reproduction--rodent Carcinogenicity--rodent Segment I: fertility and reproductive performance Subchronic dermal toxicity: 21- or 28-d Combined chronic toxicity-carcinogenicity--rodent Acute delayed neurotoxicity Skin painting--chronic	Acute inhalation toxicity Acute eye irritation--corrosivity Toxicokinetics Acute oral toxicity--nonrodent Skin sensitization--guinea pig Subchronic toxicity--nonrodent: 14- or 28-d Subchronic dermal toxicity: 21- or 28-d Subchronic oral toxicity--rodent: 90-d Teratology--rodent, rabbit	Genetic toxicity Segment I: fertility and reproductive performance Segment III: perinatal and postnatal performance Subchronic toxicity--nonrodent: 14- or 28-d Subchronic oral toxicity--nonrodent: 6- to 12-mo Neurobehavioral toxicity	Acute dermal toxicity Acute inhalation toxicity Subchronic inhalation toxicity: 90-d Chronic toxicity Segment I: fertility and reproductive performance Subchronic toxicity--nonrodent: 14- or 28-d Subchronic oral toxicity--nonrodent: 90-d Subchronic inhalation toxicity: 14- or 28-d
10-19	Subchronic inhalation toxicity: 14- or 28-d Segment III: perinatal and postnatal performance	Subchronic oral toxicity--nonrodent: 90-d Multigeneration reproduction--rodent Segment I: fertility and reproductive performance	Acute dermal irritation--corrosivity Acute eye irritation--corrosivity Multigeneration reproduction--rodent Carcinogenicity--rodent Combined chronic toxicity-carcinogenicity--rodent Human sensitization	Acute oral toxicity--nonrodent Skin sensitization--guinea pig Subchronic dermal toxicity: 21- or 28-d Neurobehavioral toxicity Multigeneration reproduction--rodent Combined chronic toxicity-carcinogenicity--rodent Skin painting--chronic Human sensitization
1-9	Subchronic toxicity--nonrodent: 14- or 28-d Neurobehavioral toxicity Human sensitization	Subchronic oral toxicity--nonrodent: 6- to 12-mo Subchronic dermal toxicity: 21- or 28-d Subchronic dermal toxicity: 90-d Combined chronic toxicity-carcinogenicity--rodent Skin painting--chronic Skin penetration	Acute dermal toxicity Acute inhalation toxicity Subchronic inhalation toxicity: 14- or 28-d Subchronic inhalation toxicity: 90-d Skin painting--chronic Implantation Skin penetration	Subchronic dermal toxicity: 90-d Segment III: perinatal and postnatal performance Implantation Skin penetration
No tests done	Subchronic oral toxicity--nonrodent: 6- to 12-mo Subchronic dermal toxicity: 90-d Subchronic eye toxicity Implantation Skin penetration	Subchronic inhalation toxicity: 14- or 28-d Subchronic inhalation toxicity: 90-d Neurobehavioral toxicity Subchronic eye toxicity Segment III: perinatal and postnatal performance Acute delayed neurotoxicity Implantation	Skin sensitization--guinea pig Subchronic dermal toxicity: 21- or 28-d Subchronic dermal toxicity: 90-d Subchronic eye toxicity Acute delayed neurotoxicity	Subchronic oral toxicity--nonrodent: 6- to 12-mo Subchronic eye toxicity Acute delayed neurotoxicity

TABLE 8 (continued)

| | Category and Subsample Size | | |
| | Chemicals in Commerce: Production Volume | | |
Percent Tested	At Least 1 Million Pounds/Year (10)	Less than 1 Million Pounds/Year (10)	Unknown/Inaccessible (20)
100	--	--	--
90-99	--	--	--
80-89	Acute parenteral toxicity	--	--
70-79	--	--	--
60-69	--	--	--
50-59	--	Acute oral toxicity--rodent Acute parenteral toxicity	Acute oral toxicity--rodent Acute dermal irritation--corrosivity
40-49	Acute oral toxicity--rodent Acute dermal irritation--corrosivity	Acute dermal irritation--corrosivity	Genetic toxicity Acute parenteral toxicity Acute eye irritation--corrosivity
30-39	Toxicokinetics	Acute dermal toxicity Genetic toxicity	--

20-29	Acute dermal toxicity
	Acute eye irritation--
	  corrosivity
	Genetic toxicity
		Acute inhalation toxicity
		Acute eye irritation--
		  corrosivity
		Subchronic oral toxicity--
		  rodent: 90-d
		Subchronic inhalation
		  toxicity: 90-d
			Acute dermal toxicity
			Acute inhalation toxicity
			Subchronic oral toxicity--
			  rodent: 14- or 28-d
			Toxicokinetics
			Skin sensitization--
			  guinea pig
			Chronic toxicity
			Human sensitization

10-19	Skin sensitization--
	  guinea pig
	Subchronic dermal
	  toxicity: 21- or 28-d
	Subchronic inhalation
	  toxicity: 14- or 28-d
	Carcinogenicity--rodent
	Skin painting--chronic
		Acute oral toxicity--
		  nonrodent
		Skin sensitization--
		  guinea pig
		Subchronic oral toxicity--
		  rodent: 14- or 28-d
		Subchronic dermal
		  toxicity: 21- or 28-d
		Subchronic inhalation
		  toxicity: 14- or 28-d
		Neurobehavioral toxicity
		Teratology--rodent,
		  rabbit
		Toxicokinetics
		Carcinogenicity--rodent
		Chronic toxicity
		Segment III: perinatal and
		  postnatal performance
		Skin painting--chronic
		Human sensitization
			Subchronic oral toxicity--
			  rodent: 90-d
			Carcinogenicity--rodent
			Subchronic dermal
			  toxicity: 21- or 28-d
			Subchronic inhalation
			  toxicity: 14- or 28-d
			Teratology--rodent,
			  rabbit
			Multigeneration repro-
			  duction--rodent
			Skin painting--chronic

1-9	--	--	Subchronic toxicity--
			  nonrodent: 14- or 28-d
			Subchronic oral toxicity--
			  6- to 12-mo
			Segment I: fertility and
			  reproductive performance
			Implantation
			Skin penetration

TABLE 8 (continued)

| | Category and Subsample Size | | |
| | Chemicals in Commerce Production Volume | | |
Percent Tested	At Least 1 Million Pounds/Year (10)	Less than 1 Million Pounds/Year (10)	Unknown/Inaccessible (20)
No tests done	Acute oral toxicity--nonrodent Acute inhalation toxicity Subchronic oral toxicity--rodent: 14- or 28-d Subchronic toxicity--nonrodent: 14- or 28-d Subchronic oral toxicity--rodent: 90-d Subchronic oral toxicity--nonrodent: 90-d Subchronic oral toxicity--nonrodent: 6- to 12-mo Subchronic dermal toxicity: 90-d Subchronic inhalation toxicity: 90-d Neurobehavioral toxicity Teratology--rodent, rabbit Multigeneration reproduction--rodent Chronic toxicity Combined chronic toxicity-carcinogenicity--rodent Subchronic eye toxicity Segment I: fertility and reproductive performance Segment III: perinatal and postnatal performance Acute delayed neurotoxicity Implantation Human sensitization Skin penetration	Subchronic toxicity--nonrodent: 14- or 28-d Subchronic oral toxicity--nonrodent: 90-d Subchronic oral toxicity--nonrodent: 6- to 12-mo Subchronic dermal toxicity: 90-d Multigeneration reproduction--rodent Combined chronic toxicity-carcinogenicity--rodent Subchronic eye toxicity Segment I: fertility and reproductive performance Acute delayed neurotoxicity Implantation Skin penetration	Acute oral toxicity--nonrodent Subchronic oral toxicity--nonrodent: 90-d Subchronic dermal toxicity: 90-d Subchronic inhalation toxicity: 90-d Neurobehavioral toxicity Combined chronic toxicity-carcinogenicity--rodent Subchronic eye toxicity Segment III: perinatal and postnatal performance Acute delayed neurotoxicity

QUALITY OF TESTING

The Committee on Toxicity Data Elements often found that studies not conforming to present testing protocol guidelines nevertheless yielded acceptable results. The committee established six classifications to describe the quality of test protocols:

● G, the highest rating, for a test protocol performed according to current reference protocol guidelines adopted by the committee.

● A, an adequate rating for a protocol that did not strictly follow current reference protocol guidelines, but was nevertheless deemed adequate for conducting a health-hazard assessment.

● IN, an inadequate rating for a protocol that was neither conducted according to guidelines nor adequate for conducting a health-hazard assessment, but was judged not to need repetition. This was used largely because other adequately conducted tests provided the necessary or closely related information or because available information was deemed to be sufficient for conducting a health-hazard assessment.

● IR, an inadequate rating for a protocol that was neither adequate nor conducted according to guidelines and that was judged to need repetition.

● C, a rating denoting indeterminable quality of a test that could not be judged. This rating was most frequently given to abstracts, review articles, and other reports that did not describe protocols fully.

● X, a notation for a test that, according to available information, was not done.

Table 9 shows the quality ratings of the 33 test types as found to have been done on the 100 substances in the subsample. The proportion of substances on which a given test type was not run (X) ranged from 37% (acute oral testing in rodents) to 100% (subchronic eye toxicity testing).

Overall, without regard for either test type or intended-use category, the Committee on Toxicity Data Elements tabulated the quality ratings from evaluations of a total of 664 tests. A tally of these quality ratings for each subsample category and for the subsample as a whole is presented in Table 10. Only about 8% of the tests met the standards of the reference protocol guidelines, and about another 19% were judged to be adequate. When more than one study of the same test type had been done, these percentages are based on the quality rating of the best study.

TABLE 9   Quality of Testing on 100 Substances in Subsample, by Test Type

	No. Substances Having Test Protocols with Indicated Quality					
Test Type	Reference Protocol Guidelines (G)	Adequate, But Not According to Reference Protocol Guidelines (A)	Inadequate, But Does Not Need Repetition (IN)	Inadequate And Needs Repetition (IR)	Adequacy Cannot Be Judged (C)	Test Not Done (X)
1. Acute oral toxicity--rodent	9	15	28	2	9	37
2. Acute oral toxicity--nonrodent	1	2	7	4	5	81
3. Acute dermal toxicity	1	7	18	2	2	70
4. Acute parenteral toxicity	2	7	21	16	15	39
5. Acute inhalation toxicity	2	6	9	3	4	76
6. Acute dermal irritation--corrosivity	5	14	11	2	5	63
7. Acute eye irritation--corrosivity	11	5	9	2	6	67
8. Skin sensitization--guinea pig	4	2	1	3	5	85
9. Subchronic oral toxicity--rodent: 14- or 28-d study	0	3	16	12	4	65
10. Subchronic toxicity--nonrodent: 14- or 28-d study	0	1	3	3	4	89
11. Subchronic oral toxicity--rodent: 90-d study	0	11	7	3	9	70
12. Subchronic oral toxicity--nonrodent: 90-d study	0	5	5	5	3	82
13. Subchronic oral toxicity--nonrodent: 6- to 12-mo study	1	1	1	2	0	95
14. Subchronic dermal toxicity: 21- or 28-d study	0	1	3	2	4	90
15. Subchronic dermal toxicity: 90-d study	0	0	1	0	1	98
16. Subchronic inhalation toxicity: 14- or 28-d study	0	2	2	3	3	90
17. Subchronic inhalation toxicity 90-d study	1	1	1	1	7	89
18. Subchronic neurotoxicity: 90-d study	0	1	2	2	2	93

92

	C1	C2	C3	C4	C5	C6
19. Teratology study--rodent, rabbit	2	2	8	12	2	74
20. Multigeneration reproduction study--rodent	0	2	2	5	3	88
21. Toxicokinetics	2	11	5	14	5	63
22. Carcinogenicity--rodent	0	6	6	8	3	77
23. Chronic toxicity	3	4	4	12	6	71
24. Combined chronic toxicity-carcinogenicity--rodent	1	1	2	0	4	92
25. Genetic toxicity	0	2	6	41	0	51
26. Subchronic eye toxicity	0	0	0	0	0	100
27. Segment I: fertility and reproductive performance	0	1	6	3	5	85
28. Segment III: perinatal and postnatal performance	1	1	0	2	4	92
29. Acute delayed neurotoxicity	0	0	2	0	1	97
30. Skin painting--chronic	0	2	4	3	2	89
31. Implantation studies	0	0	2	1	0	97
32. Human sensitization studies	5	5	1	2	2	85
33. Skin penetration studies	1	2	0	1	0	96

TABLE 10 Distribution of 664 Quality Ratings of Tests Done on 100 Substances in Subsample, by Subsample Category

Evaluation Code	Rating Description	Subsample Category and Number of Tests with Indicated Rating					
		Pesticides and Inert Ingredients of Pesticide Formulations	Cosmetic Ingredients	Drugs and Excipients in Drug Formulations	Food Additives	Chemicals in Commerce	Proportion of Tests in Whole Subsample
G	Meets current guidelines	18	6	8	5	15	8
A	Adequate, but does not meet guidelines	36	20	20	13	34	19
IN	Not adequate, but retesting not needed	53	37	29	39	35	29
IR	Inadequate and retesting needed	28	30	17	39	57	26
C	Adequacy cannot be judged	29	5	32	30	29	19
TOTAL		164	98	106	126	170	100

These data are presented for the purpose of evaluating the testing protocols; the focus is on the tests, not on the substances tested. Because the counts in Table 9 and Table 10 are not weighted to account for the different sampling rates over the seven categories, inferences as to the completeness of information on the whole select universe are not justified. Interpretation of Table 9 and Table 10 is based on the assumption that, after a decision to apply a specific test type to a specific substance, the quality of the test is independent of the reasons for doing it. The two committees also assumed that for their purpose inaccessible or unreported information did not exist. Furthermore, some studies might have been conducted according to standards higher than those indicated by a given rating, but that would not have been apparent in the reporting of those studies.

The quality of design, execution, and reporting was not uniform among the various types of tests. For some types (e.g., acute oral administration in rodents, acute dermal application, acute eye irritation and corrosivity, guinea pig skin sensitization, and subchronic dermal application for 90 d), the majority of tests were deemed not to require repetition. Tests that were more complex (e.g., teratology in rodents and rabbits, multigeneration reproductive effects in rodents, and genetic toxicity) were frequently absent and therefore were often not a part of the determination of ability to conduct a health-hazard assessment. The relationships between the year in which rated studies were conducted and their quality ratings were not determined.

The number of test types suitable for detecting substances harmful to in utero development is limited, and these few tests were rarely found for the substances examined. In general, the data most likely to exist on a substance are from tests for acute oral toxicity in adults. Tests designed to detect more chronic effects are also represented, including those involving cancer and gene mutations. Teratology studies in rodents and rabbits and multigeneration reproduction studies in rodents had been conducted infrequently, even when deemed necessary by the committee. Overall, the information available regarding toxicity to the conceptus is very limited.

Committee evaluation of an individual study protocol always included reasons for the particular rating given to the study. For the most part, these reasons were statements of a given study's specific adequacies or inadequacies with respect to its reported protocol. They were collectively tabulated for analysis to assess which deficiencies were most prevalent and which ones could be overridden in assessing the overall value of a study. The Committee on Toxicity Data Elements did not identify every reason for every test evaluation, because the development of such tabulations was an evolutionary process in the documentation of the activity. In most cases of protocol evaluation, no more than three reasons for a given rating were necessary to support the quality rating. The committee did not consider its tabulations of reasons to be appropriate for detailed quantitative analysis; nevertheless, they provide an important overview of factors that affect the quality of tests.

As the evaluations proceeded, four kinds of reasons for assigning particular ratings evolved:

- Deviations from reference protocol guidelines that, because they were minor, did not prevent a study from being used in a health-hazard assessment.

- Reasons for judging test protocols to be adequate despite their deviations from reference protocol guidelines.

- Reasons for judging test protocols to be inadequate.

- Reasons for not needing to repeat a study that had been judged to have an inadequate protocol.

The committee used this information to assess which deficiencies were most prevalent, which ones could be overridden, and which ones caused studies to be of little or no value for conducting a health-hazard assessment.

Some of the more common minor deviations from reference protocol guidelines that nevertheless resulted in a rating of adequate included the use of too few animals per dosage group, the use of only one sex, the use of too few or improper dosages, and the absence of various kinds of observations (e.g., clinical chemistry or histopathology).

Reasons for judging such tests to have been conducted adequately even though they were not performed according to reference protocol guidelines included the expectation that more information would not alter the conclusions, the sufficiency of data to evaluate toxicity or calculate an acceptable LD_{50}, and the use of dosages high enough to give positive results or exceed the limit test now prescribed in the guidelines.

Tests that were rated as inadequately conducted were often missing required observations (e.g., test-animal description, diet analysis, chemical analysis, clinical chemistry, and histopathology), had too few dosages, or lacked sufficiently detailed end points, such as data tabulation or statistical analysis of data. Occasionally, the committee recommended that these studies not be repeated, either because toxicity was sufficiently well established or because more information would have only limited value.

The first three of the six quality ratings are collectively a measure of testing that need not be repeated because its quality is adequate for conducting a health-hazard assessment, the breadth of existing information is sufficient for conducting a complete health-hazard assessment, or repetition of inadequate tests would not add much to existing information. When these first three classifications are combined, it is clear that much of the past testing has been productive. For both G and A ratings, tests were done according to reference protocol guidelines or in an otherwise acceptable fashion. Tests with G, A, or IN ratings need not be repeated. For four acute tests indicated in Table 9 (acute oral administration in rodents, acute dermal application, acute dermal irritation and corrosivity, and acute eye irritation and corrosivity), the sums of G, A, and IN ratings as a measure of testing not needed were about 83, 87, 81, and 76%, respectively. As indicated in Table 11, the percentages of four principal chronic tests (multigeneration reproduction in rodents, carcinogenicity in rodents, chronic toxicity, and combined carcinogenicity and chronic toxicity in rodents) that did not need to be repeated were consistently and substantially lower. These findings should be viewed in perspective: (1) the comparison is of simple acute tests with more complex chronic tests, (2) as indicated in Table 11, the numbers of chronic tests acutally performed were far smaller than the numbers of acute tests, and (3) the percentages are derived from small numbers of evaluated tests, particularly in the case of the chronic studies.

TESTING NEEDS

The data summarized in Table 9 are presented in detail in Tables 12 through 18 for each test performed in each category of substances examined. In these tables, one may find the percentage of the times each test was run (by substracting the value under X from 100%) for each category, as well as the percentage of the times, when the test was run, that it was given each of the five quality ratings. Upper and lower 90% confidence limits are presented with these data.

Among the 18 tests deemed by the Committee on Toxicity Data Elements to be required for pesticides and inert ingredients of pesticide formulations (see Appendix F), the proportion of substances with tests performed inadequately and needing repetition (IR + C) or not performed at all (X) ranged from 0% to 73%. For cosmetic ingredients, the corresponding range was 67-100%; for drugs and excipients in drug formulations, 25-60%; for food additives, 33-80%; and for chemicals in commerce, 45-100%. This indicates that, for each category of intended use, substantial testing and retesting remain to be performed if information gaps are to be filled. The major portion of testing need results from failure to do required tests, rather than from conducting tests in a manner inappropriate for the purpose of health-hazard assessment. If the unknown amount of information that was not available to the committee had been available, the "untested" category would be somewhat smaller than reported here.

To some extent, the apparent inadequacy of genetic-toxicity data reflects the facts that some of the test strains now required were not available until after 1976 and that the testing required involves a tier system of evaluation developed only recently. The apparent lag in testing of food additives may reflect the fact that many food additives were not generally deemed to require extensive testing until they were removed from the GRAS list in recent years.

TABLE 11   Comparison of Aggregate Quality Ratings (All G, A, and IN) for
Selected Acute Test Types and Selected Chronic Test Types

Test Type[a]	No. Substances Having at Least One Test Evaluated	Proportion with Ratings of G, A, or IN,[b] %
Acute:		
Acute oral in rodents	63	74-83-90
Acute dermal	30	76-87-95
Acute dermal irritation and corrosivity	37	70-81-91
Acute eye irritation and corrosivity	33	63-76-88
Chronic:		
Multigeneration reproduction in rodents	12	12-33-61
Carcinogenicity in rodents	23	34-52-70
Chronic toxicity	29	23-38-50
Combined carcinogenicity and chronic toxicity in rodents	8	20-50-80

[a] These tests were selected for illustrative purposes.

[b] Values are aggregate proportions of tests having quality ratings of G, A, or IN, not number of substances having these ratings. The 90% confidence limits shown are rough approximations based on the Bernoulli distribution. These proportions are based on the study of highest quality, even when other studies of the same test type were done.

Table 19 contains relative comparisons of test types that, although considered essential, were not conducted, conducted inadequately, or reported in a manner inadequate for use in a health-hazard assessment. These tests need to be conducted or repeated. In general, chronic studies, inhalation studies, and the more complex studies with specific end points (e.g., neurotoxicity, genetic toxicity, and effects on the conceptus) are most frequently needed. The gaps in toxicity information differ from one category of substances to another. To some extent, these differences may reflect the spectrum of individual tests that the Committee on Toxicity Data Elements prescribed as necessary to meet its criteria for adequacy of information in each category.

The relative comparisons in Table 19 are based on estimates for the select universe based on the sample of 675 substances and the subsample of 100 substances. Because each subsample category is small, the order is subject to substantial random errors and might change on repeating the determination of toxicity-testing needs with a new random sample.

The three greatest testing needs for pesticides and inert ingredients of pesticide formulations were in teratology, subchronic neurotoxicity, and genetic toxicity. For cosmetic ingredients, the greatest testing needs were for subchronic eye toxicity and neurobehavioral toxicity. A large variety of test types (14 test types indicated in Table 19) are needed most for drugs and excipients in drug formulations. For food additives, seven test types are needed most. For the three production categories of chemicals in commerce, eight (at least 1 million pounds per year), seven (less than 1 million pounds per year), and four (unknown or inaccessible production) test types were determined to be most needed.

HEALTH-HAZARD ASSESSMENT

The determination of toxicity-testing needs was based not only on evaluation of testing quantity and quality by the expert scientific committees. After careful review, it also involved their collective decision on the ability to conduct a health-hazard assessment with available information and on the need for additional studies or for retesting in the case of inadequate studies.

If requirements for testing were rigidly adhered to, it would be necessary to say that, in light of intended use, not one of the 100 substances in the subsample met the requirements according to the strictest standards set by the protocols chosen by the Committee on Toxicity Data Elements. However, less rigid standards permitted the committees to determine whether enough testing had been done and was of sufficiently good quality for at least a partial health-hazard assessment.

The ability to conduct complete, partial, or no health-hazard assessments of substances in each of the seven categories of the select universe is summarized in Table 20 and Figure 4. For purposes of interpretation, a complete health-hazard assessment is defined as one that provides a full estimate of hazard associated with the safe use of

TABLE 12 Quality of Testing of 3,350 Pesticides and Inert Ingredients of Pesticide Formulations in Select Universe

Test Type	Test Required[a]	Quality Ratings[b]					
		X	G	A	IN	IR	C
		Estimated Percent of Select Universe with Indicated Quality					
1. Acute oral toxicity--rodent	*	0-0-18	2-13-36	10-27-51	36-60-81	0-0-18	0-0-18
2. Acute oral toxicity--nonrodent		6-20-44	0-8-34	0-8-34	32-58-82	0-8-34	3-17-44
3. Acute dermal toxicity	*	6-20-44	0-0-22	3-17-44	56-83-97	0-0-22	0-0-22
4. Acute parenteral toxicity		10-27-51	0-9-36	0-9-36	44-73-92	0-0-24	0-9-36
5. Acute inhalation toxicity	*	14-33-58	1-10-39	9-30-61	15-40-70	1-10-39	1-10-39
6. Acute dermal irritation-corrosivity		19-40-64	1-11-43	17-44-75	17-44-75	0-0-28	0-0-28
7. Acute eye irritation-corrosivity	*	14-33-58	22-50-78	0-0-26	15-40-70	1-10-39	0-0-26
8. Skin sensitization--guinea pig	*	30-53-76	5-29-66	1-14-52	23-57-87	0-0-35	0-0-35
9. Subchronic oral toxicity--rodent: 14- or 28-d study		6-20-44	0-0-22	0-8-34	56-83-97	0-0-22	0-8-34
10. Subchronic toxicity--nonrodent: 14- or 28-d study		14-33-58	0-0-26	0-0-26	74-100-100	0-0-26	0-0-26
11. Subchronic oral toxicity--rodent: 90-d study	*	10-27-51	0-0-24	20-45-73	14-36-65	0-0-24	3-18-47
12. Subchronic oral toxicity--nonrodent: 90-d study	*	10-27-51	0-0-24	8-27-56	35-64-86	0-9-36	0-0-24
13. Subchronic oral toxicity--nonrodent: 6- to 12-mo study		49-73-90	0-0-53	0-0-53	47-100-100	0-0-53	0-0-53
14. Subchronic dermal toxicity: 21- or 28-d study	*	30-53-76	0-0-35	0-0-35	48-86-99	1-14-52	0-0-35
15. Subchronic dermal toxicity: 90-d study		49-73-90	0-0-53	0-0-53	47-100-100	0-0-53	0-0-53
16. Subchronic inhalation toxicity: 14- or 28-d study	*	36-60-81	0-0-39	1-17-58	27-67-94	0-0-39	1-17-58
17. Subchronic inhalation toxicity 90-d study		24-47-70	0-0-31	0-0-31	19-50-81	0-0-31	19-50-81
18. Neurobehavioral toxicity	*	42-67-86	0-0-45	0-0-45	34-80-99	0-0-45	1-20-66
19. Teratology study--rodent, rabbit	*	24-47-70	0-0-31	1-13-47	11-38-71	19-50-81	0-0-31
20. Multigeneration reproduction study--rodent	*	24-47-70	0-0-31	1-13-47	19-50-81	5-25-60	1-13-47
21. Toxicokinetics	*	6-20-44	3-17-44	7-25-53	3-17-44	3-17-44	7-25-53
22. Carcinogenicity--rodent	*	30-53-76	0-0-35	5-29-66	13-43-77	5-29-66	0-0-35

23. Chronic toxicity	*	10-27-51	3-18-47	0-9-36	14-36-65	3-18-47	3-18-47
24. Combined chronic toxicity-carcinogenicity-rodent	*	36-60-81	0-0-39	0-0-39	42-83-99	0-0-39	1-17-58
25. Genetic toxicity	*	6-20-44	0-0-22	0-8-34	7-25-53	39-67-88	0-0-22
26. Subchronic eye toxicity		49-73-90	0-0-53	0-0-53	47-100-100	0-0-53	0-0-53
27. Segment I: fertility and reproductive performance		36-60-81	0-0-39	0-0-39	27-67-94	1-17-58	1-17-58
28. Segment III: perinatal and postnatal performance	*	42-67-86	1-20-66	0-0-45	19-60-92	0-0-45	1-20-66
29. Acute delayed neurotoxicity		36-60-81	0-0-39	0-0-39	42-83-99	0-0-39	1-17-58
30. Skin painting--chronic		30-53-76	0-0-35	0-0-35	34-71-95	1-14-52	1-14-52
31. Implantation studies		49-73-90	0-0-53	0-0-53	47-100-100	0-0-53	0-0-53
32. Human sensitization studies		42-67-86	0-0-45	1-20-66	34-80-99	0-0-45	0-0-45
33. Skin penetration studies		49-73-90	0-0-53	0-0-53	47-100-100	0-0-53	0-0-53

a According to the standards adopted by the Committee on Toxicity Data Elements.

b X = test not done;

G = test conducted according to reference protocol guidelines;

A = test conducted adequately, but not according to reference protocol guidelines;

IN = test conducted inadequately, but retesting is not needed;

IR = test conducted inadequately, and retesting is needed;

C = test adequacy cannot be evaluated.

Percentages are presented with upper and lower 90% confidence intervals.

Value for each test type under X is a proportion of all substances in this category of the subsample.

Values in all other columns are based only on portion of category for which test was conducted.

TABLE 13 Quality of Testing of 3,410 Cosmetic Ingredients in Select Universe

Test Type	Test Required[a]	Quality Ratings[b]					
		Estimated Percent of Select Universe with Indicated Quality					
		X	G	A	IN	IR	C
1. Acute oral toxicity--rodent		14-33-58	4-20-51	1-10-39	30-60-85	1-10-39	0-0-26
2. Acute oral toxicity--nonrodent		56-80-94	0-0-63	0-0-63	14-67-98	2-33-86	0-0-63
3. Acute dermal toxicity	*	42-67-86	0-0-45	1-20-66	34-80-99	0-0-45	0-0-45
4. Acute parenteral toxicity		19-40-64	0-0-28	10-33-66	4-22-55	10-33-66	1-11-43
5. Acute inhalation toxicity		49-73-90	0-0-53	1-25-75	10-50-90	1-25-75	0-0-53
6. Acute dermal irritation-corrosivity	*	42-67-86	1-20-66	1-20-66	19-60-92	0-0-45	0-0-45
7. Acute eye irritation-corrosivity	*	49-73-90	1-25-75	10-50-90	1-25-75	0-0-53	0-0-53
8. Skin sensitization--guinea pig	*	56-80-94	2-33-86	2-33-86	0-0-63	0-0-63	2-33-86
9. Subchronic oral toxicity--rodent: 14- or 28-d study		19-40-64	0-0-28	1-11-43	34-67-90	4-22-55	0-0-28
10. Subchronic oral toxicity--nonrodent: 14- or 28-d study		56-80-94	0-0-63	0-0-63	2-33-86	2-33-86	2-33-86
11. Subchronic oral toxicity--rodent: 90-d study		49-73-90	0-0-53	0-0-53	25-75-99	0-0-53	1-25-75
12. Subchronic oral toxicity--nonrodent: 90-d study		64-87-98	0-0-78	0-0-78	0-0-78	22-100-100	0-0-78
13. Subchronic oral toxicity--nonrodent: 6- to 12-mo study		72-93-100	0-0-95	0-0-95	0-0-95	5-100-100	0-0-95
14. Subchronic dermal toxicity: 21- or 28-d study	*	64-87-98	0-0-78	0-0-78	22-100-100	0-0-78	0-0-78
15. Subchronic dermal toxicity: 90-d study	*	72-93-100	0-0-95	0-0-95	5-100-100	0-0-95	0-0-95
16. Subchronic inhalation toxicity: 14- or 28-d study		82-100-100					
17. Subchronic inhalation toxicity: 90-d study		82-100-100					
18. Neurobehavioral toxicity	*	82-100-100					
19. Teratology study--rodent, rabbit		56-80-94	0-0-63	0-0-63	2-33-86	14-67-98	0-0-63
20. Multigeneration reproduction study--rodent		64-87-98	0-0-78	3-50-97	3-50-97	3-50-97	0-0-78
21. Toxicokinetics		49-73-90	0-0-53	1-25-75	1-25-75	10-50-90	0-0-53
22. Carcinogenicity--rodent	*	36-60-81	0-0-39	6-33-73	6-33-73	1-17-58	1-17-58

23. Chronic toxicity	42-67-86	0-0-45	1-20-66	1-20-66	19-60-92	0-0-45
24. Combined chronic toxicity-carcinogenicity-rodent	72-93-100	5-100-100	0-0-95	0-0-95	0-0-95	0-0-95
* 25. Genetic toxicity	24-47-70	0-0-31	1-13-47	1-13-47	40-75-95	0-0-31
* 26. Subchronic eye toxicity	82-100-100					
27. Segment I: fertility and reproductive performance	64-87-98	0-0-78	0-0-78	3-50-97	3-50-97	0-0-78
28. Segment III: perinatal and postnatal performance	82-100-100					
29. Acute delayed neurotoxicity	82-100-100					
* 30. Skin painting--chronic	72-93-100	0-0-95	0-0-95	5-100-100	0-0-95	0-0-95
31. Implantation studies	82-100-100					
* 32. Human sensitization studies	42-67-86	0-0-45	19-60-92	0-0-45	8-40-81	0-0-45
* 33. Skin penetration studies	72-93-100	0-0-95	5-100-100	0-0-95	0-0-95	0-0-95

a According to the standards adopted by the Committee on Toxicity Data Elements.

b X = test not done;

   G = test conducted according to reference protocol guidelines;

   A = test conducted adequately, but not according to reference protocol guidelines;

   IN = test conducted inadequately, but retesting is not needed;

   IR = test conducted inadequately, and retesting is needed;

   C = test adequacy cannot be evaluated.

Percentages are presented with upper and lower 90% confidence intervals.

Value for each test type under X is a proportion of all substances in this intended-use category of the subsample without regard for whether the test was conducted. Each value under X represents the proportion of all substances in the category not having test done. Values in all other columns are based only on portion of category for which test was conducted.

TABLE 14 Quality of Testing of 1,815 Drugs and Excipients in Drug Formulations in Select Universe

Test Type	Test Required[a]	Quality Ratings[b]					
		Estimated Percent of Select Universe with Indicated Quality					
		X	G	A	IN	IR	C
1. Acute oral toxicity--rodent	*	6-20-44	0-8-34	3-17-44	25-50-75	0-0-22	7-25-53
2. Acute oral toxicity--nonrodent	*	10-27-51	0-0-24	0-9-36	35-64-86	0-9-36	3-18-47
3. Acute dermal toxicity	*	30-53-76	0-0-35	0-0-35	65-100-100	0-0-35	0-0-35
4. Acute parenteral toxicity	*	2-13-36	0-8-32	3-15-41	22-46-71	0-8-32	7-23-49
5. Acute inhalation toxicity	*	30-53-76	0-0-35	0-0-35	48-86-99	0-0-35	1-14-52
6. Acute dermal irritation-corrosivity	*	24-47-70	0-0-31	5-25-60	40-75-95	0-0-31	0-0-31
7. Acute eye irritation-corrosivity	*	24-47-70	1-13-47	1-13-47	40-75-95	0-0-31	0-0-31
8. Skin sensitization--guinea pig	*	36-60-81	0-0-39	0-0-39	61-100-100	0-0-39	0-0-39
9. Subchronic oral toxicity--rodent: 14- or 28-d study	*	14-33-58	0-0-26	0-0-26	61-90-00	0-0-26	1-10-39
10. Subchronic toxicity--nonrodent: 14- or 28-d study	*	24-47-70	0-0-31	0-0-31	53-88-99	0-0-31	1-13-47
11. Subchronic oral toxicity--rodent: 90-d study	*	14-33-58	0-0-26	4-20-51	30-60-85	0-0-26	4-20-51
12. Subchronic oral toxicity--nonrodent: 90-d study	*	14-33-58	0-0-26	4-20-51	22-50-78	4-20-51	1-10-39
13. Subchronic oral toxicity--nonrodent: 6- to 12-mo study	*	24-47-70	1-13-47	1-13-47	29-63-47	1-13-47	0-0-31
14. Subchronic dermal toxicity: 21- or 28-d study	*	36-60-81	0-0-39	0-0-39	61-100-100	0-0-39	0-0-39
15. Subchronic dermal toxicity: 90-d study	*	36-60-81	0-0-39	0-0-39	61-100-100	0-0-39	0-0-39
16. Subchronic inhalation toxicity: 28- or 14-d study	*	30-53-76	0-0-35	0-0-35	48-86-99	0-0-35	1-14-52
17. Subchronic inhalation toxicity 90-d study	*	30-53-76	0-0-35	0-0-35	48-86-99	0-0-35	1-14-52
18. Neurobehavioral toxicity	*	24-47-70	0-0-31	1-13-47	40-75-95	0-0-31	1-13-47
19. Teratology study--rodent, rabbit	*	10-27-51	0-0-24	0-0-24	35-64-86	3-18-47	3-18-47
20. Multigeneration reproduction study--rodent	*	24-47-70	0-0-31	0-0-31	40-75-95	1-13-47	1-13-47
21. Toxicokinetics	*	10-27-51	0-0-24	8-27-56	27-55-80	3-18-47	0-0-24

	X	G	A	IN	IR	C
22. Carcinogenicity--rodent	*	36-60-81	0-0-39	61-100-100	0-0-39	0-0-39
23. Chronic toxicity	*	10-27-51	0-9-36	20-45-73	0-9-36	8-27-56
24. Combined chronic toxicity--carcinogenicity--rodent	*	30-53-76	1-14-52	48-86-99	0-0-35	0-0-35
25. Genetic toxicity	*	14-33-58	0-0-26	30-60-85	15-40-70	0-0-26
26. Subchronic eye toxicity	*	36-60-81	0-0-39	61-100-100	0-0-39	0-0-39
27. Segment I: fertility and reproductive performance	*	24-47-70	0-0-31	40-75-95	0-0-31	5-25-60
28. Segment III: perinatal and postnatal performance	*	14-33-58	1-10-39	30-60-85	1-10-39	4-20-51
29. Acute delayed neurotoxicity	*	36-60-81	0-0-39	61-100-100	0-0-39	0-0-39
30. Skin painting--chronic	*	36-60-81	0-0-39	61-100-100	0-0-39	0-0-39
31. Implantation studies	*	36-60-81	0-0-39	61-100-100	0-0-39	0-0-39
32. Human sensitization studies	*	24-47-70	0-0-31	40-75-95	0-0-31	0-0-31
33. Skin penetration studies	*	30-53-76	0-0-35	48-86-99	0-0-35	0-0-35

a According to the standards adopted by the Committee on Toxicity Data Elements.

b X = test not done;

G = test conducted according to reference protocol guidelines;

A = test conducted adequately, but not according to reference protocol guidelines;

IN = test conducted inadequately, but retesting is not needed;

IR = test conducted inadequately, and retesting is needed;

C = test adequacy cannot be evaluated.

Percentages are presented with upper and lower 90% confidence intervals. Value for each test type under X is a proportion of all substances in this category of the subsample. Values in all other columns are based only on portion of category for which test was conducted.

TABLE 15 Quality of Testing of 8,627 Food Additives in Select Universe

Test Type	Test Required[a]	Quality Ratings[b]					
		X	G	A	IN	IR	C
		Estimated Percent of Select Universe with Indicated Quality					
1. Acute oral toxicity--rodent	*	14-33-58	4-20-51	4-20-51	22-50-78	0-0-26	1-10-39
2. Acute oral toxicity--nonrodent	*	49-73-90	0-0-53	0-0-53	25-75-99	1-25-75	0-0-53
3. Acute dermal toxicity		30-53-76	1-14-52	5-29-66	13-43-77	0-0-35	1-14-52
4. Acute parenteral toxicity		10-27-51	0-0-24	0-0-24	35-64-86	0-9-36	8-27-56
5. Acute inhalation toxicity		36-60-81	0-0-39	1-17-58	42-83-99	0-0-39	0-0-39
6. Acute dermal irritation-corrosivity		24-47-70	0-0-31	5-25-60	29-63-89	0-0-31	1-13-47
7. Acute eye irritation-corrosivity		19-40-64	1-11-43	1-11-43	34-67-90	0-0-28	1-11-43
8. Skin sensitization--guinea pig		49-73-90	0-0-53	0-0-53	25-75-99	1-25-75	0-0-53
9. Subchronic oral toxicity--rodent: 14- or 28-d study	*	10-27-51	0-0-24	0-0-24	20-45-73	27-55-80	0-0-24
10. Subchronic toxicity--nonrodent: 14- or 28-d study		49-73-90	0-0-53	0-0-53	25-75-99	1-25-75	0-0-53
11. Subchronic oral toxicity--rodent: 90-d study	*	36-60-81	0-0-39	1-17-58	15-50-85	6-33-73	0-0-39
12. Subchronic oral toxicity--nonrodent:90-d study	*	49-73-90	0-0-53	0-0-53	25-75-99	0-0-53	1-25-75
13. Subchronic oral toxicity--nonrodent:6- to 12-mo study		56-80-94	0-0-63	0-0-63	37-100-100	0-0-63	0-0-63
14. Subchronic dermal toxicity: 21- or 28-d study	*	42-67-86	0-0-45	0-0-45	19-60-92	0-0-45	8-40-81
15. Subchronic dermal toxicity: 90-d study		49-73-90	0-0-53	0-0-53	25-75-99	0-0-53	1-25-75
16. Subchronic inhalation toxicity: 14- or 28-d study		42-67-86	0-0-45	0-0-45	34-80-99	0-0-45	1-20-66
17. Subchronic inhalation toxicity 90-d study		36-60-81	0-0-39	1-17-58	15-50-85	1-17-58	1-17-58
18. Neurobehavioral toxicity	*	49-73-90	0-0-53	0-0-53	25-75-99	1-25-75	0-0-53
19. Teratology study--rodent, rabbit	*	30-53-76	0-0-35	1-14-52	13-43-77	13-43-77	0-0-35
20. Multigeneration reproduction study--rodent	*	49-73-90	0-0-53	0-0-53	25-75-99	0-0-53	1-25-75
21. Toxicokinetics	*	30-53-76	0-0-35	1-14-52	23-57-87	5-29-66	0-0-35
22. Carcinogenicity--rodent	*	19-40-64	0-0-28	0-0-28	17-44-75	17-44-75	1-11-43

23.	Chronic toxicity		36-60-81	0-0-39	0-0-39	15-50-85	15-50-85	0-0-39
24.	Combined chronic toxicity-- carcinogenicity--rodent	*	42-67-86	0-0-45	0-0-45	19-60-92	0-0-45	8-40-81
25.	Genetic toxicity	*	6-20-44	0-0-22	0-0-22	7-25-53	47-75-93	0-0-22
26.	Subchronic eye toxicity							
27.	Segment I: fertility and reproductive performance		42-67-86	0-0-45	0-0-45	19-60-92	1-20-66	1-20-66
28.	Segment III: perinatal and postnatal performance		49-73-90	0-0-53	0-0-53	25-75-99	1-25-75	0-0-53
29.	Acute delayed neurotoxicity		56-80-94	0-0-63	0-0-63	37-100-100	0-0-63	0-0-63
30.	Skin painting--chronic		42-67-86	0-0-45	1-20-66	19-60-93	0-0-45	1-20-66
31.	Implantation studies		56-80-94	0-0-63	0-0-63	37-100-100	0-0-63	0-0-63
32.	Human sensitization studies		42-67-86	1-20-66	0-0-45	34-80-99	0-0-45	0-0-45
33.	Skin penetration studies		49-73-90	0-0-53	0-0-53	25-75-99	1-25-75	0-0-53

a According to the standards adopted by the Committee on Toxicity Data Elements.

b X = test not done;
   G = test conducted according to reference protocol guidelines;
   A = test conducted adequately, but not according to reference protocol guidelines;
   IN = test conducted inadequately, but retesting is not needed;
   IR = test conducted inadequately, and retesting is needed;
   C = test adequacy cannot be evaluated.

Percentages are presented with upper and lower 90% confidence intervals.
Value for each test type under X is a proportion of all substances in this category of the subsample. Values in all other columns are based only on portion of category for which test was conducted.

TABLE 16 Quality of Testing of 12,860 Chemicals in Commerce in Select Universe with Production Levels of at Least 1 Million lb/yr

Test Type	Test Required[a]	Quality Ratings[b]					
		Estimated Percent of Select Universe with Indicated Quality					
		X	G	A	IN	IR	C
1. Acute oral toxicity--rodent	*	30-60-85	1-25-75	1-25-75	10-50-90	0-0-5?	0-0-53
2. Acute oral toxicity--nonrodent		74-100-100					
3. Acute dermal toxicity	*	49-80-96	0-0-78	3-50-97	0-0-78	3-50-97	0-0-78
4. Acute parenteral toxicity		4-20-51	0-0-31	0-0-31	5-25-60	29-63-89	1-13-47
5. Acute inhalation toxicity	*	74-100-100					
6. Acute dermal irritation--corrosivity	*	30-60-85	0-0-53	25-75-99	0-0-53	1-25-75	0-0-53
7. Acute eye irritation-corrosivity	*	49-80-96	22-100-100	0-0-78	0-0-78	0-0-78	0-0-78
8. Skin sensitization--guinea pig	*	61-90-99	5-100-100	0-0-95	0-0-95	0-0-95	0-0-95
9. Subchronic oral toxicity--rodent: 14- or 28-d study	*	61-90-99	0-0-95	0-0-95	5-100-100	0-0-95	0-0-95
10. Subchronic toxicity--nonrodent: 14- or 28-d study		74-100-100					
11. Subchronic oral toxicity--rodent: 90-d study	*	74-100-100					
12. Subchronic oral toxicity--nonrodent: 90-d study	*	74-100-100					
13. Subchronic oral toxicity--nonrodent: 6- to 12-mo study		74-100-100					
14. Subchronic dermal toxicity: 21- or 28-d study	*	61-90-99	0-0-95	0-0-95	0-0-95	5-100-100	0-0-95
15. Subchronic dermal toxicity: 90-d study		74-100-100					
16. Subchronic inhalation toxicity: 14- or 28-d study		61-90-99	0-0-95	0-0-95	0-0-95	5-100-100	0-0-95
17. Subchronic inhalation toxicity 90-d study		74-100-100					
18. Neurobehavioral toxicity	*	74-100-100					
19. Teratology study--rodent, rabbit	*	74-100-100					
20. Multigeneration reproduction study--rodent		74-100-100					
21. Toxicokinetics	*	39-70-91	0-0-63	2-33-86	0-0-63	14-67-98	0-0-63

	X	G	A	IN	IR	C
22. Carcinogenicity--rodent	*	61-90-99	0-0-95	0-0-95	5-100-100	0-0-95	0-0-95
23. Chronic toxicity	*	74-100-100					
24. Combined chronic toxicity-carcinogenicity-rodent		74-100-100					
25. Genetic toxicity	*	49-80-96	0-0-78	0-0-78	3-50-97	3-50-97	0-0-78
26. Subchronic eye toxicity		74-100-100					
27. Segment I: fertility and reproductive performance		74-100-100					
28. Segment III: perinatal and postnatal performance		74-100-100					
29. Acute delayed neurotoxicity		74-100-100					
30. Skin painting--chronic		61-90-99	0-0-95	0-0-95	5-100-100	0-0-95	0-0-95
31. Implantation studies		74-100-100					
32. Human sensitization studies		74-100-100					
33. Skin penetration studies		74-100-100					

a According to the standards adopted by the Committee on Toxicity Data Elements.

b X = test not done;

G = test conducted according to reference protocol guidelines;

A = test conducted adequately, but not according to reference protocol guidelines;

IN = test conducted inadequately, but retesting is not needed;

IR = test conducted inadequately, and retesting is needed;

C = test adequacy cannot be evaluated.

Percentages are presented with upper and lower 90% confidence intervals.

Value for each test type under X is a proportion of all substances in this category of the subsample.

Values in all other columns are based only on portion of category for which test was conducted.

109

TABLE 17 Quality of Testing of 13,911 Chemicals in Commerce in Select Universe with Production Levels Less Than 1 Million lb/yr

Test Type	Test Required[a]	Quality Ratings[b]					
		X	G	A	IN	IR	C
		Estimated Percent of Select Universe with Indicated Quality					
1. Acute oral toxicity--rodent	*	22-50-78	1-20-66	8-40-81	1-20-66	1-20-66	0-0-45
2. Acute oral toxicity--nonrodent		61-90-99	0-0-95	0-0-95	5-100-100	0-0-95	0-0-95
3. Acute dermal toxicity	*	39-70-91	0-0-63	0-0-63	37-100-100		
4. Acute parenteral toxicity		22-50-78	0-0-45	1-20-66	0-0-45	34-80-99	0-0-45
5. Acute inhalation toxicity	*	49-80-96	3-50-97	0-0-78	3-50-97	0-0-78	0-0-78
6. Acute dermal irritation--corrosivity	*	30-60-85	0-0-53	1-25-75	10-50-90	0-0-53	1-25-75
7. Acute eye irritation-corrosivity	*	49-80-96	0-0-78	0-0-78	3-50-97	0-0-78	3-50-97
8. Skin sensitization--guinea pig	*	61-90-99	0-0-95	0-0-95	0-0-95	0-0-95	5-100-100
9. Subchronic oral toxicity--rodent: 14- or 28-d study	*	49-80-96	0-0-78	3-50-97	3-50-97	0-0-78	0-0-78
10. Subchronic toxicity--nonrodent: 14- or 28-d study		74-100-100					
11. Subchronic oral toxicity--rodent: 90-d study	*	49-80-96	0-0-78	22-100-100	0-0-78	0-0-78	0-0-78
12. Subchronic oral toxicity--nonrodent: 90-d study	*	74-100-100					
13. Subchronic oral toxicity--nonrodent:6- to 12-mo study		74-100-100					
14. Subchronic dermal toxicity: 21- or 28-d study	*	61-90-99	0-0-95	5-100-100	0-0-95	0-0-95	0-0-95
15. Subchronic dermal toxicity: 90-d study		74-100-100					
16. Subchronic inhalation toxicity: 28- or 14-d study	*	61-90-99	0-0-95	5-100-100	0-0-95	0-0-95	0-0-95
17. Subchronic inhalation toxicity 90-d study		49-80-96	3-50-97	0-0-78	0-0-78	0-0-78	3-50-97
18. Neurobehavioral toxicity	*	61-90-99	0-0-95	0-0-95	0-0-95	5-100-100	0-0-95
19. Teratology study--rodent, rabbit	*	61-90-99	5-100-100	0-0-95	0-0-95	0-0-95	0-0-95
20. Multigeneration reproduction study--rodent	*	74-100-100					
21. Toxicokinetics	*	61-90-99	0-0-95	5-100-100	0-0-95	0-0-95	0-0-95

No.	Test	X	G	A	IN	IR	C
22.	Carcinogenicity--rodent	*	61-90-99	0-0-95	0-0-95	0-0-95	5-100-100
23.	Chronic toxicity	*	61-90-99	0-0-95	0-0-95	5-100-100	0-0-95
24.	Combined chronic toxicity-carcinogenicity-rodent		74-100-100				
25.	Genetic toxicity	*	39-70-91	0-0-63	0-0-63	37-100-100	0-0-63
26.	Subchronic eye toxicity		74-100-100				
27.	Segment I: fertility and reproductive performance		74-100-100				
28.	Segment III: perinatal and postnatal performance		61-90-99	0-0-95	0-0-95	0-0-95	5-100-100
29.	Acute delayed neurotoxicity		74-100-100				
30.	Skin painting--chronic		61-90-99	0-0-95	5-100-100	0-0-95	0-0-95
31.	Implantation studies		74-100-100				
32.	Human sensitization studies		61-90-99	0-0-95	5-100-100	0-0-95	0-0-95
33.	Skin penetration studies		74-100-100				

a According to the standards adopted by the Committee on Toxicity Data Elements.

b X = test not done;

G = test conducted according to reference protocol guidelines;

A = test conducted adequately, but not according to reference protocol guidelines;

IN = test conducted inadequately, but retesting is not needed;

IR = test conducted inadequately, and retesting is needed;

C = test adequacy cannot be evaluated.

Percentages are presented with upper and lower 90% confidence intervals.

Value for each test type under X is a proportion of all substances in this category of the subsample.

Values in all other columns are based only on portion of category for which test was conducted.

TABLE 18 Quality of Testing of 21,752 Chemicals in Commerce in Select Universe with Unknown or Inaccessible Production Levels

Test Type	Test Required[a]	Quality Ratings[b]					
		Estimated Percent of Select Universe with Indicated Quality					
		X	G	A	IN	IR	C
1. Acute oral toxicity--rodent	*	26-45-65	0-0-24	8-27-56	20-45-73	0-0-24	8-27-56
2. Acute oral toxicity--nonrodent		78-95-100	0-0-95	0-0-95	5-100-100	0-0-95	0-0-95
3. Acute dermal toxicity	*	54-75-90	0-0-45	1-20-66	8-40-81	1-20-66	1-20-66
4. Acute parenteral toxicity		39-60-78	0-0-31	0-0-31	5-25-60	5-25-60	19-50-81
5. Acute inhalation toxicity	*	54-75-90	0-0-45	1-20-66	8-40-81	1-20-66	1-20-66
6. Acute dermal irritation--corrosivity	*	30-50-70	9-30-61	1-10-39	9-30-61	1-10-39	4-20-51
7. Acute eye irritation-corrosivity	*	39-60-78	1-13-47	1-13-47	11-38-71	1-13-47	5-25-60
8. Skin sensitization--guinea pig	*	60-80-93	0-0-53	0-0-53	0-0-53	10-50-90	10-50-90
9. Subchronic oral toxicity--rodent: 14- or 28-d study	*	49-70-86	0-0-39	0-0-39	15-50-85	6-33-73	1-17-58
10. Subchronic toxicity--nonrodent: 14- or 28-d study		72-90-98	0-0-78	3-50-97	3-50-97	0-0-78	0-0-78
11. Subchronic oral toxicity--rodent: 90-d study	*	66-85-96	0-0-63	2-33-86	0-0-63	2-33-86	2-33-86
12. Subchronic oral toxicity--nonrodent: 90-d study	*	78-95-100	0-0-95	0-0-95	5-100-100	0-0-95	0-0-95
13. Subchronic oral toxicity--nonrodent: 6- to 12-mo study		78-95-100	0-0-95	0-0-95	5-100-100	0-0-95	0-0-95
14. Subchronic dermal toxicity: 21- or 28-d study	*	72-90-98	0-0-78	0-0-78	0-0-78	0-0-78	22-100-100
15. Subchronic dermal toxicity: 90-d study		86-100-100					
16. Subchronic inhalation toxicity: 14- or 28-d study	*	72-90-98	0-0-78	0-0-78	0-0-78	22-100-100	0-0-78
17. Subchronic inhalation toxicity 90-d study		86-100-100					
18. Neurobehavioral toxicity:	*	86-100-100					
19. Teratology study--rodent, rabbit	*	72-90-98	3-50-97	0-0-78	0-0-78	3-50-97	0-0-78
20. Multigeneration reproduction study--rodent	*	72-90-98	0-0-78	3-50-97	0-0-78	3-50-97	0-0-78
21. Toxicokinetics	*	54-75-90	0-0-45	1-20-66	0-0-45	34-80-99	0-0-45

No.	Test		X	G	A	IN	IR	C
22.	Carcinogenicity--rodent	*	66-85-96	0-0-63	14-67-98	0-0-63	2-33-86	0-0-63
23.	Chronic toxicity	*	60-80-93	0-0-53	1-25-75	0-0-53	10-50-90	1-25-75
24.	Combined chronic toxicity-carcinogenicity-rodent		86-100-100					
25.	Genetic toxicity	*	35-55-74	0-0-28	0-0-28	0-0-28	72-100-100	0-0-28
26.	Subchronic eye toxicity		86-100-100					
27.	Segment I: fertility and reproductive performance		78-95-100	0-0-95	5-100-100	0-0-95	0-0-95	0-0-95
28.	Segment III: perinatal and postnatal performance		86-100-100					
29.	Acute delayed neurotoxicity		86-100-100					
30.	Skin painting--chronic		72-90-98	0-0-78	0-0-78	0-0-78	22-100-100	0-0-78
31.	Implantation studies		78-95-100	0-0-95	0-0-95	0-0-95	5-100-100	0-0-95
32.	Human sensitization studies		60-80-93	10-50-90	0-0-53	0-0-53	0-0-53	10-50-90
33.	Skin penetration studies		78-95-100	0-0-95	5-100-100	0-0-95	0-0-95	0-0-95

a According to the standards adopted by the Committee on Toxicity Data Elements.

b X = test not done;

G = test conducted according to reference protocol guidelines;

A = test conducted adequately, but not according to reference protocol guidelines;

IN = test conducted inadequately, but retesting is not needed;

IR = test conducted inadequately, and retesting is needed;

C = test adequacy cannot be evaluated.

Percentages are presented with upper and lower 90% confidence intervals.

Value for each test type under X is a proportion of all substances in this category of the subsample.

Values in all other columns are based only on portion of category for which test was conducted.

TABLE 19 Relative Comparison of Testing Need by Intended-Use Category of Select Universe[a]

	Category and Subsample Size		
Pesticides and Inert Ingredients of Pesticide Formulations (15)	Cosmetic Ingredients (15)	Drugs and Excipients in Drug Formulations (15)	Food Additives (15)
Neurobehavioral toxicity	Neurobehavioral toxicity	Acute inhalation toxicity	Acute inhalation toxicity--nonrodent
Teratology--rodent, rabbit	Subchronic eye toxicity	Skin sensitization--guinea pig	Subchronic oral toxicity--nonrodent: 90-d study
Genetic toxicity	Subchronic dermal toxicity: 90-d	Subchronic dermal toxicity: 90-d	Subchronic oral toxicity--nonrodent: 6- to 12-mo study
Subchronic inhalation toxicity: 14- or 28-d	Skin painting--chronic	Subchronic inhalation toxicity: 14- or 28-d	Neurobehavioral toxicity
Multigeneration reproduction--rodent	Skin penetration	Subchronic inhalation toxicity: 90-d	Multigeneration reproduction--rodent
Carcinogenicity--rodent	Skin sensitization--guinea pig	Multigeneration reproduction--rodent	Combined chronic toxicity-carcinogenicity--rodent
Combined chronic toxicity-carcinogenicity--rodent	Subchronic dermal toxicity: 21- or 28-d	Carcinogenicity--rodent	Genetic toxicity
Acute delayed neurotoxicity	Genetic toxicity	Genetic toxicity	
		Subchronic eye toxicity	
Subchronic dermal toxicity: 21- or 28-d	Human sensitization	Segment I: fertility and reproductive performance	Subchronic oral toxicity--rodent: 90-d study
		Acute delayed neurotoxicity	Teratology--rodent, rabbit
	Acute eye irritation--corrosivity	Skin painting--chronic	Carcinogenicity--rodent
	Carcinogenicity--rodent	Implantation	

114

Skin sensitization--
  guinea pig
Toxicokinetics
Chronic toxicity

Acute inhalation toxicity

Acute eye irritation--
  corrosivity
Subchronic oral toxicity--
  rodent: 90-d

Subchronic oral toxicity--
  nonrodent: 90-d

Acute dermal toxicity

Acute dermal toxicity
Acute dermal irritation--
  corrosivity

Acute dermal toxicity
Subchronic toxicity
  nonrodent: 14- or 28-d
Subchronic oral toxicity--
  nonrodent: 90-d
Subchronic oral toxicity--
  nonrodent: 6- to 12-mo
Neurobehavioral toxicity
Teratology--rodent,
  rabbit
Chronic toxicity
Combined chronic toxicity--
  carcinogenicity--rodent
Segment III: perinatal and
  postnatal performance

Acute oral toxicity--
  nonrodent
Acute dermal irritation--
  corrosivity
Acute eye irritation--
  corrosivity
Subchronic oral toxicity--
  rodent: 90-d
Human sensitization
Acute oral toxicity--
  rodent
Acute parenteral toxicity
Subchronic oral toxicity--
  rodent: 14- or 28-d

Subchronic oral toxicity--
  rodent: 14- or 28-d
Toxicokinetics

Acute oral toxicity--
  rodent

115

TABLE 19 (continued)

Category and Subsample Size

Chemicals in Commerce by Production Level

At Least 1 Million Pounds/Year (10)	Less Than 1 Million Pounds/Year (10)	Level Unknown/ Inaccessible (20)
Acute inhalation toxicity	Skin sensitization-- guinea pig	Skin sensitization-- guinea pig
Subchronic oral toxicity-- rodent: 90-d	Subchronic oral toxicity-- nonrodent: 90-d	Subchronic dermal toxicity: 21- or 28-d
Subchronic oral toxicity-- nonrodent: 90-d	Neurobehavioral toxicity	Neurobehavioral toxicity
Subchronic dermal toxicity: 21- or 28-d	Multigeneration reproduction--rodent	Genetic toxicity
Neurobehavioral toxicity	Carcinogenicity--rodent	
Teratology--rodent, rabbit	Chronic toxicity	Subchronic oral toxicity-- rodent: 90-d
Multigeneration reproduction--rodent	Genetic toxicity	Subchronic oral toxicity-- nonrodent: 90-d
Chronic toxicity		Teratology--rodent, rabbit
	Acute eye irritation-- corrosivity	Multigeneration reproduction--rodent
Acute dermal toxicity	Subchronic dermal toxicity: 21- or 28-d	Toxicokinetics
Skin sensitization-- guinea pig	Teratology--rodent, rabbit	Chronic toxicity
Subchronic oral toxicity-- rodent: 14- or 28-d	Toxicokinetics	
Toxicokinetics		Carcinogenicity--rodent
Carcinogenicity--rodent	Acute inhalation toxicity	
Genetic toxicity	Subchronic oral toxicity-- rodent: 14- or 28-d	Acute dermal toxicity
	Subchronic oral toxicity-- rodent: 90-d	Acute parenteral toxicity
Acute eye irritation-- corrosivity		Subchronic oral toxicity-- rodent: 14- or 28-d
	Acute dermal toxicity	
Acute dermal irritation-- corrosivity	Acute dermal irritation-- corrosivity	Acute eye irritation-- corrosivity
Acute oral toxicity-- rodent	Acute oral toxicity-- rodent	Acute dermal irritation-- corrosivity

aIn each column, tests are arranged in order of relative need from greatest to least, with groupings indicating equal need. For each test type, relative need is based on the number of substances in each category that required the test according to the standards adopted by the Committee on Toxicity Data Elements, but on which the test was not done (X), was done inadequately and needed repetition (IR), or could not be evaluated (C).

116

TABLE 20 Ability to Conduct Health-Hazard Assessment of Substances in Seven Categories of Select Universe

Category	No. of Substances in Category	Estimated Percentages[a] of Substances in Category with:			Less than Minimal Toxicity Information[b]	No Toxicity Information[c]
		Minimal Toxicity Information				
		Complete Health-Hazard Assessment Possible	Partial Health-Hazard Assessment Possible	No Health-Hazard Assessment Possible		
Pesticides and inert ingredients of pesticide fomulations	3,350	3-10-19	13-24-34	0-2-7	16-26-37	26-38-50
Cosmetic ingredients	3,410	0-2-7	7-14-21	5-10-17	12-18-24	48-56-64
Drugs and excipients in drug formulations	1,815	9-18-29	8-18-28	0-3-7	15-36-46	15-25-35
Food additives	8,627	2-5-11	8-14-19	0-1-4	26-34-42	38-46-54
Chemicals in commerce:						
$\geq$1 million lb/yr	12,860	0-0-16	4-11-18	4-11-18	0-0-2	72-78-84
$\leq$1 million lb/yr	13,911	0-0-17	5-12-16	5-12-20	0-0-2	70-76-82
Unknown/inaccessible	21,752	0-0-7	5-10-15	4-8-13	0-0-2	76-82-88

a Estimated percentages are given with upper and lower 90% confidence limits.
b Partial health-hazard assessment of some substances with less than minimal toxicity information may be possible.
c On basis of search strategy used to find minimal toxicity information (Chapter 2).

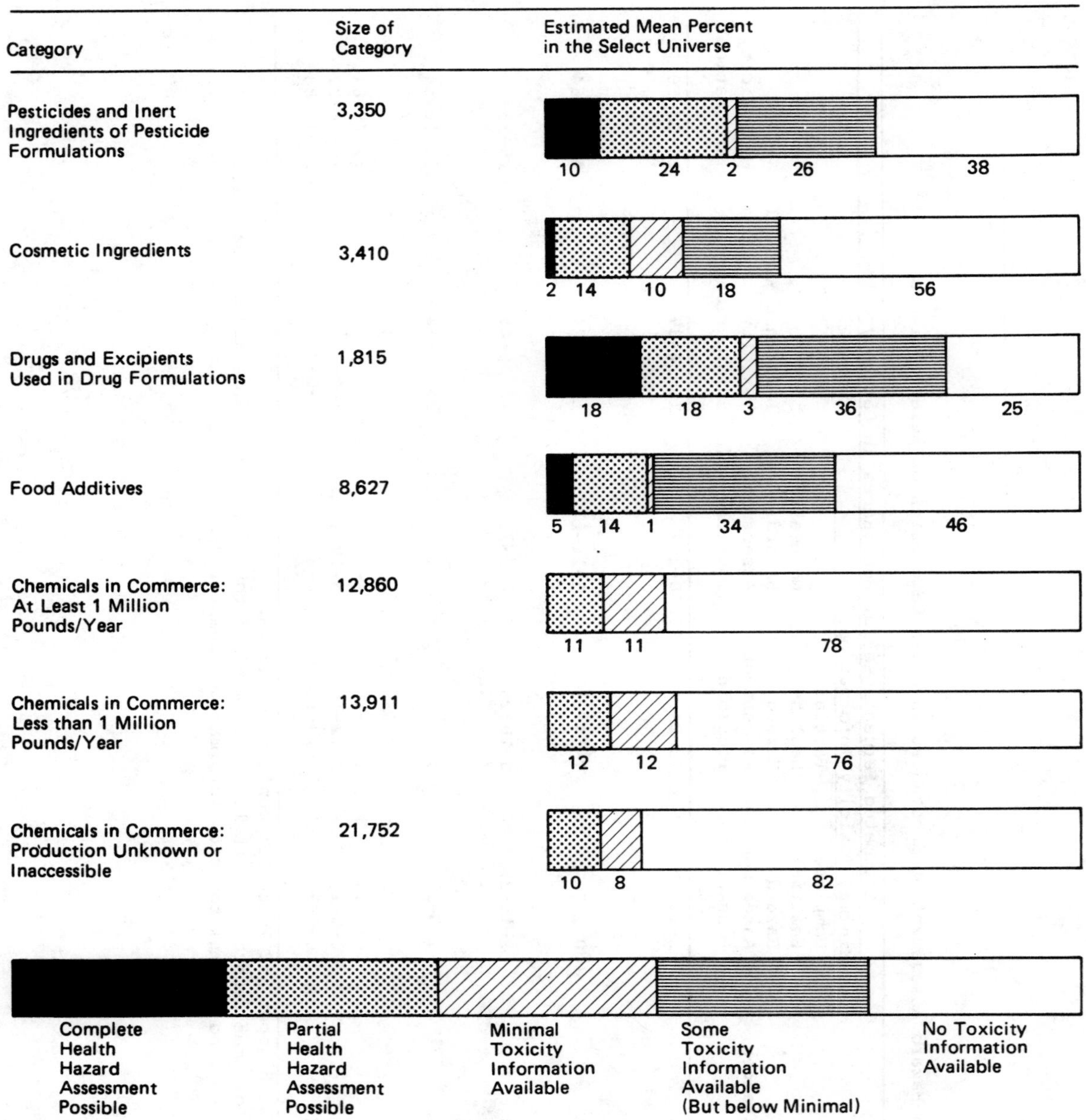

Category	Size of Category	Estimated Mean Percent in the Select Universe
Pesticides and Inert Ingredients of Pesticide Formulations	3,350	10 24 2 26 38
Cosmetic Ingredients	3,410	2 14 10 18 56
Drugs and Excipients Used in Drug Formulations	1,815	18 18 3 36 25
Food Additives	8,627	5 14 1 34 46
Chemicals in Commerce: At Least 1 Million Pounds/Year	12,860	11 11 78
Chemicals in Commerce: Less than 1 Million Pounds/Year	13,911	12 12 76
Chemicals in Commerce: Production Unknown or Inaccessible	21,752	10 8 82

| Complete Health Hazard Assessment Possible | Partial Health Hazard Assessment Possible | Minimal Toxicity Information Available | Some Toxicity Information Available (But below Minimal) | No Toxicity Information Available |

FIGURE 2   Ability to conduct health-hazard assessment of substances in seven categories of select universe.

a substance. A partial health-hazard assessment is defined as one that has a limited characterization of the hazard associated with the safe use of a substance. Therefore, a partial health-hazard assessment has a broad range extending from very limited (e.g., acute toxicity by one route of administration) to almost complete (e.g., full acute-toxicity and chronic-toxicity evaluation except for inadequate neurobehavioral-toxicity determination).

The percentages of the select universe in Table 20 and Figure 4 are estimates based on analysis of the subsample of 100 substances, all of which met the prescribed requirements for minimal toxicity information. Results of this analysis indicate not only the percentage of substances in each of the seven categories that have sufficient testing of adequate quality to conduct a health-hazard assessment, but also the percentage that would require additional testing according to the standards adopted by the Committee on Toxicity Data Elements if an assessment were to be performed. It should be remembered that the requirement for minimal toxicity information (see Chapter 2) varied among the seven categories. Chemicals in commerce were considered to have met the requirement if any one of the five tests used to define minimal toxicity information were done. Therefore, chemicals in commerce that had less than minimal toxicity information as determined by the search strategy used had, by definition, no information.

In general, proportionately more testing has been undertaken on pesticides and inert ingredients of pesticide formulations and on drugs and excipients in drug formulations. In these two categories, 36% and 39% of substances met the requirements for minimal toxicity information, respectively; the Committee on Toxicity Data Elements judged it possible to make at least a partial health-hazard assessment of 94% and 92%, respectively, of the substances with minimal toxicity information.

Cosmetic ingredients and food additives have been somewhat less thoroughly tested. Minimal-toxicity-information requirements were met by about 26% and 20% of substances in these categories, respectively, and at least a partial health-hazard assessment was judged possible for about 62% and 95%, respectively, of the substances with minimal toxicity information.

In contrast, only about 20% of the substances in each of the three production categories of chemicals in commerce had minimal toxicity information; at least a partial health-hazard assessment was judged possible for about 50% of the substances with minimal toxicity information in each of the three categories. Virtually all the substances in the three subsample categories of chemicals in commerce with minimal toxicity information required additional toxicity testing if a complete health-hazard assessment were to be performed. Chemicals in commerce with indicated 1977 production of at least 1 million pounds have been tested no more often or more adequately than substances with 1977 production of less than 1 million pounds.

It is evident that, even in categories in which the greatest amount of testing has been done, there is still a great deal to do. For the chemicals in commerce, this arises in part from the facts that regulation of these substances has come about only recently and that the regulatory requirements for testing have generally been less than those for substances in the other categories. The three categories of chemicals in commerce also do not specifically reflect the substances of greatest commercial use. Exposure information may be present to a greater degree on such substances than on those selected for the sample.

INTERPRETATION AND ANALYSIS OF PHYSICOCHEMICAL AND EXPOSURE DATA

The committees attempted to relate the quantity and quality of toxicity testing to several factors that seem to be logical determinants of testing, such as breadth of known exposure, expected trends in exposure, physicochemical properties and chemical fate of the substances, and strength of evidence of toxicity in humans, including the severity of reported chronic human toxicity. In addition, the committees sought information on occupational and environmental exposure and attempted to relate this information to the extent and quality of toxicity testing. These efforts were limited to substances in the subsample.

As the dossiers for the 100 substances in the subsample were examined, it became evident that characterization of the substances with respect to each of these factors, if it was possible at all, was based on sparse information. Most available information was on the physicochemical properties of the substances; the least was on exposure. However, no comprehensive method of gathering the needed information could be identified, and, in the end, the principal basis for characterizing exposure information was the knowledge and expertise of the committee members. For example, a judgment about widespread exposure was made on the basis of available information--such as the number of persons exposed in occupational or environmental settings, production volumes, environmental stability, or number of drug prescriptions sold--or on the individual experience and personal knowledge of committee members. It is possible that exposure information is more complete on substances of greatest commercial importance than on other substances.

Information on physicochemical properties and potential exposure was tabulated and tallied as present or absent, but its quality and quantity were not assessed. Therefore, conclusions drawn from these data should be interpreted conservatively. A multiple-choice summary checklist (see Appendix L) for each substance in the subsample of 100 was completed by the Committee on Toxicity Data Elements after review of all information in the dossier. These data were then examined for correlations among amount and quality of testing, potential for exposure, knowledge of physicochemical properties, and concern for potentially adverse human health effects.

Because the samples were small and sampling in the several categories of substances was not proportional, averages or percentages based on totals could be misleading, especially for larger groupings of substances, and particularly if they are taken to be firm estimates. Furthermore, because the numbers are small, detailed inferences to the select universe are not appropriate. Nonetheless, the findings will give the reader a reasonable impression, based on the subsample of 100 substances, of how the data are arrayed.

The following observations are apparent from the committees' analysis:

● Of the 100 substances in the subsample, 42 were known to have widespread exposure. An additional 14 had limited exposure potential, which, however, would be intensive for specific groups.

● For 20 of the 100 substances, physicochemical data led to a high concern about potential adverse human health effects. For 32 substances, there was moderate concern.

● The amount of testing that had been performed was not related to the committee's judgment that the chemicals warranted additional concern on the basis of physicochemical information.

Of the seven categories, more information was generally available on the four that had specific, regulated uses than on the three categories of chemicals in commerce. The committees believe that this might be the result of two factors: In contrast with the other four categories, chemicals in commerce have not been subject to regulation until recently and thus toxicity testing has been required for a shorter time; and the four categories that have been subject to regulation for a longer time are also those to which humans are intentionally exposed and which would therefore be of greater concern to toxicologists. The committee members recognize that these reasons are interdependent.

Table 21 summarizes the extent of availability of various kinds of data on substances on which a partial or complete health-hazard assessment could be conducted. Within the seven categories, data on solubility or partition coefficient, physical state, manufacturing process, and melting or boiling point were least available for substances in the three categories of chemicals in commerce.

Other chemical data that were available could assist in conducting a health-hazard assessment. Chemical-reactivity data were most available for the pesticides and inert ingredients of pesticide formulations (73%), drugs and excipients in drug formulations (67%), food additives (47%), the three production categories of chemicals in commerce (20-45%), and cosmetic ingredients (40%). Chemical-reactivity data were considered to be available if there was any information on hydrolysis, photochemical changes, potential for absorption or desorption, or nonbiologic systems, including shelf-life and oxidation and reduction potential.

121

TABLE 21 Availability of Specific Information on Substances in Seven Categories of Select Universe

Type of Data	Proportion of Substances on Which Data are Available, %	
	Pesticides and Inert Ingredients of Pesticide Formulations, Cosmetic Ingredients, Drugs and Excipients in Drug Formulations, and Food Additives	Chemicals in Commerce (Three Categories)
Intended use	93-100	90-100
Production	67-100	65-100
Solubility or partition coefficient	93-100	60-80
Physical state	87-100	50-90
Manufacturing process	87-100	40-70
Melting or boiling point	67-80	40-70

Of all the types of chemical information sought, data on the overall availability of analytic methods were least available (0-27%). Analytic methods were considered to be available if there was any information on the pure substance only in nonbiologic systems, on chemical reactions in biologic systems, or on evaluation in environmental media (air, water, and food) or direct human substrates (urine, blood, other body fluids, and tissues). Data elements concerning use, production processes and volumes, chemical bioavailability, and exposure related to the workplace, the general environment, and intended use are necessary to assess the relation between exposure potential and overall health hazard. Data on intended-use exposure or, where appropriate, indications of uses other than those intended (e.g., drug abuse) were more readily available for pesticides and inert ingredients of pesticide formulations (67%), drugs and excipients in drug formulations (100%), cosmetic ingredients (67%), and food additives (60%) than for chemicals in commerce (0-30%). Data elements involving bioavailability were available for 73% of pesticides and inert ingredient of pesticide formulations and for 20% of the low-volume chemicals in commerce. Information on bioavailability was considered available if there was any information on environmental stability and turnover rates; biodegradation; excretion and elimination; occurrence in air, water, tissues, or food chains; or bioaccumulation. Data related to occupational and environmental exposure potentials frequently were absent, ranging from 45% to 100% absent in the seven categories.

The availability of data concerning occupational or intended-use exposure had little or no relation to whether the literature contained information on bioavailability or environmental exposure. Pertinent data concerning the determination of exposure were scanty, even though exposure assessment is inherent in any approach to health-hazard assessment.

Although the committees' standards for judging the presence of various types of physicochemical charateristics were liberal, the proportions of substances without such data are high. The hazards associated with human exposure to chemicals depend on the nature, frequency, and intensity of exposure, as well as on the toxicity of the chemicals. The major activity of the committees, however, was devoted to an analysis of the quality and quantity of the toxicity data base, and the bulk of this report describes the analytic process and the results of the analysis. Although the committees acknowledge the importance of both exposure and toxicity data in hazard evaluation, they recognized that it would not be possible to use an approach that assembles an exposure data base amenable to the kinds of systematic analysis applied to the toxicity data base. The following are among the many factors that limit the application of such an approach:

● There are few legal reporting requirements for human exposure to chemicals. Even data on production volumes of substances and numbers of people involved in manufacture, distribution, use, and waste disposal are limited.

● There is little incentive for voluntary reporting of either production or exposure data in the open literature or in accessible agency files, and the few data available are often reported in forms that limit their comparability.

● Environmental-concentration data collected for compliance monitoring--such as coal-mine dust content, ambient-air concentrations of criteria pollutants, and concentrations of pesticide residues in foods--are for specific substances that were not in the random sample selected for this study. Furthermore, data collected for compliance monitoring may be of limited value in evaluating population exposures.

● Little is known about physical processes and procedures that affect the exposure potential for uses other than those intended. For example, the intensity of occupational exposure is strongly influenced by the choice of process and control equipment, and the intensity of environmental exposure is strongly influenced by the selection of waste-disposal technique, chemical reactivity, and degree of biodegradability.

In view of the great importance of exposure data and indexes of hazard assessment and the nearly complete absence of such data, the committees recommend that planning begin for the development of much more extensive, detailed, and accurate data bases than now exist for exposure assessments. The committees recognize that this will require a substantial effort and allocation of resources. It will also depend on further interagency collaboration and communication.

6

SUMMARY AND CONCLUSIONS

A select universe of 65,725 listings of substances was created from lists of pesticides and registered inert ingredients of pesticide formulations, cosmetic ingredients, drugs and excipients in drug formulations, food additives, and chemicals in commerce as listed in the Inventory of the Toxic Substances Control Act. A sample of 675 substances chosen by a systematic stratified random process was selected from the 65,725 entries. A random subsample of 100 substances on which there was at least a prescribed minimum of toxicity information was then selected from the random sample. The subsample contained representatives of seven categories of substances: (1) 15 pesticides and inert ingredients of pesticide formulations, (2) 15 cosmetic ingredients, (3) 15 drugs and excipients in drug formulations, (4) 15 food additives, and 40 chemicals in commerce, which were divided into (5) 10 with 1977 production of at least 1 million pounds, (6) 10 with 1977 production of less than 1 million pounds, and (7) 20 whose 1977 production volume was unknown or inaccessible because of manufacturers' claims of confidentiality.

The findings in this report apply to a total universe of approximately 53,500 unique substances, inasmuch as approximately 20% of the 65,725 listed substances were duplicates. A detailed dossier describing the adequacy of toxicity testing for conducting a health-hazard assessment was prepared for each of the 100 substances in the subsample. The analyses in this report are based on evaluations of testing adequacy for the subsample and information on the availability, nature, and extent of minimal toxicity information on the sample of 675 substances as prescribed by the Committee on Toxicity Data Elements.

Of the seven categories, drugs and excipients in drug formulations have been tested most adequately. For about 18% of the substances in this category, sufficiently complete health-hazard assessments can be conducted that no further toxicity testing would be required. Chemicals in commerce, as enumerated in the Inventory of the Toxic Substances Control Act, have had least adequate testing. On no substance in the three production categories of chemicals in commerce examined by the committees is information sufficient to permit a complete health-hazard assessment. Partial assessments could be made for 10-37% of the substances in the select universe. Even though the subsample in each category was small, so that the confidence limits of these estimated mean percentages form a wide range, it is obvious that, overall, most of the substances have not been adequately tested.

Adequacy of testing is, in large part, related to the intended use and regulatory history of a substance. Thus, pesticides and inert ingredients of pesticide formulations and drugs and excipients in drug formulations, which have the longest history of required testing, are the most fully tested substances. Substances that are ingested or otherwise applied to humans have in general been subjected to more testing than those to which there may be no intended or expected human exposure. Complete health-hazard assessments appear to require further testing of 82% of the drugs and excipients in drug formulations, 90% of the pesticides and inert ingredients of pesticide formulations, 95% of the food additives, 98% of the cosmetic ingredients, and essentially all the substances in the three production categories of chemicals in commerce.

The committees concluded that lack of availability of information could not be used to judge adequacy of testing for the purposes of this report. The estimated percentages expressing the judgment that further testing is required exclude proprietary data, particularly on drugs, that were not available to the committees and would presumably not be available to any other independent body assessing the state of toxicity information.

Toxicity tests, when conducted, have been of sufficiently high quality to obviate repetition or further testing. Approximately 80% of the acute-toxicity tests (based on the acute-toxicity test of highest quality for a given substance when more than one study was done) were judged not to need repetition by the committees. Only 33-50% of the chronic, long-term tests were considered not to need repetition. The test types that were done least frequently included those which are long, difficult, and complex.

The committees did not evaluate or categorize the results of testing, such as the frequency of finding specific kinds of carcinogenicity, neurotoxicity, or genetic toxicity. Thus, no conclusions can be presented on the toxicity of the 100 substances examined in detail. Such categorization would require a different kind of review, as well as a sample of much more than 100 substances to determine reliable estimates of proportions of the select universe known to have specific health effects.

Assessment of human health hazards requires information on the nature and extent of exposure, as well as on toxicity. For 36 of the 100 substances in the subsample, no data were available from which the committees could determine the extent of exposure, and, for 75 of the substances in the subsample, no information was available from which trends in exposure could be estimated.

REFERENCES

Interagency Regulatory Liaison Group, Testing Standards and Guidelines Work Group. 1981a. Recommended Guidelines for Acute Dermal Toxicity Test. Washington, D.C.: Interagency Regulatory Liaison Group. 12 pp.

Interagency Regulatory Liaison Group, Testing Standards and Guidelines Work Group. 1981b. Recommended Guideline for Acute Eye Irritation Testing. Washington, D.C.: Interagency Regulatory Liaison Group. 12 pp.

Interagency Regulatory Liaison Group, Testing Standards and Guidelines Work Group. 1981c. Recommended Guideline for Acute Oral Toxicity Testing in Rodents. Washington, D.C.: Interagency Regulatory Liaison Group. 11 pp.

Interagency Regulatory Liaison Group, Testing Standards and Guidelines Work Group. 1981d. Recommended Guidelines for Teratogenicity Studies in the Rat, Mouse, Hamster, or Rabbit. Washington, D.C.: Interagency Regulatory Liaison Group. 12 pp.

Interagency Regulatory Liaison Group, Epidemiology Work Group. 1981e. Guidelines for documentation of epidemiologic studies. Am. J. Epidemiol. 114:609-613

National Research Council, Committee for the Working Conference on Principles of Protocols for Evaluating Chemicals in the Environment. 1975. Principles for Evaluating Chemicals in the Environment. Washington, D.C.: National Academy of Sciences. 454 pp.

National Research Council, Committee for the Revision of NAS Publication 1138. 1977a. Principles and Procedures for Evaluating the Toxicity of Household Substances. Washington, D.C.: National Academy of Sciences. 130 pp.

National Research Council, Safe Drinking Water Committee. 1977b. Drinking Water and Health. Vol. 1. Washington, D.C.: National Academy of Sciences. 939 pp.

National Research Council, Safe Drinking Water Committee. 1980. Drinking Water and Health. Vol. 3. Washington, D.C.: National Academy of Sciences. 415 pp.

Organisation for Economic Co-operation and Development. 1979. Short-Term and Long-Term Toxicology Groups. Final Report. 185 pp. (unpublished)

Organisation for Economic Co-operation and Development. 1981. Guidelines for the Testing of Chemicals. Paris: Organisation for Economic Co-operation and Development. c. 700 pp.

U.S. Environmental Protection Agency. 1979. Toxic Substances Control Act Chemical Substances Inventory. Vol. II. User Guide and Indices to the Initial Inventory. Substance Main Index. Washington, D.C.: U. S. Government Printing Office. 799 pp.

U.S. Environmental Protection Agency. 1980. Toxic Substances Control Act Chemical Substances Inventory. Cumulative Supplement. Washington, D.C.: U.S. Government Printing Office. 535 pp.

APPENDIX A

SAMPLE OF 675 SUBSTANCES AND SUBSAMPLE OF 100 SUBSTANCES FROM THE "SELECT UNIVERSE"

The 675 substances in the sample and their CAS Registry numbers (where available) are listed below in randomly ordered sequence within the following seven categories: pesticides and inert ingredients of pesticide formulations, cosmetic ingredients, drugs and excipients in drug formulations, food additives, and the three production categories of chemicals in commerce as listed in the Inventory of the Toxic Substances Control Act. The chemical names are formatted as they appeared in the lists from which they were selected, with a few minor changes for clarity. Substances that were selected for the subsample of 100 are noted with an asterisk (*). The selection process to form the subsample in each category ceased where a solid line appears; it indicates that the required number for the subsample in the given category had been found. An equal sign (=) is used to indicate a word broken for spacing purposes where no hyphen normally occurs.

PESTICIDES AND INERT INGREDIENTS OF PESTICIDE FORMULATIONS

	Chemical	CAS number
	Ammonium ligninsulfonate	8061-53-8
	Diethylaminoethanolamine	Unavailable
*	[2,2,2-Trichloro-1-hydroxyethyl) dimethylphosphonate]	52-68-6
	Alkyldimethylbenzylammonium chloride	8001-54-4
	Carbamic acid, dipropylthio-, S-tert-butyl ester	2212-63-7
	Soap bark	Unavailable
*	1,2,4-Thiadiazole, 5-ethoxy-3-(trichloromethyl)-	2593-15-9
	1-Butanesulfonothioic acid, S-(chloromethyl) ester	16008-31-4
*	Phenol, 4-(di-2-propenylamino)-3,5-dimethyl-, methyl= carbamate (ester)	6392-46-7
	Sulfonated oleic acid, potassium salt	Unavailable
*	2H-1,3,5-Thiadiazine-2-thione, tetrahydro-3,5-dimethyl-	533-74-4
	Ethanol, 2-butoxy-, phosphate (3:1)	78-51-3
	Sodium decyldiphenyletherdisulfonate	36445-71-3
	Trichlorobenzyl chloride	1344-32-7
	Benzenecarbothioamide, 2,5-dichloro-	69622-81-7
*	Citric acid, trisodium salt	68-04-2
	Benzenamine, compound with 1,3,5-trinitro- benzene (1:1)	3101-79-9
*	Ethylene thiourea	96-45-7
	Acetic acid, (2,4-dichlorophenoxy)-, compound with 1,1',1"-nitrilotris[2-propanol]	32341-80-3
*	4,4'-Bipyridinium, 1,1'-dimethyl-, dichloride	1910-42-5
	Oxirane, methyl-, polymer with oxirane, monobutyl ether	9038-95-3
*	Potassium iodate	7758-05-6

Chemical	CAS Number
* Pyridine	110-86-1
Phosphorodithioic acid, S-(chloromethyl) O,O-, diethyl ester	24934-91-6
* 2,3,5-Trichloro-4-(propylsulfonyl)pyridine	38827-35-9
* p-Benzoquinone	106-51-4
Glycine, N-(2-[bis(carboxymethyl)amino]ethyl)- N-(2-hydroxyethyl)-, trisodium salt	139-89-9
2-Propanamine, sulfate	60828-92-4
* p-Nitrophenyldimethylthionophosphate	297-97-2
2,4-Dichlorophenoxyacetic acid, alkylamine salt	Unavailable
* Sodium acetate	127-09-3
Agrobacterium radiobacter	Unavailable
Carbamic acid, dimethyl-, 3-methyl-1-phenyl- 1H-pyrazol-5-yl ester	87-47-8
Carbonic acid, methyl 2-(1-methylheptyl)-, 4,6-dinitrophenyl ester	5386-68-5
* Phosphorodithioic acid, O,O-dimethyl ester, S-ester with N-(mercaptomethyl)phthalimide	732-11-6
* C.I. Pigment green 21 (Copper acetoarsenite, solid)	12002-03-8
2,5-Cyclohexadiene-1,4-dione, 2,3,5,6-tetrachloro-	118-75-2
Mercury, (acetato-O) (methylphenyl)-	1300-78-3
Copper hydroxide [Cu(OH)$_2$]	20427-59-2
Acetic acid, (2,4-dichlorophenoxy)-, methyl-2-(methyl- 2-[methyl-2-(2-methylpropoxy)ethoxy]-ethoxy)ethyl ester	53535-28-7
Phenarsazine, 10,10'-oxybis[5,10-dihydro-	4095-45-8
Alkyldimethylbenzylammonium chloride	8001-54-5
Benzoic acid, 2-hydroxy-, compound with 4-chloro= benzenamine (1:1)	53404-66-3
N-(2-[2-Hydroxyethoxypoly(ethyleneoxy)polypropylene= oxy)]propyl)hexanamide	Unavailable
Heptadecenylimidazoline	Unavailable
Benzenamine, ar,ar-dichloro-	27134-27-6
Sulfuric acid, zinc salt (1:1), monohydrate	7446-19-7
Sodium pentaborate	Unavailable
1H-Imidazo[4,5-b]pyridine, 6-chloro-2-(trifluoromethyl)-	13577-71-4
[1,1'-Biphenyl]-2-ol, ammonium salt	52704-98-0

COSMETIC INGREDIENTS

Poloxamine 1301	11111-34-5
Sucrose benzoate/sucrose acetate isobutyrate	Unavailable
Acetylated lanolin ricinoleate	977055-85-8
PEG-70 hydrogenated lanolin	68648-27-1

COSMETIC INGREDIENTS

	Chemical	CAS Number
*	Maleic acid	110-16-7
	Pareth-91-8	68439-46-3
	Methylpropylcellulose	977057-25-2
	Safflower glyceride	977058-10-8
	PEG-30 glyceryl oleate	68889-49-6
	Octadecene/maleic anhydride copolymer	25266-02-8
*	FD & C Red No. 40	25956-17-6
	Ammonium phosphate	7722-76-1
	PPG-8-ceteth-10	9087-53-0
*	4,4'-Isopropylidenediphenol	80-05-7
*	Ethyl linolenate	1191-41-9
	Allantoin calcium pantothenate	4207-41-4
	Nonoxynol-8	26027-38-3/
		37205-87-1
*	Calcium acetate	62-54-4
	Sucrose benzoate	12738-64-6
	Laureth-3	3055-94-5
	Potassium oleate	143-18-0
*	PEG-100 stearate	9004-99-3
	Cocamine oxide	61788-90-7
	Sodium myristyl sulfate	1191-50-0
*	Tetrasodium EDTA	64-02-8
	Cetearyl alcohol	8005-44-5
	Dimethyl cocamine	61788-93-0
*	p-Cresol	106-44-5
*	DM Hydantoin	77-71-4
	Isosteareth-6 carboxylic acid	Unavailable
*	Dehydroacetic acid	520-45-6
	Spinach extract	Unavailable
	Benzophenone-11	1341-54-4
*	PEG-200	25322-68-3
*	Guanidine carbonate	593-85-1
	PPG-2 methyl ether	13429-07-7/
		37286-64-9
	D & C Orange No. 5 zirconium lake	977054-31-1
	Hydrogenated tallow amine oxide	61788-94-1
*	Sodium bromate	7789-38-0
	Barium sulfide	21109-95-5
	Oleth-15	9004-98-2/
		25190-05-0
	Phloroglucinol	108-73-6
	Zinc myristate	16260-28-8
	Acetylated glycol stearate	Unavailable
	Vinylpyrrolidone/vinyl acetate/itaconic	68928-72-3
	Sorbitan triisostearate	54392-27-7
	PEG-14 oleate	9004-96-0

Chemical	CAS Number
Honey extract	Unavailable
Quaternium-8	977066-07-1
* Trisodium EDTA	150-38-9
PEG-45 stearate	9004-99-3
Acetylated hydrogenated tallow glycerides	977063-59-4
Trioleth-8 phosphate	977058-53-9
* Zinc carbonate	3486-35-9
Glyceryl tri-C_{10-18} acids	Unavailable
Potassium iodide	7681-11-0
Benzophenone-7	85-19-8
Ethylene brassylate	105-95-3
Sodium monodiethylaminopropyl cocoaspartate	977068-51-1
Nonoxynol-4	7311-27-5
Ceteareth-17	977063-70-9
Ammonium myreth sulfate	27731-61-9
Poloxamine 908	11111-34-5
Isostearamidopropyl dimethylamine lactate	55852-15-8
Ditridecyl sodium sulfosuccinate	2673-22-5
Styrene/maleic anhydride copolymer	9011-13-6
Grape juice	977064-74-6
FD & C Green No. 3	2353-45-9
Corn poppy extract	Unavailable
PEG-8 oleate	9004-96-0/
	977055-26-7
N-Phenyl-p-phenylenediamine	101-54-2
Oleth-4-phosphate	39464-69-2
Rue oil	8014-29-7
Decyl mercaptomethylimidazole	977064-18-8
Urocanic acid	104-98-3
PPG-20 lanolin ether	68439-53-2
Isopropylamine	75-31-0
Quaternium-3	68989-01-5
PEG-7 hydrogenated castor oil	61788-85-0
Hydrogenated soybean oil	8016-70-4
C_{11-13} isoparaffin	977068-15-7
Magnesium stearate	557-04-0
Dimethiconol	31692-79-2
PPG-4-ceteth-5	9087-53-0
D & C Orange No. 4	633-96-5
Tetramethylammonium chloride	75-57-0
Sodium bicarbonate	144-55-8
Sorbitan sesquiisostearate	977067-59-6
PEG-32	25322-68-3
Ascorbyl dipalmitate	28474-90-0
Lauramine oxide	1643-20-5
Hydrogenated animal glyceride	Unavailable

Chemical	CAS Number
PEG-7 glyceryl cocoate	977064-68-8
Cocamidopropylamine oxide	68155-09-9
TEA-oleamido PEG-2 sulfosuccinate	Unavailable
Pareth-45-11	Unavailable
Nonyl nonoxynol-10 phosphate	Unavailable
Stearoxytrimethylsilane	Unavailable
Solvent yellow 44	2478-20-8
Alginic acid	9005-32-7

DRUGS AND EXCIPIENTS IN DRUG FORMULATIONS

	Chemical	CAS Number
*	Acetylcysteine	616-91-1
	Linear tridecyl benzene sulfonate	Unavailable
*	Dichlorophen	97-23-4
*	Diphenidol hydrochloride	3254-89-5
*	Lactose	63-42-3
	Benzalkonium chloride	8001-54-5
*	Methylergonovine maleate	57432-61-8
	Flavor anise	Unavailable
	Chlorphenoxamine hydrochloride	562-09-4
*	Spironolactone	52-01-7
*	Norethindrone acetate	51-98-9
*	Methamphetamine hydrochloride	300-42-5
	Amylopectin	9037-22-3
	Flavor mint	Unavailable
*	Phytonadione	84-80-0
*	Cottonseed oil	8001-29-4
	Potassium formate	590-29-4
	Polyoxyethylene propylene	Unavailable
	Ethyl vanillin	121-32-4
	Triethanolamine polypeptide oleate condensate	Unavailable
*	Amantadine hydrochloride	665-66-7
	Sulfoxone sodium	144-75-2
*	Carboxypolymethylene	9007-20-9
*	Peanut oil	8002-03-7
	Insulin suspension, isophane, purified pork	Unavailable
	Bromodiphenhydramine hydrochloride	1808-12-4
*	Pyridoxine hydrochloride	58-56-0
	Calcium phosphate, tribasic	12167-74-7
*	Gelatin	9000-70-8
	Promalgen type G	Unavailable
	Mullein leaf	Unavailable
	Sodium chromate, CR-51	10039-53-9
	Phenacemide	63-98-9

DRUGS AND EXCIPIENTS IN DRUG FORMULATIONS, continued

Chemical	CAS Number
Guanidine hydrochloride	50-01-1
Orphenadrine hydrochloride	341-69-5
Deferoxamine mesylate	70-51-9
Cetyl alcohol	124-29-8
Clonazepam	1622-61-3
Testosterone	58-22-0
Epinephrine	51-43-4
Sennoside A	Unavailable
Thyrotropin	9002-7-5
Opalux AS 8010-A (Black)	Unavailable
Ipodate sodium	1221-56-3
Undecoylium chloride-iodine	1338-54-1
Flavor sherry, imitation	Unavailable
Wax, white	8006-40-4/ 8012-89-3
Hydrolose	9004-64-2
Mebutamate	64-55-1
Sodium phosphate, monobasic	7558-80-7

FOOD ADDITIVES

Chemical	CAS Number
Mannose	31103-86-3
3,5-Dimethyltetrahydro-1,3,5-thiadiazine-2-thione	533-74-4
p-Acetamidobenzoic acid	556-08-1
Vanadium tetrachloride	7632-51-1
Ethyl 3-hydroxybutyrate	5405-41-4
2,7-Dinitroso-1-naphthol	977014-63-3
Fennel	977001-13-0
* Norharman	244-63-3
Ionone, gamma	79-76-5
Triethylamine hydrochloride	554-68-7
* Cupric sulfate, anhydrous	7758-98-7
Ammonium thiocyanate	1762-95-4
Yeast extract, Baker's	8013-01-2
Dimethylphenylpiperazinium iodide	54-77-3
Sulfide ion	18496-25-8
Allyl nonanoate	7493-72-3
Geranium oil	8000-46-2
Benzyl thiocyanate	3012-37-1
Polyvinyl ethyl ether	25104-37-4
Elemene, alpha-	5951-67-7
Methyl isobutyrate	547-63-7
* Jasmine absolute	8031-01-4

	Chemical	CAS Number
*	Calcium stearate	1592-23-0
	Pentaerythritol tetrakis(3-mercaptopropionate)	7575-23-7
	Propyl 2-furanacrylate	623-22-3
	Butter fat	977018-87-3
	CI Fluorescent Brightener #109	61951-68-6
	Soybean mill feed	977030-55-9
	Cobalt(2+) caprylate	1588-79-0
	Tetramethyl tin	594-27-4
	Chromous oxide	12018-00-7
*	1-Monostearin	123-94-4
	Asafetida oil	977017-80-3
*	Hydrazine hydrate	7803-57-8
	DI-Dodecyl tin oxide	2273-48-5
	Molybdic acid	11099-00-6
	Celery seed extract	Unavailable
	Diethylene glycol dibenzoate	120-55-8
	p-Menth-1-en-9-ol	18479-68-0
*	Sodium lauryl sulfate	151-21-3
	Guanidoethyl cellulose	9069-21-0
	Lipase, animal	977033-78-5
	Silicon	7440-21-3
	2-Ethylhexyl 9,10-epoxystearate	141-38-8
	1,4-Dihydroxy-9,10-anthraquinone	81-64-1
	Phytoene	540-04-5
	Isoamyl isobutyrate	2050-01-3
	2-Tridecanone	593-08-8
	N-tert-Butylacrylamide	107-58-4
*	Riboflavin supplement	977030-53-7
*	Acenaphthylene	208-96-8
	Mannide monoleate	25339-93-9
*	Di-(2-methoxyethyl) phthalate	117-82-8
*	Diethylene glycol	111-46-6
*	Linseed oil	8001-26-1
	Artichoke leaf	Unavailable
	N-Stearoylsarcosine	142-48-3
	Ethyl 2-methylbutyrate	7452-79-1
	Chromium hydroxide	12626-43-6
	Xylyl sulfone	27043-27-2
	p-Cymen-8-ol	1197-01-9
	Molybdate Orange	12656-85-8
	Ammonium isovalerate	7563-33-9
	Feculose starch acetate	977033-03-6
	Rhynchosia pyramidalis	977030-08-2
	Cobalt tallate	61789-52-4

FOOD ADDITIVES, continued

Chemical	CAS Number
* Sodium laureth-3 sulfate	13150-00-0
* Silica	7631-86-9
Elaidic acid	112-79-8
2-tert-Butyl-4-ethylphenol	96-70-8
Allyl isovalerate	2835-39-4
1-Methylpiperazine	109-01-3
* Calcium saccharin	6381-91-5
Polyvinyl chloride	9002-86-2
Isoamyl cinnamate	7779-65-9
Benzyl phenylacetate	102-16-9
Sulfasomidine	[515-64-0]
Butirosin sulfate	51022-98-1
Guaiaretic acid	500-40-3
Soybean hull, ground	977032-85-1
Norbixin	542-40-5
Triethyl lead	5224-23-7
Propyl phenol	31019-46-2
Humulus	977001-58-3
Phthalocyanine	574-93-6
Cupric hydroxide	20427-59-2
Cedarwood oil terpene	68608-32-2
C.I. Disperse Orange #3	730-40-5
1,4-Dianilinoanthraquinone	2944-12-9
Dimethylol melamine	5001-80-9
Tetrakis(hydroxymethyl)phosphonium chloride	124-64-1
2-Ethylhexyl mercaptoacetate	7659-86-1
Pentaerythritol monostearate	78-23-9
Itaconic acid-methyl methacrylate copolymer	27155-24-4
2,6,6-Trimethyl-2-cyclohexen-1-one	20013-73-4
Geranial	141-27-5
3,4,5,6-Dibenzacridine	224-53-3
Ion-exchange membrane	Unavailable
Methyl hydrogen siloxane	63148-57-2
Valproic acid	99-66-1

CHEMICALS IN COMMERCE
Production at Least 1 Million Pounds/Year

Chemical	CAS Number
7-Oxabicyclo[4.1.0]heptan-2-one, 6-methyl-3- (1-methylethyl)-	5286-38-4
2-Pyrazolin-5-one, 1-(p-aminophenyl)-3-ethoxy-	4105-91-3
Bismuth, compound with gadolinium (1:1)	12010-44-5

CHEMICALS IN COMMERCE
Production at Least 1 Million Pounds/Year, continued

Chemical	CAS Number
Benzene, (2-iodoethyl)-	17376-04-4
Molybdenum phosphide (MoP)	12163-69-8
D-Glucose, enzyme-hydrolyzed	68921-30-2
Thiazole, 2-(2-methylpropyl)-	18640-74-9
1,2-Benzenedicarboxylic acid, bis(2-ethylhexyl) ester, polymer with 1,3-diisocyanatomethyl benzene, methyloxirane, and 1,2,3-propanetriol	68492-79-5
Amines, N,N,N'-trimethyl-N'-tallow alkyl= trimethylenedi-	68783-25-5
2-Propenoic acid, butyl ester, polymer with ethenyl acetate and 2-hydroxyethyl-2-propenoate	65776-73-0
Poly(oxy-1,2-ethanediyl), alpha,alpha'-[(ethyloctadecyliminio)di-2,1-ethanediyl]= bis[omega-hydroxy-, ethyl sulfate	42845-62-5
1,7-Naphthalenedisulfonic acid, 4-[(2,4-dichloro= benzoyl)amino]-5-hydroxy-6-[(2-methoxyphenyl)= azo]-, disodium salt	6416-33-7
* Benzenesulfinic acid, 4-chloro-	100-03-0
1,2-Benzenediamine, N-methyl-, dihydrochloride	25148-68-9
Bismuth hydroxide	10361-43-0
* Ethanol, 2-[(2-[(2-aminoethyl)amino]ethyl)= amino]-	1965-29-3
Phenol, 4,4'-(3H-2,1-benzoxathiol-3-ylidene)= bis(2,5-dimethyl-, S,S-dioxide	125-31-5
Benzenethiol, 4-dodecyl-, hydrogen phosphoro= dithioate, zinc salt	65045-85-4
* Isoquinoline, 1,2,3,4-tetrahydro-	91-21-4
* Pentanamide, N,N-dimethyl-	6225-06-5
2-Propenamide, N-(hydroxymethyl)-, polymer with 1,3-butadiene and 2-propenenitrile	26603-98-5
* 9H-Fluorene, 2-nitro-	607-57-8
Tannins, salts with 2-[3-(1,3-dihydro-1,3,3-trimethyl-2H-indol-2-ylidene)-1-propenyl]-1,3,3-trimethyl-3H-indolium	68957-25-5
Benzenepropanoic acid, 4-(bis[2-(benzoyloxy)= ethyl]amino)-alpha,beta-dicyano-, ethyl ester	65151-61-3
Silane, (3-isocyanatopropyl)trimethoxy)-	15396-00-6
Benzenesulfonic acid, 3-[(ethoxycarbonyl)amino]-, monosodium salt	71215-93-5
Hexanedioic acid, polymer with methyloxirane polymer with oxirane ether with oxybis=[propanol] (2:1)	63549-52-0
Octanoic acid, mixed esters with triethylene glycol hexanoate	68130-48-3
Benzene, 1-iodo-3-nitro-	645-00-1

137

Chemical	CAS Number
Calcium hydroxide, reaction products with iron oxide (Fe$_2$O$_3$) and magnesium hydroxide	68411-13-2
1-Propanaminium, N-ethyl-N,N-dimethyl-3-[(1-oxo= eicosyl)amino]-, ethyl sulfate	67846-22-4
8-Oxa-3,5-dithia-4-stannaundecan-1-ol, 4,4-dimethyl-9-oxo-,propanoate	67905-21-9
Poly(oxy-1,2-ethanediyl), alpha-[2,4-bis= (2-phenyl-1-propenyl)phenyl]-omega-hydroxy-	72088-88-1
2-Naphthalenesulfonic acid, 7-amino-5-[(4-[(2-bromo-1-oxo-2-propenyl)amino]-2-[(4-methyl-3-sulfophenyl)sulfonyl]phenylazo)]-, disodium salt	70210-02-5
Acetamide, N,N'-1,3-propanediylbis-, N-[3-C$_{20-30}$-(alkyloxy)propyl] derivatives	70528-81-3
Benzene, 1,1'-[1,2-ethanediylbis(thio)]bis-	622-20-8
2-Naphthalenesulfonic acid, 6-[(2,6-dimethylphenyl)= amino]-4-hydroxy-	23973-67-3
1-Propanamine, 2-chloro-N,N-dimethyl-, hydrochloride	4584-49-0
Ethanone, 1-(2,4,5-triethoxyphenyl)-	63213-29-6
Acetonitrile, 2,2',2'',2'''-(1,2-ethanediyl= dinitrilo)tetrakis-	5766-67-6
* Carbamic acid, (4-chlorophenyl)-, 1-methylethyl ester	2239-92-1
2,4,6(1H,3H,5H)-Pyrimidinetrione, 5-phenyl-1-(phenylmethyl)-	72846-00-5
Yttrium oxide sulfate, ytterbium-doped	68585-88-6
Didymium (rare earth mixture)	8006-73-3
Ethanol, 2,2'-oxybis-, polymer with alpha-hydro-omega-hydroxypoly(oxy-1,4-butanediyl) and 1,1'-methylenebis[4-isocyanatobenzene]	64078-69-9
* Benzene, 1,2,3,5-tetramethyl-	527-53-7
Phenol, isooctyldinitro-	37224-61-6
1,3,-Naphthalenedisulfonic acid, 7-[(4-[(4-[(4-aminobenzoyl)amino]-2-methyl= phenyl)azo]-2-methylphenyl)azo]-, disodium salt	6949-09-3
Hexanedioic acid, dimethyl ester, polymer with N,N'-bis(2-aminoethyl)-1,2-ethanediamine and dimethyl pentanedioate	72175-31-6
Bicyclo[3.1.0]hex-2-ene, 2-methyl-5-(1-methylethyl)-	2867-05-2
Benzo[a]phenoxazin-7-ium, 9-(dimethylamino)-, chloride	966-62-1
* Acetic acid, chloro-, 2-phenylethyl ester	7476-91-7
Antimony phosphide (SbP)	25889-81-0
Poly(oxy-1,2-ethanediyl), alpha-hydro-omega-hydroxy-, ether with 2-[(2-hydroxyethyl)amino]-2-(hydroxy= methyl)-1,3-propanediol (4:1)	72269-66-0

Chemical	CAS Number
Iron, complexes with diazotized 2-amino-4,6-dinitro= phenol monosodium salt coupled with diazotized 4-amino-5-hydroxy-2,7-naphthalenedisulfonic acid, diazotized 4-amino-3-methylbenzenesulfonic acid, diazotized 4-nitrobenzenamine, and resorcinol	71662-50-5
Glycerides, tallow di-	68553-08-2
1,3-Isobenzofurandione, polymer with 1,2-ethane= diol, 2,5-furandione, and 2,2'-oxybis[ethanol]	28679-80-3
2-Naphthalenesulfonic acid, 5-[bis(methylsulfonyl)= amino]-1-[(methylsulfonyl)oxy]-, sodium salt	58596-06-8
2-Butenedioic acid, (E)-, polymer with 1,3-butadiene, ethenylbenzene, (1-methylethenyl)benzene, methyl 2-methyl-2-propenoate, and 2-propenenitrile	69898-51-7
Phosphonic acid, (1,6-hexanediylbis[nitrilobis= (methylene)])tetrakis-, hexammonium disodium salt	68298-90-8
Vanadic acid ($H_4V_2O_7$), tetracesium salt	55343-67-4
Cyclohexanone, 2,6-dimethyl-4-(3-methylbutyl)-	71820-43-4
Oxirane, methyl-, polymer with oxirane, mono= (hydrogen sulfate), tridecyl ether	70850-89-4
Oils, menhaden, polymers with benzoic acid, glycerol, and isophthalic acid	68458-39-9
Benzenesulfonic acid, 2,5-dichloro- 4-[4-([3-([3-([1-(2,5-dichloro-4-sulfophenyl)- 4,5-dihydro-5-oxo-1H-pyrazol-4-yl]azo)benzoyl]= [phenylmethyl]amino)-4-methylphenyl]azo)- 4,5-dihydro-5-oxo-1H-pyrazol-1-yl]-, disodium salt	71050-54-9
1H-Purine-2,6,8(3H)-trione, 7,9-dihydro-, calcium salt	827-37-2
Vanadium silicide (V_3Si)	12039-76-8
Formaldehyde, polymer with methylphenol and 1,3,5,7-tetraazatricyclo[3.3.1.1^{3,7}]decane	68845-06-7
Coal, sulfonated	69013-20-3
2,5-Hexanediol, 2,5-dimethyl-	110-03-2
1(3H)-Isobenzofuranone, 3,3-bis[4-(sulfooxy)= phenyl]-, dipotassium salt	52322-16-4
Benzene, 1,2,4-trichloro-3-(chloromethyl)-	1424-79-9
Ethanone, 2-(acetyloxy)-1,2-diphenyl-	574-06-1
1H-Isoindol-1-one, 3-amino-	14352-51-3
Poly(difluoromethylene), alpha-hydro-= omega-[(phosphonooxy)methyl]-	72987-44-1
Poly[oxy(methyl-1,2-ethanediyl)], alpha-hydro- omega-[(([3[(([2-(1-aziridinyl)ethoxy]carbonyl)= amino]methylphenyl)amino]carbonyl)oxy]-, ether with 2-ethyl-2-(hydroxymethyl)-1,3-propanediol (3:1)	68015-74-7

Chemical	CAS Number
[1,1'-Binaphthalene]-,8,8'-dicarboxylic acid	29878-91-9
1H-Isoindole-5-carboxylic acid, 2,3-dihydro-1,3-dioxo-	20262-55-9
Terpenes and terpenoids, Litsea cubela-oil, hydrogenated	68608-36-6
(1,1'-Bicyclohexyl)-2-carboxylic acid, 4',5-di=hydroxy-2',3-dimethyl-5',6-bis[(1-oxo-2-pro=penyl)oxy]-, methyl ester	67952-52-7
* Ethene, (2,2,2-trifluoroethoxy)-	406-90-6
Acetamide, N-[2-(acetyloxy)ethyl]-	16180-96-4
Chromate(3-), (3-hydroxy-4-[(2-hydroxy-1-naphthalenyl)=azo]-7-nitro-1-naphthalenesulfonato[3-])=(4-hydroxy-3-[([2-hydroxy-5-([4-(phenylazo)=phenyl]azo)phenyl]methylene)amino]benzene=sulfonato[3-])-, trisodium	72479-29-9
* Stannane, difluorodimethyl-	3582-17-0
Butanedioic acid, bis(2-mercaptoethyl) ester	60642-67-3
Starch, 2,3-dialdehydo	9047-50-1
2-Propenoic acid, 2-methyl-, 2-ethyl-2([(2-methyl-1-oxo-2-propenyl)oxy]methyl)-1,3-propanediyl ester, polymer with ethenylbenzene, chloromethylated, trimethylamine-quaternized	68908-37-2
Benzene, 1,2-dichloro-4-(trichloromethyl)-	13014-24-9
Cyclopentanone, 2-(3-methyl-2-butenyl)-	2520-60-7
1,3-Benzenedicarbonyl dichloride, 5-hydroxy-	61842-44-2
Acetic acid, (4-formylphenoxy)-	22042-71-3
Ethanaminium, N,N,N-trimethyl-2-[(2-methyl-1-oxo-2-propenyl)oxy]-, chloride, polymer with 2-propenamide and N,N,N-trimethyl-2-[(2-methyl-1-oxo-2-propenyl)oxy]ethanaminium methyl sulfate	68227-15-6
4(1H)-Pyrimidinone, 6-hydroxy-	1193-24-4
Phenol, 4,4'-(1-methylethylidene)bis-, polymer with N,N'-bis(2-aminoethyl)-1,2-ethanediamine and (chloromethyl)oxirane, nonylphenol-modified	68951-48-4
Naphthalenesulfonic acid, dibutyl-, ammonium salt	68379-06-6
1,4-Benzenedicarboxylic acid, dimethyl ester, polymer with dimethyl pentanedioate, 1,6-hexanediol, and 2,2'-oxybis[ethanol]	71519-81-8
Resin acids and rosin acids, hydrogenated, esters with diethylene glycol	68648-51-1
Benzenesulfonic acid, 5-chloro-2,4-dinitro-	56961-56-9
Cyclohexane, isocyanato-	3173-53-3
Triethylenetetramine, polymer with ethylene oxide	31510-83-5
Pyrazineethanethiol	35250-53-4
1,4-Benzenedicarboxylic acid, polymer with 1,3-dihydro-1,3-dioxo-5-isobenzofurancarboxylic acid, hexanedioic acid and 1,2-propanediol	70729-94-1

CHEMICALS IN COMMERCE
Production at Least 1 Million Pounds/Year, continued

Chemical	CAS Number
Fatty acids, castor-oil, polymers with cottonseed-oil fatty acids, dehydrated castor-oil fatty acids, glycerol, phthalic anhydride and soya fatty acids	68525-89-3
2-Propenoic acid, 2-methyl-, 3-(trimethoxysilyl)= propyl ester, polymer with N-(1,1-dimethyl-3-oxo= butyl)-2-propenamide, ethenyl acetate and 2-ethylhexyl 2-propenoate	67785-57-3
Iron, complexes with 2-ethylhexanoic acid and tall-oil fatty acids	68187-36-0
Uranium bromide (UBr$_4$)	13470-20-7
Hexanoyl chloride	142-61-0
Benzeneethanol, alpha-butyl-, acetate	40628-77-1
Cyclotrisiloxane, hexamethyl-, polymer with (1,1-dimethylethyl)ethenylbenzene and (1-methyl= ethenyl)benzene	66836-92-8
1,2-Benzenedicarboxylic acid, compound with benzenamine (1:1)	50930-79-5
Imidodisulfuric acid, ammonium salt	27441-86-7
Benzenamine, N,N-dimethyl-4-([4-(methylamino)= phenyl]methyl)-	53477-27-3
Zinc, chloro([2,2',2''-nitrilotris= (ethanolato)](1-)-N,O,O',O'')-	33520-38-6
Uranium fluoride (UF$_5$)	13775-07-0
Safflower oil, polymer with glycerol and TDI	68072-09-3
Poly(oxy-1,2-ethanediyl), alpha,alpha'-[(docosyl= imino)di-2,1-ethanediyl]bis[omega-hydroxy-	38796-84-8
Cyclopentanecarboxylic acid, 1-amino-	52-52-8
Benzenediazonium, 5-[(butylamino)sulfonyl]-2-methoxy-, (T-4)-tetrachlorozincate(2-) (2:1)	62778-15-8
3-Butenal, 2,3-dimethyl-4-(2,6,6-trimethyl-2-cyclo= hexen-1-yl)-	68140-49-8
Fatty acids, C$_{5-10}$, esters with polypentaerythritol	68915-66-2
2-Propenoic acid, 2-methyl-, C$_{7-18}$-alkyl esters, polymer with 2-(methyl[(gamma-omega-perfluoro-C$_{8-14}$-alkyl)sulfonyl]amino)ethyl 2-methyl-2-propenoate	68988-55-6
Alcohols, C$_{6-12}$, ethoxylated	68439-45-2
2-Propenoic acid, 2-chloro-, methyl ester	80-63-7
Benzene, 1-methyl-4-phenoxy-	1706-12-3
1-Propanaminium, N-ethyl-N,N-dimethyl-3-[(2-methyl-1-oxo-2-propenyl)amino]-, ethyl sulfate	70942-19-7

141

	1-Hexene, 6-chloro-	928-89-2
	Formaldehyde, polymer with phenol and 3a,4,7,7a-tetrahydro-4,7-methano-1H-indene	29862-25-7
	4,7-Methano-1H-indenecarboxaldehyde, octahydro-	30772-79-3
	Hexadecanoic acid, octadecyl ester	2598-99-4
	Poly(oxy-1,2-ethanediyl), alpha-(1-oxo-2-propenyl)-omega-methoxy-	32171-39-4
	2-Propenoic acid, 2-methyl-, 2-hydroxypropyl ester, homopolymer	25703-79-1
	1H-Benzimidazole, 5-chloro-2-methyl-	2818-69-1
	Silane, bicyclo[2.2.1]hept-2-yltrichloro-	18245-29-9
	Pyridinium, 1-[2-(p-[(2-cyano-4-nitrophenyl)=azo]-N-ethylanilino)ethyl]-, chloride	23258-43-7
*	2-Furanmethanol, tetrahydro-, phosphate (3:1)	10427-00-6
	m-Dioxane, 2,5,5-trimethyl-2-propyl-	5421-99-8
	Silicic acid (H4SiO4), tetraphenyl ester	1174-72-7
	Benzamide, 4-methoxy-3-nitro-N-phenyl-	97-32-5
	Indol-3-ol, dihydrogen phosphate (ester), disodium salt	3318-43-2
*	1-Propanaminium, N,N,N-tripropyl-, bromide	1941-30-6
	Heptanoic acid, anhydride	626-27-7
	Nitrous acid, 3-methylbutyl ester	110-46-3
	Dextran, hydrogen sulfate	9042-14-2
	Azulene, 1,2,3,4,5,6,7,8-octahydro-1,4-dimethyl-7-(1-methylethylidene)-, (1S-cis)-	38-84-6
	Silane, 1,4-phenylenebis[chlorodimethyl-	1078-97-3
	Formic acid, rubidium salt	3495-35-0
*	Butanedioic acid, tetrafluoro-	377-38-8
	Glycols, polyethylene, hydrogen sulfate, eicosyl ether, sodium salt	26636-38-4
*	Silane, dichloroethenylethyl-	10138-21-3
*	2,8,9-Trioxa-5-aza-1-silabicyclo[3.3.3]undecane, 1-methyl-	2288-13-3
	Methanediamine, dihydrochloride	57166-92-4
*	Ichthammol	8029-68-3
	Benzenediazonium, 4,4'-(1,2-ethenediyl)bis=[3-sulfo-, dichloride	13954-62-6
	D-Arabinitol	488-82-4
	Phosphate(1-), hexafluoro-, ammonium	16941-11-0
	Platinum, dichloro[1,2,5,6-eta)-1,5-cyclooctadiene]-	12080-32-9
*	Terbium oxide	12738-76-0
	1H-Benzotriazolecarboxylic acid	60932-58-3
	Hexanoic acid, decyl ester	52363-43-6
	Docosanoic acid, 3-hydroxy-2,2-bis(hydroxymethyl)=propyl ester	53161-46-9
	Ethanol, 2,2,2-trifluoro-, 4-methylbenzenesulfonate	433-06-7
	Ethanol, 2-methoxy-, sodium salt	3139-99-9

Chemical	CAS Number
Benzene, 1-(chloromethyl)-2,4-dimethyl-	824-55-5
2-Butenoic acid, 2-methyl-, butyl ester, (Z)-	7785-64-0
1,6-Hexanediamine, N,N'-dibutylidene-	1002-91-1
Benzenesulfonamide, N,4-dimethyl-	640-61-9
2-Naphthalenecarboxylic acid, 5-(acetylamino)- 1-hydroxy-	63133-78-8
9,10-Anthracenedione, 1,4-bis[(2,6-diethyl= phenyl)amino]-	20241-74-1
* Ethanesulfonyl chloride, 2-chloro-	1622-32-8
1H-1,2,4-Triazol-3-amine, 5-(methylthio)-	45534-08-5
Benzamide, N-hydroxy-N-phenyl-	304-88-1
Phosphonic acid, dodecyl-, diethyl ester	4844-38-6
Iridium oxide	1312-46-5
9-Octadecenoic acid (Z)-, methyl ester, sulfurized, copper-treated	61788-34-9
2-Propenoic acid, ethyl ester, polymer with ethenylbenzene, formaldehyde, and 2-propenamide	28650-65-9
1H-3a,7-Methanoazulene, octahydro-3,8,8-trimethyl- 6-methylene-, [3R-(3alpha,3abeta,7beta,= 8aalpha)]-	546-28-1
1,2-Propanediol, 3-[(2-hydroxyethyl)thio]-	1468-40-2
Lead ruthenium oxide	37194-88-0
2-Butenedioic acid (Z)-, mono(2-ethylhexyl) ester, polymer with ethenyl acetate and 2-ethylhexyl 2-propenoate	61909-78-2
2,3b-Methano-3bH-cyclopenta[1,3]cyclopropa= [1,2]benzene-4-methanol, octahydro- 7,7,8,8-tetramethyl-	59056-64-3
Benzenemethanesulfonic acid, alpha,4-dihydroxy- 3-methoxy-, monosodium salt	19473-05-3
1,2-Ethanediamine, N,N'-bis(1,1-dimethylethyl)-	4062-60-6
2-Propenal, 2-methyl-3-[2-(1-methylethyl)phenyl]-	6502-23-4
9,12-Tetradecadien-1-ol, (Z,E)-	51937-00-9
Hexanoic acid, 2,2,3,3,4,4,5,6,6,6-decafluoro- 5-(trifluoromethyl)-	15899-29-3
1,3-Benzenedicarboxylic acid, polymer with 2,5-furandione, 1,6-hexanediol, and 1,2-propanediol	42133-48-2
Butanamide, N-(3-aminophenyl)-3-oxo-, mono= hydrochloride	59994-21-7
Acetic acid, sec-octyl ester	54515-77-4
Trisiloxane, 1,1,1,3,5,5,5-heptamethyl- 3-[(trimethylsilyl)oxy]-	17928-28-8
Benzenepropanol, 4-(1,1-dimethylethyl)-beta-methyl-	56107-04-1
Benzoic acid, 2-hydroxy-, strontium salt (2:1)	526-26-1

Chemical	CAS Number
2-Propenoic acid, diester with butanediol	31442-13-4
Acetamide, 2-chloro-2,2-difluoro-	354-28-9
Docosanoic acid, octadecyl ester	24271-12-3
3-Cyclohexene-1-carboxaldehyde, dimethyl-	27939-60-2
2-Propenoic acid, 3-(1H-imidazol-4-yl)-	104-98-3
Propanedinitrile, ([3-chloro-4-(octadecyl= oxy)phenyl]hydrazono)-,	41319-88-4
Phenol, 4,4'-(2-pyridinylmethylene)bis-, diacetate (ester)	603-50-9
1-Butanesulfonic acid, 4-[(4-aminophenyl)butyl= amino]-	35079-64-2
Benzoic acid, 4-[1-([(5-[(4-[2,4-bis(1,1-dimethyl= propyl)phenoxy]-1-oxobutyl)amino]-2-chloro= phenyl)amino]carbonyl)-3,3-dimethyl-2-oxo= butoxy]-, methyl ester	63217-24-3
Oils, cascarilla	8007-06-5
* 1,3-Cyclopentadiene, 1,2,3,4,5,5-hexachloro-	77-47-4
1,2-Benzenediamine, 4-nitro-, dihydrochloride	6219-77-8
1-Nonanamine	112-20-9
* Hexanoic acid, 2-propenyl ester	123-68-2
Furan, 2,2,3,3,4,4,5-heptafluorotetrahydro- 5-(nonafluorobutyl)-	335-36-4
9-Octadecenoic acid, (Z)-, potassium salt	143-18-0
Thiocyanic acid, ammonium salt	1762-95-4
Styrax balsam	8046-19-3
4(1H)-Pyrimidinone, 2,3-dihydro-6-propyl-2-thioxo-	51-52-5
Bicyclo[3.1.1]heptane, 2,6,6-trimethyl-	473-55-2
Benzenesulfonic acid, 2,5-dichloro-4-[4,5-dihydro- 3-methyl-5-oxo-4-(phenylazo)-1H-pyrazol-1-yl]-, sodium salt	6359-97-3
Benzoic acid, 2-hydroxy-, methyl ester	119-36-8
1-Octen-3-ol	3391-86-4
2-Naphthalenesulfonic acid, 8-hydroxy-5,7-dinitro-	483-84-1
1,8-Naphthalenedicarboxylic acid, dipotassium salt	1209-84-3
Barium peroxide	1304-29-6
2-Propenoic acid, octadecyl ester	4813-57-4
9,10-Anthracenedione, 1,4-diamino-	128-95-0
2-Propenoic acid, homopolymer, sodium salt	9003-04-7
Chromate(3-), bis[5-([5-chloro-2-hydroxyphenyl]= azo)-6-hydroxy-2-naphthalenesulfonato(3-)]-, disodium hydrogen	6408-02-2
9-Octadecenamide, N-(2-aminoethyl)-N-(2-hydroxy= ethyl)-, (Z)-	93-81-2
Benzenesulfonamide, 4-amino-	63-74-1
Orange flower water	8030-28-2

CHEMICALS IN COMMERCE
Production Less than 1 Million Pounds/Year, continued

Chemical	CAS Number
9,10-Anthracenedione	84-65-1
Benzene, fluoro-	462-06-6
1(3H)-Isobenzofuranone, 6-(dimethylamino)- 3,3-bis[4-(dimethylamino)phenyl]-	1552-42-7
Benzaldehyde	100-52-7
Butanamide, N-(2-methoxyphenyl)-3-oxo-	92-15-9
Oils, citronella	8000-29-1
9H-Carbazole, 9-ethyl-3-nitro-	86-20-4
C.I. Pigment Yellow 35	8048-07-5
Phosphoric acid, monosodium salt	7558-80-7
2-Propen-1-ol, 2-methyl-	513-42-8
2,7-Naphthalenedisulfonic acid, 4-amino-5-hydroxy- 3-[(4-nitrophenyl)azo]-6-(phenylazo)-, disodium salt	1064-48-8
Benzeneacetic acid, alpha-amino-, (+)-	2835-06-5
Lead	7439-92-1
Poly(oxy-1,2-ethanediyl), alpha-(isooctylphenyl)- omega-hydroxy-	9004-87-9
Benzenamine, N,N-diethyl-4-nitro-	2216-15-1
Benzeneacetic acid, 1-methylethyl ester	4861-85-2
2,4-Pentanediol, 2-methyl-	107-41-5
Dammar	9000-16-2
2-Imidazoline-1-ethanol, 2-(8-heptadecynyl)-, monoacetate (salt)	3388-72-5
3,4-Pyridinedimethanol, 5-hydroxy-6-methyl-, hydrochloride	58-56-0
L-Serine	56-45-1
Vitamin B$_{12}$	68-19-9
2-Propenoic acid, 2-methyl-, dodecyl ester	142-90-5
Benzenesulfonic acid, 2,2'-(1,2-ethenediyl)bis= (5-[(4-[bis(2-hydroxyethyl)amino]-6-(phenyl= amino)-1,3,5-triazin-2-yl)amino]-, disodium salt	4193-55-9
1,3-Propanediol, 2-[(benzoyloxy)methyl]- 2-methyl-, dibenzoate	4196-87-6
Octadecanoic acid, 12-hydroxy-, methyl ester	141-23-1

CHEMICALS IN COMMERCE
Production Unknown or Inaccessible

Chemical	CAS Number
2-Propanone, 1-(2-furanyl)-	6975-60-6
1-Hexadecen-3-ol, 3,7,11,15-tetramethyl-	505-32-8
Phenol,dodecyl-, lead(2+) salt	68586-21-0
Poly(oxy-1,2-ethanediyl), alpha-[2,4-bis= (1-methylpropyl)phenyl]-omega-hydroxy-	67970-22-3

145

Chemical	CAS Number
2,7-Naphthalenedisulfonic acid, 4-amino-6-([4-([(4-[(2,4-diaminophenyl)azo]phenyl)=amino]sulfonyl)phenyl]azo)-5-hydroxy-3-[(4-nitrophenyl)azo]-	72089-20-4
Fatty acids, tall-oil, polymers with glycerol, maleic anhydride, phthalic anhydride, and soybean oil	68015-41-8
* Benzoic acid, 2-hydroxy-5-[(4-nitrophenyl)azo]-, monosodium salt	1718-34-9
Safflower oil, polymer with conjugated safflower oil, glycerol, methyl methacrylate, penta=erythritol, phthalic anhydride, and styrene	68083-08-9
Propanoic acid, 3,3'-thiobis-, diethyl ester	673-79-0
Distillates (petroleum), solvent-dewaxed light paraffinic	64742-56-9
Poly(oxy-1,2-ethanediyl), alpha-(carboxymethyl)-omega-hydroxy-, C_{12-13}-alkyl ethers	70750-17-3
Trisiloxane, 1,1,1,5,5,5-hexamethyl-3-phenyl-3-[(trimethylsilyl)oxy]-	2116-84-9
2-Propenoic acid, 2-methyl-, methyl ester, polymer with ethenylbenzene, 2-propenenitrile, and 2-propenoic acid	38684-13-8
Quaternary ammonium compounds, bis(hydrogenated tallow alkyl)dimethyl, methyl sulfates	61789-81-9
Manganese alloy, base, Mn_{65-68}, Fe_{10-23}, Si_{12-21}, $C_{0.5-3}$, $P_{0-0.2}$ (ASTM A483)	12743-28-1
2-Naphthalenesulfonic acid, 6-hydroxy-5-[(2-methoxy-5-methyl-4-sulfophenyl)azo]-, disodium salt	25956-17-6
* Germanium	7440-56-4
2-Propenenitrile, polymer with 1,3-butadiene and ethenylbenzene, ammonium salt	67952-85-6
Phenol, 4-(2,4-dichlorophenoxy)-	40843-73-0
* Lithium, (1-methylpropyl)-	598-30-1
Oils, menhaden, polymers with p-tert-butylphenol, formaldehyde, glycerol, pentaerythritol, phthalic anhydride, and rosin	68553-68-4
6-Octenoic acid, 3,7-dimethyl-, methyl ester	2270-60-2
* Propanal, 2-methyl-	78-84-2
Ethanaminium, N-[4-([4-(diethylamino)phenyl]=[4-(ethylamino)-1-naphthalenyl]methylene)-2,5-cyclohexadien-1-ylidene]-N-ethyl-, trihydroxypentatriacontaoxo[phosphato(3-)]=dodecamolybdate(4-) (4:1)	69070-64-0

Chemical	CAS Number
* 1,3-Propanediol, 2-methyl-2-[(nitrooxy)methyl]-, dinitrate (ester)	3032-55-1
* 2,6,10,14,18,22-Tetracosahexaene, 2,6,10,= 15,19,23-hexamethyl-, (all-E)-	111-02-4
* Benzene, 1,1'-oxybis[2,3,4,5,6-pentabromo-	1163-19-5
2-Propenoic acid, polymer with ethenylbenzene, ethyl 2-propenoate, formaldehyde, and 2-propenamide	67846-51-9
2,7-Naphthalenedisulfonic acid, 4-hydroxy- 3-([4-([2-(sulfooxy)ethyl]sulfonyl)phenyl]= azo)-, tripotassium salt	72187-37-2
1-Penten-3-one, 1-(2,6,6-trimethyl-2-cyclo= hexen-1-yl)-	127-42-4
Fatty acids, C_{12-18}, polymers with adipic acid, C_{14-18} fatty acids, 1,6-hexanediol, isodecanol, and propylene glycol	71060-65-6
* Bicyclo[7.2.0]undec-4-ene, 4,11,11-trimethyl- 8-methylene-, [1R-(1R*,4E,9S*)]-	87-44-5
Vanadic acid, ammonium salt	11115-67-6
2-Propanethiol	75-33-2
1-Octacosanol	557-61-9
* Carbonic acid, nickel(2+) salt (1:1)	3333-67-3
* Ferrocene	102-54-5
Soybean oil, polymer with isophthalic acid and trimethylolethane	66070-63-1
Tin hydroxide	12054-72-7
* C.I. Pigment Green 7	1328-53-6
Zeolites, calcium-iron-magnesium-vanadium- containing	68918-02-5
Glycine, N-phenyl-, monosodium salt	10265-69-7
* tert-Dodecanethiol	25103-58-6
2-Anthracenesulfonic acid, 1-amino-9,10-dihydro- 4-[(4-[(methylamino)methyl]phenyl)amino]- 9,10-dioxo-, monosodium salt	67905-55-9
Benzenemethanol, alpha-ethynyl-alpha-methyl-	127-66-2
Fatty acids, tall-oil, polymers with dipropylene glycol, maleic anhydride, and pitch	68459-12-1
* Acetaldehyde, chloro-	107-20-0
Fatty acids, tall-oil, compounds with N-cyclohexyl- N-methylcyclohexanamine	68188-05-6
Benzenesulfinic acid, methyl-, bis[4-(dimethyl= amino)phenyl]methyl ester	29061-52-7
* Zinc, bis(2,4-pentanedionato-O,O')-, (T-4)-	14024-63-6
1H-Benzotriazole, sodium salt	15217-42-2

CHEMICALS IN COMMERCE
Production Unknown or Inaccessible, continued

Chemical	CAS Number
Poly(oxy-1,2-ethanediyl), alpha,alpha',alpha'',= alpha'''-(1,4-phenylenebis[methylene(octadecyl= nitrilio)di-2,1-ethanediyl])tetrakis= [omega-hydroxy-, dichloride	68140-77-2
Leach solutions, copper, spent	69012-76-6
Oils, walnut, polymers with glycerol and phthalic anhydride	68553-87-7
Amides, C16-18 and C18-unsaturated, N,N-bis= (hydroxyethyl)	68603-38-3
2-Propenoic acid, 3,3'-(1,4-phenylene)bis-	16323-43-6
Lanthanum iodide (LaI3)	13813-22-4
* 2,5-Cyclohexadiene-1,4-dione, dioxime	105-11-3
Cyclohexene, 1-ethenyl-	2622-21-1
Ethane, 1,1-dichloro-1,2,2,2-tetrafluoro-	374-07-2
* Hydroxylamine, sulfate (2:1)	10039-54-0
2-Naphthalenecarboxanilide, 3-hydroxy- 4-[(4-methoxy-2-nitrophenyl)azo]-	4154-63-6
Benzene, trichloro-, polymer with 1,4-dichloro= benzene and sodium sulfide (Na2S)	72276-00-7
1,4-Benzenedimethanamine	539-48-0
1H-Pyrazole-3-carboxylic acid, 1-(3-aminophenyl)- 4-[(2-methoxy-4-[(3-sulfophenyl)azo]phenyl)= azo]-, disodium salt	68227-66-7
Hexanoic acid, 2-ethyl-, cobalt(2+) salt	136-52-7
Urea, polymer with formaldehyde and 1,3,5,7-tetra= azatricyclo-[3.3.1.1^3,7]decane, butylated ethylated	69898-36-8
2-Propenoic acid, 4-(1-methyl-1-phenylethyl)phenyl ester	54449-74-0
* 2-Butene, 1,4-dibromo-, (E)-	821-06-7
Benzene, 1-chloro-4-(methylthio)-2-nitro-	1199-36-6
8-Quinolinol, 7-C12-16-alkyl derivatives	68511-63-7
Phenol, 2,2'-methylenebis[4-chloro-	97-23-4
Ethanesulfonic acid, 2-[cyclohexyl(1-oxo= tetradecyl)amino]-, sodium salt	63217-16-3
Silicon(1+), tris(2,4-pentanedionato-O,O')-, (OC-6-11)-, hexafluoroantimonate(1-)	67251-37-0
3H-Pyrazol-3-imine, 2,4-dihydro-5-methyl-2-phenyl-	6401-97-4
Poly(oxy-1,2-ethanediyl), alpha-sulfo- omega-(9,12-octadecadienyloxy)-, sodium salt, (Z,Z)-	65086-42-2
Benzenediazonium, 5-methoxy-4-methyl- 2-([2-(methylsulfonyl)-4-nitrophenyl]azo)-, sulfate (1:1)	72152-44-4

148

Chemical	CAS Number
2-Naphthalenesulfonic acid, 4-hydroxy- 3-[(4-methoxy-2-sulfophenyl)azo]- 7-(methylamino)-	51838-10-9
* Gallium oxide	12024-21-4
Hexanedioic acid, polymer with 5-amino- 1,3,3-trimethylcyclohexanemethanamine, 1,3-diiso=cyanatomethylbenzene, and 1,2-propanediol	68212-31-7
Acetic acid, mercapto-, 2,2-bis([(mercapto= acetyl)oxy]methyl)-1,3-propanediyl ester	10193-99-4
* Benzenesulfonic acid, dodecyl-, sodium salt	25155-30-0
Soybean oil, polymer with benzoic acid, isophthalic acid, and phthalic anhydride	68554-84-7
Safflower oil, polymer with p-tert-butylbenzoic acid, glycerol, lauric acid, and phthalic anhydride	68474-53-3
Acetic acid, (3-methylbutoxy)-, 2-propenyl ester	67634-00-8
Fatty acids, coco, polymers with isophthalic acid, neopentyl glycol, and trimellitic anhydride	68525-97-3
1,4-Diisocyanatobenzene	104-49-4
* Tungsten carbide	12070-12-1
Benzene, 2-methoxy-1-(1-methoxyethoxy)-4- (2-propenyl)-	68213-85-4
Amides, coco	61789-19-3
Oxirane, methyl-, polymer with oxirane, mono= (hydrogen, sulfate), sodium salt	67426-60-2
Benzeneacetic acid, 5-methyl-2-(1-methylethyl)= cyclohexyl ester, [1R-(1alpha,2beta,5alpha)]-	26171-78-8
3,6,9,12,15,18,21,24-Octaoxaheptacosanoyl fluoride, 2,4,4,5,7,7,8,10,10,11,13,13,14,= 16,16,17,19,19,20,22,22,23,25,25,26,26,= 27,27,27,-nonacosafluoro-2,5,8,11,14,17,20,= 23-octakis(trifluoromethyl)-	13140-26-6
3-Thiazolidineethanol, 2-imino-alpha-phenyl-, 4-methylbenzenesulfonate (salt) (1:1)	18126-02-8
Aluminum hydroxide	21645-51-2
Amines, di-C$_{14-18}$-alkyl	68037-98-9
Zinc, diamminebis(nitrato-O,O')-	33363-00-7
Poly[oxy(methyl-1,2-ethanediyl)], alpha,alpha',= alpha'',alpha'''-(1,2-ethanediylbis[nitrilo= bis(methyl-2,1-ethanediyl)])tetrakis= [omega-hydroxy-, sodium salt	68567-72-6
Phosphoramidic acid, dibutyl-, diethyl ester	67828-17-5
Amines, bis(hydrogenated tallow alkyl)	61789-79-5

Chemical	CAS Number
Starch, polymer with 2-propenenitrile, hydrolyzed, sodium salt	68442-56-8
Coconut oil, polymer with ethylene glycol, pentaerythritol, and phthalic anhydride	66070-89-1
Manganese nitride	12033-07-7
Thiols, C_{4-20}, gamma-omega-perfluoro	68140-19-2
1,3-Benzenedicarboxylic acid, 5-sulfo-, 1,3-bis(2-hydroxyethyl) ester, monosodium salt	24019-46-3
Fatty acids, tallow, lithium salts	64755-02-8
Amines, C_{10-16}-alkyldimethyl	67700-98-5
1,3,5-Triazine-2,4,6-triamine, polymer with formaldehyde, hydrochloride	62412-64-0
Octanoic acid, 2-butyl-	27610-92-0
Cyclopentanone, 2-hexylidene-	17373-89-6
Fatty acids, tall-oil, reaction products with 2-[(2-aminoethyl)amino]ethanol, quaternized with sulfate diethyl	70955-34-9
Benzenediazonium, 4-(diethylamino)-, chloride, compound with zinc chloride	17409-47-1
Titanium nitride	25583-20-4
Quaternary ammonium compounds, bis(hydrogenated tallow alkyl)dimethyl, chlorides	61789-80-8
1,3,5-Triazine-2,4,6-triamine, polymer with formaldehyde, isobutylated	68002-21-1
Hexanedioic acid, polymer with 1,6-hexanediol	25212-06-0
Naphtho[1,2-d]thiazol-2-amine	40172-65-4
Formaldehyde, polymer with oxybis[propanol] and phenol, methylated	68441-82-7
Oils, cod, polymerized, oxidized	68082-76-8
Fatty acids, C_{6-12}	67762-36-1
2-Propenoic acid, 2-methyl-, methyl ester, polymer with butyl 2-propenoate	25852-37-3
Poly[oxy(methyl-1,2-ethanediyl)], alpha-hydro-= omega-hydroxy-, polymer with 1,1'-methylene= bis[4-isocyanatobenzene]	9048-57-1
Hexanedioic acid, polymer with 2-ethyl-2-(hydroxy= methyl)-1,3-propanediol and 2,2'-oxybis[ethanol]	28183-09-7
Bacillus megaterium	68038-67-5
Ethanedioic acid, bis(3,4,6-trichloro-2-[(pentyl= oxy)carbonyl]phenyl) ester	30431-54-0

APPENDIX B

TESTING FOR VARIOUS SITUATIONS OF CHEMICAL USE AND GENERAL EXPOSURE TO DIRECT AND INDIRECT FOOD ADDITIVES

(Legend of symbols and reference notations follows Appendix G)

Toxicity Tests	Direct Food Additives (Including Colors)	Indirect Food Additive — Virtually Nil Migration (<0.05 ppm)	Indirect Food Additive — Insignificant Migration (0.05–1 ppm)	Indirect Food Additive — Significant Migration (>1 ppm)	General Exposure — Occupational	General Exposure — Environmental
1. Acute oral toxicity--rodent	X	X	X	X	X	X
2. Acute oral toxicity--nonrodent	-	*	-	-	-	-
3. Acute dermal toxicity	-	-	-	-	X	X
4. Acute parenteral toxicity	-	-	-	-	-	-
5. Acute inhalation toxicity	-	-	-	-	X	X
6. Acute dermal irritation--corrosivity	-	-	-	-	X	X
7. Acute eye irritation--corrosivity	-	-	-	-	X	X
8. Skin sensitization--guinea pig	-	X	-	-	X	X
9. Subchronic oral toxicity--rodent: 14- or 28-d	-	-	-	-	-	-
10. Subchronic toxicity--nonrodent: 14- or 28-d	-	-	X	-	*	*
11. Subchronic oral toxicity--rodent: 90-d	-	-	-	-	*	*
12. Subchronic oral toxicity--nonrodent: 90-d	X	-	X	X	X	*
13. Subchronic oral toxicity--nonrodent: 6- to 12-mo	*	-	-	*	-	-
14. Subchronic dermal toxicity: 21- or 28-d	-	-	-	-	-	-
15. Subchronic dermal toxicity: 90-d	-	-	-	-	X	*
16. Subchronic inhalation toxicity: 28- or 14-d	-	-	-	-	*	*
17. Subchronic inhalation toxicity: 90-d	-	-	-	-	X	*
18. Neurobehavioral toxicity (see Appendix I)	X	-	-	-	-	-

151

APPENDIX B (continued)

| Toxicity Tests | Direct Food Additives (Including Colors) | Indirect Food Additives | | | General Exposure | |
| | | Insignificant Migration | | Significant migration (>1 ppm) | Occupational | Environmental |
		Virtually Nil Migration (<0.05 ppm)	(0.05-1 ppm)			
19. Teratology study--rodent, rabbit	X	-	X	X	X	X
20. Multigeneration reproduction study--rodent	X	-	*	-	X	*
21. Toxicokinetics	*	*[c]	-	*	*	*
22. Carcinogenicity--rodent	X[a]	*[c]	*[c]	X[a]	X	*
23. Chronic toxicity	-	-	-	-	X	*
24. Combined chronic toxicity-carcinogenicity--rodent	X[b]	-	-	X[b]	-	-
25. Genetic toxicity (see Appendix J)	X	X	X	X	X	X
26. Subchronic eye toxicity	-	-	-	-	-	-
27. Segment I: fertility and reproductive performance	-	-	-	-	-	-
28. Segment III: perinatal and postnatal performance	-	-	-	-	-	-
29. Acute delayed neurotoxicity	-	-	-	-	-	-
30. Skin painting--chronic	-	-	-	-	-	-
31. Implantation studies	-	-	-	-	-	-
32. Human sensitization studies	-	-	-	-	-	-
33. Skin penetration studies	-	-	-	-	-	-

APPENDIX C

TESTING FOR VARIOUS SITUATIONS OF CHEMICAL USE AND GENERAL EXPOSURE TO ORAL OR PARENTERAL DRUGS OR COLOR ADDITIVES FOR SUTURES

(Legend for symbols and reference notations follows Appendix G)

Toxicity Tests	Period of Oral or Parenteral Use				Color Additive for Sutures	General Exposure	
	Several Days	Up to 2 Weeks	Up to 3 Months	6 Months to Unlimited		Occupational	Environmental
1. Acute oral toxicity--rodent	x^f	x^f	x^f	x^f	-	x	x
2. Acute oral toxicity--nonrodent	x^f	x^f	x^f	x^f	-	-	-
3. Acute dermal toxicity	-	-	-	-	-	x	x
4. Acute parenteral toxicity	x^f	x^f	x^f	x^f	-	-	-
5. Acute inhalation toxicity	-	-	-	-	-	x	x
6. Acute dermal irritation--corrosivity	-	-	-	-	-	x	x
7. Acute eye irritation--corrosivity	-	-	-	-	-	x	x
8. Skin sensitization--guinea pig	-	-	-	-	-	x	x
9. Subchronic oral toxicity--rodent: 14- or 28-d	x^g	$*^g$	$*^g$	-	-	-	-
10. Subchronic toxicity--nonrodent: 14- or 28-d	x	$*^g$	$*^g$	-	-	-	-
11. Subchronic oral toxicity--rodent: 90-d	-	x^g	x^h	$*^g$	-	*	*
12. Subchronic oral toxicity--nonrodent: 90-d	-	x^g	x^h	$*^g$	-	*	*
13. Subchronic oral toxicity--nonrodent: 6- to 12-mo	-	-	-	x^j	-	x	*
14. Subchronic dermal toxicity: 21- or 28-d	-	-	-	-	-	-	-
15. Subchronic dermal toxicity: 90-d	-	-	-	-	-	-	-
16. Subchronic inhalation toxicity: 28- or 14-d	-	-	-	-	-	*	*
17. Subchronic inhalation toxicity: 90-d	-	-	-	-	-	x	*
18. Neurobehavioral toxicity (see Appendix I)	-	x	x	x	-	-	-

153

APPENDIX C (continued)

Toxicity Tests	Period of Oral or Parenteral Use				Color Additive for Sutures	General Exposure	
	Several Days	Up to 2 Weeks	Up to 3 Months	6 Months to Unlimited		Occupational	Environmental
19. Teratology study--rodent, rabbit	X	X	X	X	-	X	X
20. Multigeneration reproduction study--rodent	X	X	X	X	-	X	*
21. Toxicokinetics	*	*	*	*	-	*	*
22. Carcinogenicity--rodent	-	-	-	*	*c	X	*
23. Chronic toxicity	-	-	-	X^i	-	X	*
24. Combined chronic toxicity--carcinogenicity--rodent	-	-	-	*	-	-	-
25. Genetic toxicity (see Appendix J)	X	X	X	X	-	X	X
26. Subchronic eye toxicity	-	-	-	-	-	-	-
27. Segment I: fertility and reproductive performance	X	X	X	X	-	-	-
28. Segment III: perinatal and postnatal performance	X	X	X	X	-	-	-
29. Acute delayed neurotoxicity	*	*	-	-	-	-	-
30. Skin painting--chronic	-	-	-	-	-	-	-
31. Implantation studies	-	-	-	-	X^d	-	-
32. Human sensitization studies	-	-	-	-	-	-	-
33. Skin penetration studies	-	-	-	-	-	-	-

154

APPENDIX D

TESTING FOR VARIOUS SITUATIONS OF CHEMICAL USE AND GENERAL EXPOSURE TO INHALATION AND VETERINARY DRUGS

(Legend of symbols and reference notations follows Appendix G)

Toxicity Tests	Inhalation (General Anesthetics)	Veterinary Drugs	General Exposure Occupational	General Exposure Environmental
1. Acute oral toxicity--rodent	-	x^o	x	x
2. Acute oral toxicity--nonrodent	-	x^o	-	-
3. Acute dermal toxicity	-	-	x	x
4. Acute parenteral toxicity	*	-	-	-
5. Acute inhalation toxicity	-	-	x	x
6. Acute dermal irritation--corrosivity	-	-	x	x
7. Acute eye irritation--corrosivity	-	-	x	x
8. Skin sensitization--guinea pig	-	-	x	x
9. Subchronic oral toxicity--rodent: 14- or 28-d	-	-	x	x
10. Subchronic toxicity--nonrodent: 14- or 28-d	-	-	-	-
11. Subchronic oral toxicity--rodent: 90-d	-	x^o	*	*
12. Subchronic oral toxicity--nonrodent: 90-d	-	x^o	*	*
13. Subchronic oral toxicity--nonrodent: 6- to 12-mo	-	-	x	*
14. Subchronic dermal toxicity: 21- or 28-d	-	-	-	-
15. Subchronic dermal toxicity: 90-d	-	-	x	*
16. Subchronic inhalation toxicity: 28- or 14-d	x^1	-	*	*
17. Subchronic inhalation toxicity: 90-d	-	-	x	*
18. Neurobehavioral toxicity (see Appendix I)	-	-	x	x

APPENDIX D (continued)

Toxicity Tests	Inhalation (General Anesthetics)	Veterinary Drugs	General Exposure Occupational	General Exposure Environmental
19. Teratology study--rodent, rabbit	X	-	X	X
20. Multigeneration reproduction study--rodent	X	-	X	*
21. Toxicokinetics	*	-	*	*
22. Carcinogenicity--rodent	-	-	X	*
23. Chronic toxicity	-	-	X	*
24. Combined chronic toxicity-carcinogenicity--rodent	-	-	-	-
25. Genetic toxicity (see Appendix J)	X	-	X	X
26. Subchronic eye toxicity	-	-	-	-
27. Segment I: fertility and reproductive performance	X	-	-	-
28. Segment III: perinatal and postnatal performance	X	-	-	-
29. Acute delayed neurotoxicity	-	-	-	-
30. Skin painting--chronic	-	-	-	-
31. Implantation studies	-	-	-	-
32. Human sensitization studies	-	-	-	-
33. Skin penetration studies	-	-	-	-

156

APPENDIX E

TESTING FOR VARIOUS SITUATIONS OF CHEMICAL USE AND GENERAL EXPOSURE TO COSMETICS AND DERMAL, VAGINAL-RECTAL, OVER-THE-COUNTER, AND OPHTHALMIC DRUGS

(Legend of symbols and reference rotations follows Appendix G)

Toxicity Tests	Cosmetics	Dermal Drugs	Vaginal-Rectal Drugs	Over-the-Counter Drugs	Ophthalmic Drugs	General Exposure Occupational	General Exposure Environmental
1. Acute oral toxicity--rodent	-	-	*[f]	x[f]	*[f]	X	X
2. Acute oral toxicity--nonrodent	-	-	*[f]	x[f]	*[f]	-	-
3. Acute dermal toxicity	X	X	-	x[f]	-	X	X
4. Acute parenteral toxicity	-	-	-	x[f]	-	-	-
5. Acute inhalation toxicity	-	-	x[h]	x[f]	-	X	X
6. Acute dermal irritation--corrosivity	X	X	x[h]	*[f]	-	X	X
7. Acute eye irritation--corrosivity	X	-	-	*[f]	-	X	X
8. Skin sensitization--guinea pig	X	X	-	*[f]	-	X	X
9. Subchronic oral toxicity--rodent: 14- or 28-d	-	-	-	-	-	X	X
10. Subchronic toxicity--nonrodent: 14- or 28-d	-	-	-	-	-	-	-
11. Subchronic oral toxicity--rodent: 90-d	-	-	x[n]	*	-	*	*
12. Subchronic oral toxicity--nonrodent: 90-d	-	-	x[n]	-	-	*	*
13. Subchronic oral toxicity--nonrodent: 6- to 12-mo	-	-	-	-	-	X	*
14. Subchronic dermal toxicity: 21- or 28-d	X	X	-	-	-	-	-
15. Subchronic dermal toxicity: 90-d	*	x[k]	-	*	-	X	*
16. Subchronic inhalation toxicity: 28- or 14-d	-	-	-	*	-	*	*
17. Subchronic inhalation toxicity: 90-d	-	-	-	*	-	X	*
18. Neurobehavioral toxicity (see Appendix I)	*	-	*	X	-	-	-

157

APPENDIX E (continued)

Toxicity Tests	Cosmetics	Dermal Drugs	Vaginal-Rectal Drugs	Over-the-counter Drugs	Ophthalmic Drugs	General Exposure Occupational	General Exposure Environmental
19. Teratology study--rodent, rabbit	-	X	X	X	X	X	X
20. Multigeneration reproduction study--rodent	-	X	X	X	X	X	*
21. Toxicokinetics	*c	*	*	*	*	*	*
22. Carcinogenicity--rodent	-	-	-	*i	-	X	*
23. Chronic toxicity	-	-	-	X	-	X	*
24. Combined chronic toxicity-carcinogenicity--rodent	-	-	-	*	-	-	-
25. Genetic toxicity (see Appendix J)	X	X	X	X	X	X	X
26. Subchronic eye toxicity	X	-	-	-	X^m	-	-
27. Segment I: fertility and reproductive performance	-	X	X	X	X	-	-
28. Segment III: perinatal and postnatal performance	-	X	X	X	X	-	-
29. Acute delayed neurotoxicity	-	*	-	-	-	-	-
30. Skin painting--chronic	X	*	-	-	-	-	-
31. Implantation studies	-	-	-	-	-	-	-
32. Human sensitization studies	X^e	*	-	-	-	-	-
33. Skin penetration studies	*	-	-	-	-	-	-

APPENDIX F

TESTING FOR VARIOUS SITUATIONS OF CHEMICAL USE AND GENERAL EXPOSURE TO PESTICIDES

(Legend of symbols and reference notations follows Appendix G)

Toxicity Tests	Pesticides	General Exposure Occupational	General Exposure Environmental
1. Acute oral toxicity--rodent	X	X	X
2. Acute oral toxicity--nonrodent	-	-	-
3. Acute dermal toxicity	X	X	X
4. Acute parenteral toxicity	-	-	-
5. Acute inhalation toxicity	X	X	X
6. Acute dermal irritation--corrosivity	X	X	X
7. Acute eye irritation--corrosivity	X	X	X
8. Skin sensitization--guinea pig	X	X	X
9. Subchronic oral toxicity--rodent: 14- or 28-d	-	X	X
10. Subchronic toxicity--nonrodent: 14- or 28-d	-	-	-
11. Subchronic oral toxicity--rodent: 90-d	X	*	*
12. Subchronic oral toxicity--nonrodent: 90-d	X	*	*
13. Subchronic oral toxicity--nonrodent: 6- to 12-mo	-	X	*
14. Subchronic dermal toxicity: 21- or 28-d	X	-	-
15. Subchronic dermal toxicity: 90-d	-	X	*
16. Subchronic inhalation toxicity: 28- or 14-d	X	*	*
17. Subchronic inhalation toxicity: 90-d	-	X	*
18. Neurobehavioral toxicity (see Appendix I)	X	X	X

APPENDIX F (continued)

Toxicity Tests	Pesticides	General Exposure Occupational	General Exposure Environmental
19. Teratology study--rodent, rabbit	X	X	X
20. Multigeneration reproduction study--rodent	X	X	*
21. Toxicokinetics	X	*	*
22. Carcinogenicity--rodent	x^p	X	*
23. Chronic toxicity	x^p	X	*
24. Combined chronic toxicity--carcinogenicity--rodent	x^q	-	-
25. Genetic toxicity (see Appendix J)	X	X	X
26. Subchronic eye toxicity	-	-	-
27. Segment I: fertility and reproductive performance	-	-	-
28. Segment III: perinatal and postnatal performance	-	-	-
29. Acute delayed neurotoxicity	*	X	X
30. Skin painting--chronic	-	-	-
31. Implantation studies	-	-	-
32. Human sensitization studies	-	-	-
33. Skin penetration studies	-	-	-

APPENDIX G

TESTING FOR VARIOUS SITUATIONS OF CHEMICAL USE AND GENERAL EXPOSURE TO OTHER MARKETABLE CHEMICALS

(Legend of symbols and reference notations follows Appendix G)

Toxicity Tests	Other Marketable Chemicals	General Exposure Occupational	General Exposure Environmental
1. Acute oral toxicity—rodent	X	X	X
2. Acute oral toxicity—nonrodent	–	–	–
3. Acute dermal toxicity	X	X	X
4. Acute parenteral toxicity	–	–	–
5. Acute inhalation toxicity	X	X	X
6. Acute dermal irritation—corrosivity	X	X	X
7. Acute eye irritation—corrosivity	X	X	X
8. Skin sensitization—guinea pig	X	X	X
9. Subchronic oral toxicity—rodent: 14- or 28-d	X	X	X
10. Subchronic toxicity—nonrodent: 14- or 28-d	–	–	–
11. Subchronic oral toxicity—rodent: 90-d	X	*	*
12. Subchronic oral toxicity—nonrodent: 90-d	X	*	*
13. Subchronic oral toxicity—nonrodent: 6- to 12-mo	–	X	*
14. Subchronic dermal toxicity: 21- or 28-d	X	–	–
15. Subchronic dermal toxicity: 90-d	–	X	*
16. Subchronic inhalation toxicity: 28- or 14-d	–	*	*
17. Subchronic inhalation toxicity: 90-d	–	X	*
18. Neurobehavioral toxicity (see Appendix I)	X	X	X

APPENDIX G (continued)

Toxicity Tests	Other Marketable Chemicals	General Exposure	
		Occupational	Environmental
19. Teratology study--rodent, rabbit	X	X	X
20. Multigeneration reproduction study--rodent	X	X	*
21. Toxicokinetics	X	*	*
22. Carcinogenicity--rodent	X	X	*
23. Chronic toxicity	X	X	*
24. Combined chronic toxicity-carcinogenicity--rodent	-	-	-
25. Genetic toxicity (see Appendix J)	X	X	X
26. Subchronic eye toxicity	-	-	-
27. Segment I: fertility and reproductive performance	-	-	-
28. Segment III: perinatal and postnatal performance	-	-	-
29. Acute delayed neurotoxicity	-	X	X
30. Skin painting--chronic	-	-	-
31. Implantation studies	-	-	-
32. Human sensitization studies	-	-	-
33. Skin penetration studies	-	-	-

162

Symbols and reference notations for Appendixes B-G

X = Appropriate test; list of checked tests will be considered as minimal necessary for evaluation of adequate data base.

* = If indicated by available data or information; it will be responsibility of reviewing toxicologist to examine data from listed necessary tests and then judge whether these additional tests also will be necessary.

a = Use rodent other than rat.

b = Use only rat.

c = If carcinogenicity is suspected.

d = Lifetime--rat if carcinogenicity is suspected; short-term for absorbable sutures; ocular for ophthalmic sutures.

e = Do repeat patch test with photosensitization test.

f = Acute toxicity should be determined in three or four species by appropriate route(s) of intended use.

g = By appropriate route of intended use.

h = Will also require up to 6-mo study by appropriate route.

i = 18- or 24-mo study.

j = 12-mo study.

k = May require up to 6 mo on intact skin.

l = Four species; 3 h/d (5 d/wk) under conditions to be used clinically.

m = One species; duration commensurate with clinical use.

n = Duration and number of applications determined by intended use.

o = Additional studies appropriate to duration and route of intended use; some studies required in target species; if target species is food-producing animal, see direct food additives in Appendix B.

p = Or perform test 24.

q = Or perform tests 22 and 23.

APPENDIX H

REFERENCE PROTOCOLS FOR TOXICITY TESTING

Test	Reference Source[a]
1. Acute oral toxicity--rodent	IRLG, 1981c
2. Acute oral toxicity--nonrodent	OECD, 1981, 401:1-7;[b] when rabbit is used as nonrodent, fewer than 10 (5 per sex) at each dosage will be acceptable; for dogs or other large nonrodents, ascending-dose study will be acceptable
3. Acute dermal toxicity	IRLG, 1981a
4. Acute parenteral toxicity	OECD, 1981, 401:1-7; guidelines for acute oral toxicity should be followed, but administration will be by intravenous, intramuscular, subcutaneous, or intraperitoneal routes
5. Acute inhalation toxicity	NRC, 1977a
6. Acute dermal irritation--corrosivity	OECD, 1981, 404:1-6
7. Acute eye irritation--corrosivity	IRLG, 1981b
8. Skin sensitization--guinea pig	OECD, 1981, 406:1-9;
9. Subchronic oral toxicity--rodent: 14- or 28-d study	OECD, 1981, 407:1-9
10. Subchronic toxicity--nonrodent 14- or 28-d study	OECD, 1981, 407:1-9
11. Subchronic oral toxicity--rodent: 90-d study	OECD, 1981, 408:1-10
12. Subchronic oral toxicity--nonrodent: 90-d study	OECD, 1981, 409:1-9

Test	Reference sources
13. Subchronic oral toxicity--nonrodent: 6- to 12-mo study	OECD, 1981, 409:1-9
14. Subchronic dermal toxicity: 21- or 28-d study	OECD, 1981, 410:1-8
15. Subchronic dermal toxicity: 90-d study	OECD, 1981, 411:1-10
16. Subchronic inhalation toxicity: 28- or 14-d study	NRC, 1977a
17. Subchronic inhalation toxicity: 90-d study	NRC, 1977a
18. Subchronic neurotoxicity: 90-d study	OECD, 1979, pp. 106-109
19. Teratology study--rodent, rabbit	IRLG, 1981d
20. Multigeneration reproduction study-- rodent	U.S. Environmental Protection Agency, 1978
21. Toxicokinetics	OECD, 1981, 415:1-5
22. Carcinogenicity--rodent	OECD, 1981, 451:1-19
23. Chronic toxicity	OECD, 1981, 452:1-15
24. Combined chronic toxicity- carcinogenicity--rodent	OECD, 1981, 453:1-16
25. Genetic toxicity	OECD, 1979, pp. 114-116
26. Subchronic eye toxicity	U.S. Department of Health, Education, and Welfare, 1973[c]
27. Segment 1: fertility and reproductive performance	U.S. Department of Health, Education, and Welfare, 1973[c]
28. Segment III: perinatal and postnatal performance	U.S. Department of Health, Education, and Welfare, 1973[c]

29. Acute delayed neurotoxicity	U.S. Environmental Protection Agency, 1978
30. Skin painting--chronic	OECD, 1981, 451:1-15
31. Implantation studies	Guidelines for chronic toxicity (test 23) should be followed with test material implanted, rather than administered in diet or parenterally
32. Human sensitization studies	Marzulli and Maibach, 1980
33. Skin penetration studies	Marzulli et al., 1969

---

a   IRLG = Interagency Regulatory Liaison Group; OECD = Organisation for Economic Co-operation and Development; NRC = National Research Council.
b   Penultimate version of OECD guidelines.
c   Further descriptions of segments I and III, the Food and Drug Administration Bureau of Drugs requirements for reproduction studies, may be found in Collins (1978).

REFERENCES

Collins, T. F. X. 1978. Reproduction and teratology guidelines. J. Environ. Path. and Toxicol. 2:141-147.

Interagency Regulatory Liaison Group, Testing Standards and Guidelines Work Group. 1981a. Recommended Guidelines for Acute Dermal Toxicity Test. Washington, D.C.: Interagency Regulatory Liaison Group. 12 pp.

Interagency Regulatory Liaison Group, Testing Standards and Guidelines Work Group. 1981b. Recommended Guideline for Acute Eye Irritation Testing. Washington, D.C.: Interagency Regulatory Liaison Group. 12 pp.

Interagency Regulatory Liaison Group, Testing Standards and Guidelines Work Group. 1981c. Recommended Guideline for Acute Oral Toxicity Testing in Rodents. Washington, D.C.: Interagency Regulatory Liaison Group. 11 pp.

Interagency Regulatory Liaison Group, Testing Standards and Guidelines Work Group. 1981d. Recommended Guidelines for Teratogenicity Studies in the Rat, Mouse, Hamster, or Rabbit. Washington, D.C.: Interagency Regulatory Liaison Group. 12 pp.

Marzulli, F. N., W. C. Brown, and H. I. Maibach. 1969. Techniques for studying skin penetration. Toxicol. Appl. Pharmacol. (Suppl.) 3:76-83.

Marzulli, F. N., and H. I. Maibach. 1980. Contact allergy: Predictive testing of fragrance ingredients in humans by Draize and Maximization methods. J. Environ. Pathol. and Toxicol. 3:235-245.

National Research Council, Committee for the Revision of NAS Publication 1138. 1977. Principles and Procedures for Evaluating the Toxicity of Household Substances. Washington, D.C.: National Academy of Sciences. 130 pp.

Organisation for Economic Co-operation and Development. 1979. Short-Term and Long-Term Toxicology Groups. Final Report. 185 pp. (unpublished)

Organisation for Economic Co-operation and Development. 1981. Guidelines for the Testing of Chemicals. Paris: Organisation for Economic Co-operation and Development. c. 700 pp.

U.S. Department of Health, Education, and Welfare, Food and Drug Administration. 1973. Introduction to Total Drug Quality. DHEW Publication No. (FDA) 74-3006. Washington, D.C.: U.S. Department of Health, Education, and Welfare. 101 pp.

U.S. Environmental Protection Agency. 1978. Proposed guidelines for registering pesticides in the United States; Hazard evaluation: Humans and domestic animals. Fed. Reg. 43(163):37336-37403.

APPENDIX I

REFERENCE PROTOCOL GUIDELINES
FOR NEUROBEHAVIORAL-TOXICITY TESTS

Neurobehavioral-toxicity testing requires both morphologic and behavioral assessment. With current methods, the effects of some chemicals are most reliably detected by measuring morphologic changes, whereas other chemicals produce behavioral changes without, as yet, the identification of any morphologic basis. Because methods for neurobehavioral-toxicity testing are still being developed and validated, a high degree of standardization in such testing is inappropriate at this time. Nevertheless, a scheme proposed by a committee of the National Research Council (1977) gave specific and detailed suggestions of protocols for neurobehavioral-toxicity testing. The scheme, presented here, is a reasonable starting point for neurobehavioral-toxicity testing, and the suggested protocols were used as the reference protocol guideline in this study.

The 1977 committee recommended that both conditioned and unconditioned behaviors be studied. For unconditioned behavior, it suggested that circadian patterns of spontaneous motor activity (SMA)[*] be measured. SMA in rodents has been widely studied to document behavioral changes produced by chemicals.

This procedure does not require any special training of personnel. For conditioned behavior, it suggested that performance maintained by schedules of food presentation be used as a measure. Some types of schedule-controlled responding[**] require little training of the animal, but provide an estimate of the effects of a substance on fairly complex behavioral processes. Reiter et al. (1981) have found that measurements

---

[*] Spontaneous motor activity (SMA) is a measurement of an animal's body movement under a given set of environmental conditions. SMA may be detected by a variety of techniques, including photocell-beam interruption and changes in a capacitance field. It is important that measurements be objective and unbiased. Use of automatic instrumentation is recommended for the measurement of SMA.

[**] Schedule-controlled responding is maintained by reinforcement. For example, animals are trained to make specific responses to specific stimuli to obtain food or water or to avoid shock. The relationship of the response to delivery of such reinforcers is called the schedule of reinforcement, and the behavior may be referred to as schedule-controlled responding (Ferster and Skinner, 1957).

169

of SMA and schedule-controlled behavior are two of the most sensitive tests for determining the behavioral effects of pesticides. A battery of sophisticated techniques for more detailed evaluation of substances for neurobehavioral effects is gradually being developed (Geller et al., 1979).

Measurement of morphologic changes in the nervous system requires in situ perfusion of the animal and the use of contemporary methods of nervous system preparation for examination by light microscopy and perhaps electron microscopy. Spencer et al. (1980) have developed procedures for tissue preparation based on glutaraldehyde fixation that permit classification of neurotoxins according to their target sites (e.g., soma and myelin). Their system permits assessment of the location, type, and degree of neurotoxic damage. For the neurotoxins studied thus far, there is a good correlation between the type of neurotoxic damage observed in animal models and the type of neurotoxic damage produced by the same substances in humans.

In addition to the behavioral and morphologic testing recommended for the screening of neurotoxins, a host of other approaches to neurotoxicity testing are under development. Among these are tests involving tissue cultures, electrophysiology, and neurochemistry. The current stage of development of some of these approaches has been outlined briefly by Spencer et al. (1980).

PRINCIPLES AND DESCRIPTION OF TESTING
FOR NEUROBEHAVIORAL EFFECTS

Many substances damage the nervous system (Spencer and Schaumburg, 1980) and thereby produce a variety of behavioral changes. Although no substance or series of substances can now be recommended as a reference for neurotoxicity testing, positive controls are always desirable. Agents with established neurobehavioral toxicity should be tested repeatedly to establish the sensitivity of the measurement system when new substances with potential neurobehavioral toxicity are under investigation.

A range of doses of a substance should be administered for acute-toxicity testing of neurobehavioral effects, including doses large enough to produce obvious effects. Fractions of the LD_{50} should be used, and the substances should be administered by the route of likely human exposure. Effects of the substances on SMA and schedule-controlled behavior should then be determined. Later, the animals should be sacrificed for histologic examination by light microscopy and possibly electron microscopy. For subchronic studies, three doses, including the highest dose that did not affect behavior or produce morphologic changes on acute administration, should be administered repeatedly.

Healthy adult animals should be randomly assigned to treatment and control groups. In some instances, behavior is exceedingly stable, and pre-exposure behavior provides control observations for the exposure period. Rats are the preferred animal, although mice are also appropriate. Standard breeds and strains should be used, and at least six animals should be studied at each dose.

SMA must be measured under carefully controlled conditions (National Research Council, 1977). For example, strain of animal, feeding and watering schedules, time of day of testing, and conditions of temperature, humidity, and lighting must be specified. The device used to measure SMA must also be described in detail, although no particular method has yet been recommended.

For measurement of schedule-controlled responding, conditions must also be specified in detail. Again, the species, body weight, feeding and watering schedules, etc., must be specified. The reinforcement schedule should be described in sufficient detail to allow others to repeat the experiment under nearly identical conditions.

A reasonable time should be allowed to elapse, after the substance is administered, for absorption and distribution before behavior is measured. Animals should be tested daily for at least 1 wk after acute administration to determine the consequences of delayed neurotoxicity, such as that observed with alkyl tin compounds. Similar principles should be followed for subchronic testing. Where possible, it is desirable to have histologic examinations and behavioral testing of the same animals.

Additional animals exposed to the substance, but not tested for behavioral effects, can be added to the design as necessary to increase group size for time-response effects. Methods similar to those described by Spencer et al. (1980) are appropriate.

Tests of statistical significance should be applied to the data. Reproductions of histologic slides can often be used to illustrate types of neurotoxicity, although tables are generally necessary to document the frequency with which the illustrated events occurred. Specificity of neurobehavioral toxicity may be expressed as the ratio of the LD_{50} to the neurobehaviorally toxic dose. Neurobehavioral-toxicity evaluation and its relative importance should include a comparison with other toxicity data on the substance. For example, neurotoxic effects manifested only by a dose 10 times higher than that required for carcinogenicity would be of little regulatory interest.

The test report should include dose-effect data for behavior and morphology, a description of the location and type of neural lesions, the time course of the development of behavioral and morphologic changes, statistical treatment of the results, and an indication of the degree to which the toxicity is reversible.

Detection of either morphologic changes or behavioral changes caused by effects on the central nervous system can be considered as evidence of neurobehavioral toxicity that may be extrapolated to humans. The degree to which such extrapolation is valid is under investigation. More sophisticated understanding of specific types of behavioral toxicity would require a series of tests that is beyond the present protocol. When morphologic changes have been observed, the selection of later behavioral tests should be strongly influenced by the type, degree, and location of the morphologic changes.

DETERMINATION OF ADEQUACY
OF NEUROBEHAVIORAL-TOXICITY TESTING

Because no reference protocol guidelines for neurobehavioral-toxicity testing have been established and because much evidence of neurobehavioral toxicity exists in qualitative descriptions of the adverse effects of substances, the Committee on Toxicity Data Elements has established a series of criteria for judging the adequacy of neurobehavioral-toxicity testing, including observations of effects.

The neurobehavioral-toxicity data base (see Figure I-1) should contain studies on function (both conditioned and unconditioned behavior) and morphology (neuropathology). The committee regarded neurobehavioral-toxicity testing to be (1) adequate, according to reference protocol guidelines, only if all three study types followed the procedures described here; (2) adequate, but not according to reference protocol guidelines, if all three study types were conducted adequately, but not according to the procedures described here; and (3) inadequate if at least one of the three study types was missing or was conducted inadequately. The need to repeat studies conducted inadequately depends on the amount and quality of existing neurobehavioral-toxicity information on a substance. For example, documentation of extensive neural lesions (a morphologic response) might make behavioral testing (a functional response) unnecessary for regulatory purposes, because the neuropathologic effects already constitute a hazard. These criteria allowed the committee to determine whether neurobehavioral effects were observed with adequate or inadequate procedures.

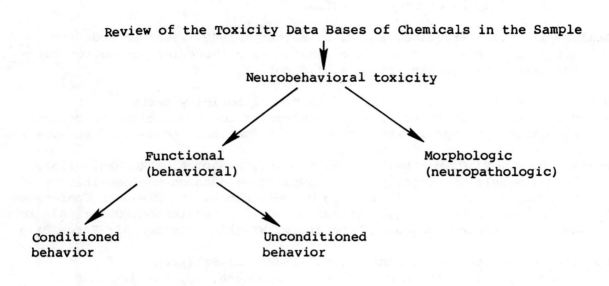

Review of the Toxicity Data Bases of Chemicals in the Sample

Neurobehavioral toxicity

Functional (behavioral)

Morphologic (neuropathologic)

Conditioned behavior

Unconditioned behavior

FIGURE I-1   Studies necessary for judging adequacy of neurobehavioral-toxicity testing.

REFERENCES

Ferster, C. B., and B. F. Skinner. 1957. Schedules of Reinforcement.
    New York: Appleton-Century-Crofts.

Geller, I., W. C. Stebbins, and M. J. Wayner. 1979. Test methods for
    definition of effects of toxic substances on behavior and neuromotor
    function. Neurobehav. Toxicol. 1:Suppl. 1.

National Research Council. 1977. Behavioral toxicity tests,
    pp. 111-118. In Principles and Procedures for Evaluating the Toxicity of
    Household Subtances. Washington, D.C.: National Academy of Sciences.

Reiter, L. W., R. C. MacPhail, P. H. Ruppert, and D. A. Eckerman. 1981.
    Animal models of toxicity: Some comparative data on the sensitivity of
    behavioral tests, pp. 11-23. In Proceedings of the Eleventh Conference on
    Environmental Toxicology. AFAMRL-TR-80-125. Patterson, Ohio: Air Force
    Aerospace Medical Research Laboratory, Wright-Patterson Air Force Base.

Spencer, P. S., and H. H. Schaumburg. 1980. Classification of
    neurotoxic disease. A morphological approach, pp. 92-99. In P. S.
    Spencer and H. H. Schaumburg, Eds. Experimental and Clinical
    Neurotoxicology. Baltimore: Williams and Wilkins.

Spencer, P. S., M. C. Bischoff, and H. H. Schaumburg. 1980. Neuro-
    pathological methods for the detection of neurotoxic disease, pp. 743-757.
    In P. S. Spencer and H. H. Schaumburg, Eds. Experimental and Clinical
    Neurotoxicology. Baltimore: Williams and Wilkins.

APPENDIX J

REFERENCE PROTOCOL GUIDELINES
FOR GENETIC-TOXICITY TESTS

The OECD (Organisation for Economic Co-operation and Development, 1981) has recommended a set of guidelines for the assessment of genetic toxicity. Government agencies in the United States are attempting to make their approaches compatible with those guidelines, and the Committee on Toxicity Data Elements accepted the guidelines as the standard against which genetic-toxicity studies were to be measured.

The OECD testing strategy has two stages. The first stage is the generation of a minimal set of data on previously untested substances: one test for gene mutation (Ames Salmonella/liver microsome reverse-mutation assay or Escherichia coli reverse-mutation assay) and one test for chromosomal damage (rodent micronucleus assay or in vitro chromosomal-aberration assay). OECD-approved protocols have been published for the conduct of these four assays. For a substance to be considered in compliance with the OECD guidelines, it must have been tested under at least one approved protocol for each of the two end points.

The second stage of the OECD strategy is additional testing of substances that produced positive results in the first step, as indicated in Figure J-1. Protocols have been drafted for several additional in vitro and in vivo assays. Some of these assays focus on both gene mutations and chromosomal damage. The protocols for the additional tests are in draft form and have not been approved by the OECD working groups.

Once the protocols have been approved, a substance identified as positive in one or both of the Stage 1 tests will require additional testing in a Stage 2 assay or assays measuring the end point(s) found positive in Stage 1. Such a substance will not be viewed as having been adequately tested without assessment in at least one Stage 2 test.

Using this framework for assessing the adequacy of genetic-toxicity testing protocols, the Committee on Toxicity Data Elements established the following criteria for its use in determining adequacy of genetic-toxicity testing within the subsample of 100 substances:

● A substance studied for each Stage 1 end point in tests conducted according to the OECD protocol guidelines was considered as having met the committee's standard for the genetic-toxicity reference protocol guidelines (G).

● A substance studied for each Stage 1 end point in one test conducted according to the OECD protocol guidelines and in the other not according to the OECD protocol guidelines, but nevertheless regarded by the committee as adequate, was considered to have met the committee's standard for an adequate (A) study of genetic toxicity.

175

● A substance studied for each Stage 1 end point in tests not conducted according to the OECD protocol guidelines, but nevertheless regarded by the committee as adequate, was considered to have met the committee's standard for an adequate (A) study of genetic toxicity.

● A substance studied for only one Stage 1 end point in tests conducted according to the OECD protocol guidelines, or nevertheless regarded by the committee as adequate, was considered to be inadequately tested (I). Tests must be conducted for both Stage 1 end points for a substance to be regarded as adequately tested.

● A substance studied for only one Stage 1 end point in tests conducted according to an acceptable standard (not necessarily in accordance with the OECD protocol guidelines) and also studied in other adequately conducted mutagenicity tests not prescribed in Stage 1 was considered to be inadequately tested (I). Nevertheless, further genetic-toxicity testing might not be required by the committee.

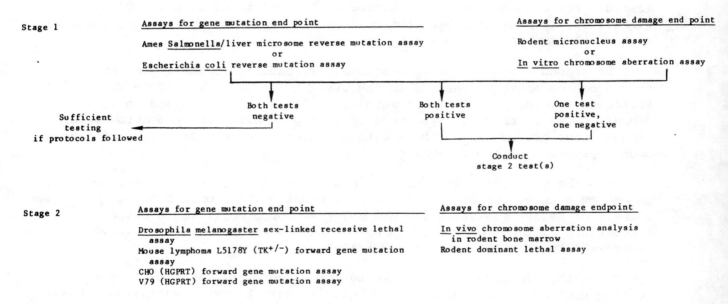

FIGURE J-1 End points and tests defining them in the two-stage assessment of genetic toxicity.

REFERENCE

Organisation for Economic Co-operation and Development. 1981. Guidelines for the Testing of Chemicals. Paris: Organisation for Economic Co-operation and Development. 700 pp.

APPENDIX K

CONCEPTUAL ISSUES CONCERNING THE INTERPRETATION OF RESULTS OF STUDIES OF REPRODUCTIVE AND DEVELOPMENTAL TOXICITY

Interpretation of studies of reproductive and developmental toxicity required an awareness of potential confounding factors, such as interspecies differences in toxicokinetics and developmental biology. These factors notwithstanding, tests for adverse effects on reproduction are not the same as tests for adverse effects on development (e.g., in utero growth retardation, functional decrements evident after birth, death of the conceptus, and production of live or dead terata).

Some overlap of effects can be expected in studies designed to evaluate reproductive or developmental effects. A multigeneration study may be designed primarily to assess effects on reproduction, but litter size, for instance, may be reduced, not only by diminished fertility, but also through death or resorption of malformed concepti, thereby indicating a substance's ability to disrupt development. Similarly, a Segment II study,* designed primarily to reveal developmental effects, may show that a substance induces abortion by affecting uterine function. This is not evidence that it directly disrupts the development of the embryo. Thus, in evaluation of the adequacy of results of a Segment II study, data from other studies may be used. If groups in the Segment II study are too small to provide useful results, a multigeneration study can sometimes be used as an aid to determine whether the developmental effects of a substance have been adequately tested. If results with a restricted number of animals in a Segment II study are consistent with those of a multigeneration study, the Segment II report may be judged as an adequate assessment of developmental effects. A substance's developmental toxicity might range from the apparent absence of effect to a clearly defined and obvious dose-response relationship. Even if such dose-related findings do not prove to be statistically significant (because of the use of small samples), they may be biologically important in light of data from related studies.

---

* In a Segment II study, two species of pregnant animals are exposed only during organogenesis, and fetuses are examined 1 day before parturition to determine developmental toxicity.

177

APPENDIX L

MAJOR COMPONENTS OF A DOSSIER

A dossier consisted of (1) a synopsis containing a substance's physicochemical characteristics, manufacturing processes, production, intended and other uses, chemical fates, exposure potentials, and all retrieved information deemed by the NRC staff to be relevant; (2) a list or lists of the adequacy ratings of required toxicity tests that were performed; (3) summary ratings for the quality of all information in the dossier; and (4) a data sheet detailing physicochemical, use, and exposure information. These four components are given below.

SYNOPSIS

Chemical name, CAS Registry number, and intended-use category
Physicochemical properties, manufacturing processes, production, uses,
     chemical fate, and exposure potential
   Physicochemical properties
   State(s)
   Solubility (fat, water)
   Chemical uniqueness
   Manufacturing process
   Production level
   Consumption level
   Uses
      Intended
      From other than originating lists
      Other than those indicated from sampling lists
   Statement of chemical fate
   Exposure potential
      In intended-use setting (dose, duration, frequency, route)
      In occupational setting (dose, duration, frequency, route)
      In environmental setting (dose, duration, frequency, route)
   Roadblocks to getting above information
   Level of concern based on all the above information
Synopsis of the health-effects data base
   Human
   Animal
   Summary of the toxicity data base
      Toxicologic uniqueness (e.g., reversibility of effect)
      For required tests
      For tests not required
   Roadblocks to getting toxicity information
Adequacy of the data base
   Analysis of individual toxicity studies
   Analysis of the complete toxicity data base
   Analysis of the complete data base

Adequacy Ratings of Required Toxicity Tests That Were Performed for a Pesticide or Inert Ingredient of Pesticide Formulation[a]

Necessary Toxicity Tests[b]	Ref. Notation[c]	Intended-Use Exposure
1. Acute oral toxicity--rodent		
3. Acute dermal toxicity		
5. Acute inhalation toxicity		
7. Acute eye irritation-corrosivity		
8. Skin sensitization--guinea pig		
11. Subchronic oral toxicity--rodent: 90-d		
12. Subchronic oral toxicity--nonrodent: 90-d		
14. Subchronic dermal toxicity: 21- or 28-d		
16. Subchronic inhalation toxicity: 28- or 14-d		
18. Neurobehavioral toxicity		
a. Functional: conditioned behavior		
b. Functional: unconditioned behavior		
c. Morphological		
19. Teratology--rodent, rabbit		
20. Multigeneration reproduction--rodent		
21. Toxicokinetics		
22. Carcinogenicity--rodent	q	
23. Chronic toxicity	q	
24. Combined chronic toxicity-carcinogenicity--rodent	r	
25. Genetic toxicity		
a. Ames reverse mutation (his-)		
b. E. coli reverse mutation (trp-)		
c. Rodent micronucleus		
d. In vitro chromosomal aberration		
e. Drosophila SLRL		
f. Mouse lymphoma (TK) forward mutation		
g. CHO (HGPRT) forward mutation		
h. V79 (HGPRT) forward mutation		
i. In vivo chromosomal analysis in bone marrow		
j. Rodent dominant-lethal assay		
29. Acute delayed neurotoxicity	*	

Other tests not required but performed:

---

[a] This is a sample summary for a pesticide or inert ingredient of a pesticide formulation in intended-use setting. Similar summaries were prepared for occupational and environmental settings and for all other intended-use categories for the substance being evaluated. Toxicity tests necessary for these settings are described in Appendix F. Measures of adequacy are according to the criteria described in Chapter 4.

[b] Test numbers are those used in the list of 33 test types in Appendix F.

[c] These and other reference notations are described in the legend of symbols and reference notations following Appendix G.

Summary Ratings for the Quality of All Information in a Dossier

1. Breadth of known exposure

   a. Widespread
   b. Limited, but intensive to specific groups
   c. Limited
   d. Rare
   e. Unknown

2. Trend in per capita exposure

   a. Increasing
   b. Decreasing
   c. Stable
   d. Unknown

3A. Level of concern resulting from physicochemical properties and chemical fate[a]

   a. High
   b. Moderate
   c. Low
   d. Unknown

3B. Level of concern resulting from physicochemical properties and exposure potential[a]

   a. High
   b. Moderate
   c. Low
   d. Unknown

---

[a] Ranking substances according to level of concern is based on chemical and physical data for only two exposure groups--adult males and adult females--not the conceptus. Chemical and physical data are generally not applicable to fetal development, nor is molecular structure indicative of a potential for reproductive/developmental toxicity. For instance, alkylating agents have been shown to injure embryos, yet they are of low priority for reproductive/ developmental toxicity testing, because they are no more hazardous to embryos than to adults. Even without testing, reproductive/developmental toxicity will be evident only at or very near the toxic dose for adults.

4. Strength of evidence of toxicity in humans

    a. High
    b. Medium
    c. Low or absent

5. Severity of chronic human toxicity

    a. Serious
    b. Moderate
    c. Minor
    d. Unknown (no evidence of toxicity
         or evidence of no toxicity)

6. Evidence of inaccessible toxicity data

    a. Yes
    b. No

7. Availability of appropriate test battery

	Intended Use	Occupational	Environmental
a. Complete			
b. Incomplete			
c. Essentially no results available			

8. Quality of available tests

    a. Most high
    b. Mixed
    c. Most low
    d. None available

9. Is all available information sufficient to allow an assessment of the hazard?

	Intended Use	Occupational	Environmental
a. Yes			
b. Limited assessment is possible			
c. No assessment is possible			

Physicochemical, Use, and Exposure Information

I. Chemical Identification

    A. Nomenclature
    B. Structure
    C. Chemical Abstracts Service (CAS) Registry number
    D. Defined purity
    E. Impurities/contaminants (identity and quantity)

II. Production, Use, and Waste Disposal

    A. Origin
        1. Starting materials
        2. Manufacturing processes
    B. Products
        1. Major products (percent yield)
        2. Magnitude of byproducts (number and volume)
    C. Wastes
        1. Identity
        2. Quantity
        3. Disposal methods
        4. Estimated unintentional losses (quantity)
        5. Physical state (gas, liquid, solid)
    D. Production
        1. Yearly volume (lb/yr)
        2. Intended/registered uses
        3. Estimated unintentional losses (quantity)
        4. Physical state (gas, liquid, solid)

III. Physical Characteristics

    A. Boiling point
    B. Density or specific gravity
    C. Vapor pressure
    D. Particle size
    E. Water solubility
    F. Organic (fat) solubility
    G. Complexity (single compound, family, mixture of families)
    H. pH
    I. Dissociation constant

IV. Chemical Reactivity in Nonbiologic Systems

    A. Shelf life/chemical stability
    B. Potential for oxidation/reduction
    C. Potential for hydrolysis
    D. Potential for photochemical changes
    E. Potential adsorption/desorption

V. Availability of Analytic Methods for Detection

    A. Evaluation of neat compound in nonbiologic systems
    B. Evaluation of chemical reactions in biologic systems
    C. Evaluation of environmental media only (air, water, food), not to include human exposure
    D. Evaluation of direct human substrates (urine, blood, other fluids, tissues, etc.)

VI. Chemical Bioavailability

    A. Environmental stability, turnover rates
    B. Biodegradation, excretion, elimination
    C. Occurrence in air, water, tissues, and/or food chain
    D. Bioaccumulation

VII. Exposure

    A. Production processes
      1. Open
      2. Closed
    B. Occupational exposure
      1. Number of people exposed:
        a. Production and direct use
        b. Other
      2. Monitoring
      3. Route of exposure
      4. Comments[a]
    C. Environmental exposure
      1. Number of people exposed
      2. Monitoring
      3. Comments[a]
    D. Intentional exposure
      1. Route
      2. Number of people exposed
        a. Conditions of recommended use
        b. Abuse
      3. Monitoring
      4. Comments[a]

---

[a] Information regarding special groups (age, sex, and ethnic groups, both demographic and geographic).

APPENDIX M

PROCEDURES, RATIONALE, AND RESULTS OF DATA IDENTIFICATION AND ACQUISITION

TYPES OF DATA COLLECTED

To estimate the potential for human exposure, the Committee on Toxicity Data Elements sought information on manufacturing processes, intended uses, production and consumption, and chemical and physical properties of each substance. In addition, potential exposure routes were identified, where possible, and durations and frequencies of exposures were estimated. The committee recognized that the potential for exposure may vary for each manufacturing process, intended use, or grade (purity) of the substance. For each substance, therefore, searches were made for information on the following subjects:

● The chemistry of the substance, including its chemical and generic nomenclature and Chemical Abstracts Service (CAS) Registry number, purity, identification of contaminants, and physical and chemical properties that influence its potential toxicity (fat and water solubility, molecular weight, and volatility).

● Manufacturing processes, including pathways of chemical synthesis, and production volumes.

● Uses, both industrial and consumer, including information on transportation, storage, and disposal.

● Chemical fate (including environmental pathways that lead to secondary human exposure), persistence, and bioavailability.

● Exposure of humans, including route, duration, dose, and frequency of human exposure and rate of absorption.

● Toxicity, particularly data with the potential to predict hazards to human health, including results of studies in mammalian and nonmammalian animal models.

● Epidemiologic and clinical studies.

SOURCES OF DATA

Resources were not sufficient to search every possible source of remotely relevant information. Data banks were selected after consideration of the likelihood of redundancy with other data sources, the time required to retrieve citations and collect cited materials and review them for relevance, and the accessibility of the identified data.

Five general sources of data were used in the initial search:

● On-line data bases, in which chemical information and citations of toxicologic and related biomedical publications are mechanically stored.

● Secondary sources, such as textbooks and technical review documents, which generally contain citations to primary sources that could be reviewed if necessary.

● The collection of toxicologic documents maintained since 1957 by the Toxicology Information Center (TIC) of the NRC Board on Toxicology and Environmental Health Hazards.

● Government agency files (both confidential and open to public access) on sample substances that are regulated.

● Manufacturers, trade associations, and large industrial consumers of the substances in the sample.

ACCESS TO DATA

The Committee on Toxicity Data Elements encountered several roadblocks when searching for data. These difficulties affected the completeness of the data bases compiled for each substance and, as a consequence, assessment of their adequacy. The following discussion is important, because these impediments to data access not only have affected this study, but also can be expected to influence future data evaluation and analysis.

Some information (especially that important for exposure estimation) has been classified as confidential and is stored in the files of government agencies or manufacturing companies. The committee questions whether the secrecy of data pertinent to human exposure is in the public interest or even in the narrower interests of industry. Data compiled in technical reports or bulletins that have not been published in the open literature are also inaccessible. Some primary data sources no longer exist or cannot be found, because they have been filed in large, poorly indexed record repositories.

Some kinds of data that are important to the determination of the adequacy of toxicity testing protocols often are not reported. Included in this category are detailed identification of chemical contaminants or substances used in mixtures; the species, strain, age, sex, breeding source, breeding techniques, diet, and water supply of the animal tested; and the length of time between exposure and observed effect or between exposure and a conclusion that no effect occurred. Negative results (findings of no toxic effects) often are not reported at all. Such data may be important to the assessment of data-base adequacy, but may be unavailable or even unknown to the organization that holds them.

AUTOMATED STORAGE AND RETRIEVAL DATA BASES

The committee anticipated that on-line data bases would provide most of the needed toxicity references, as well as the information on uses of chemicals. However, there is only limited coverage of data published before the mid-1960s, when the first large-scale automated data bases were developed.

The primary set of data bases chosen for this study collectively covers a broad spectrum of chemical and toxicologic information. Table M-1 lists these data bases and briefly describes their scope. The numbers of records listed are those at the time the search was initiated. As indicated in the table, some of the data bases grew during the year that the searches were conducted.

The committee also used a secondary set of data bases that are more specialized data bases created for comprehensive coverage of specific topics. Specialized collections with relevance to the committee's task are described in Table M-2.

The amount and type of data retrieved by on-line searching of bibliographic files depend in great measure on the search strategy used. In the retrieval of computerized information, the selection process is a stepwise removal of unwanted data. At each step, the operator judges whether the information is relevant or not. Information designated as not useful is "discarded" from the process. For this project, the operators were conservative in their selection process, because it seemed better to retrieve irrelevant information and remove it manually than to overlook important citations. The NRC personnel in charge of the on-line retrieval consistently tended to keep more information than committee members believed to be relevant.

The committee recognized the need for uniformity in the search for data on each substance. For some compounds, very little information was found, even through a search of all the data bases. For others, thousands of indexed documents were identified by the computer as having

TABLE M-1 Automated Data Bases Used to Obtain Information on 100 Chemicals in Sample

Name	Data-Base Type	Approximate No. Records	Period Covered[a]	Subjects
CHEMLINE	Chemical dictionary	416,000 (September 1981) 530,754 (July 1982)	NA	Chemical names, synonyms, CAS Registry numbers, citation file locations, and molecular formulas
SANSS (CIS)	Chemical dictionary	180,000 (September 1981)	NA	Same as CHEMLINE with structure display
CHEMNAME	Chemical dictionary	737,000 (September 1981) 1,180,000 (June 1982)	NA	Same as CHEMLINE with substructure searching
Toxicology Data Bank	Secondary source material reviewed by peer group	1,100 (September 1981) 3,555 (July 1982)	NA	Profile, including uses; potential for poisoning, fire, explosion, and toxicity; pharmacology; metabolism; manufacturers' information; physical properties; import, export, and production data (records completed for only a few chemicals)
RTECS	Primary and secondary source	48,569 (September 1981) 53,425 (July 1982)	NA	NIOSH-compiled data with references to primary toxicologic literature; lists regulations
Chemical Abstracts	Bibliographic file	3,686,000 (September 1981) 5,462,000 (June 1982)	1967–	U.S. and foreign data on physical, analytic, and applied chemistry; properties and reactions; use; and toxicity
Chemical Industry Notes	Bibliographic file	230,000 (September 1981)	1974–	Industry-oriented information on process methods and production volumes; some information on uses

Database	Type	Records	Date	Coverage	Description
TOXLINE and TOXBACK	Compendium of bibliographic files	1,167,000 1,299,028	(September 1981) (July 1982, purged of some duplicates)	1965–	Eleven data bases were drawn together to create TOXLINE; U.S. and foreign literature on toxicologic subjects (different files on teratology, mutagenicity, pharmacology, environmental pollution, air pollution, pesticides, epidemiology, and chemical-biologic activity) with several general toxicology files
CANCERLINE	Bibliographic file	264,682 306,611	(September 1981) (July 1982)	1963–	U.S. and foreign literature on carcinogenicity, tumor initiation, antitumor activity, and other topics

a NA = not applicable.

189

TABLE M-2 Specialized Data Bases

Category of Substance	Name	Approximate No. Records[a]	Period Covered	Subjects
Cosmetic ingredients	MEDLINE	3,500,000	1966–	Biomedical and clinical literature citations and research information
	BIOSIS	2,735,000	1969–	Biomedical and clinical literature citations and research information
Drugs and excipients in drug formulations	MEDLINE	3,500,000	1966–	Biomedical and clinical literature citations and research information
	BIOSIS	2,735,000	1969–	Biomedical and clinical literature citations and research information
	International Pharmaceutical Abstracts	48,000	1970–	Clinical and animal studies in pharmacology; research and theoretical works on practical drug use
Pesticides and inert ingredients of pesticide formulations	AGRICOLA	1,150,000	1970–	Biologic and agricultural science
	Commonwealth Agricultural Bureaux	635,000	1970–	Biologic and agricultural science
	Environmental Periodical Bibliography	108,000	1973–	General human ecology, nutrition and health
Food additives	MEDLINE	3,500,000	1966–	Biomedical and clinical literature citations and research information
	BIOSIS	2,735,000	1969–	Biomedical and clinical literature citations and research information
	FOODS ADLIBRA	57,400	1974–	Food packaging, food technology, nutrition, and toxicology

a Numbers of records given for each data base reflect numbers at beginning of searching task. During period of searching, substantial numbers of entries were added.

190

some potentially relevant information. In the latter situation, time constraints precluded review of every document. For this reason, review articles were provided to committee members, who were asked to identify citations that applied to each of the toxicity tests required for a substance's category of intended use. Those primary documents were then collected. When some required toxicity data elements were not uncovered in this way, the data bases were searched again with a focus on the missing elements. For this purpose, a list of search terms was developed for each of the major toxicity subjects: carcinogenicity, mutagenicity, reproductive/developmental toxicity, organ and system toxicity, and metabolic effects (toxicokinetics). This procedure eliminated many citations that were not relevant or were already identified.

This strategy of data compilation included feedback from committee members after they examined data on a substance. On the basis of their experience, they identified what data might be missing. If the result of a search seemed inappropriate in any way, a followup search was conducted.

The greatest difficulty encountered in the use of the powerful searching capabilities of automated files was specific identification of substances that were not clearly defined. Some chemical terms on the list constituting the "select universe" signify materials with uniform compositions, but others describe generic sets of compounds or single compounds that may occur in a variety of physical forms. Still others are chemical mixtures or biologic products that vary in composition from source to source or sample to sample. A few of the chemical names were vague or ambiguous.

OTHER SOURCES

Textbooks, technical reports, reviews, abstracts, and patent applications provided only limited toxicologic information, but were excellent sources of information on chemical and physical properties, intended uses, and manufacturing processes. To obtain a consistent base of data, a standard search pattern was instituted. One source of information was the computerized toxicity data bases. In addition to abstracts of research papers, abstracts of patent applications were found by computer searches. These often contained information on potential uses. Helpful information was also obtained from the files of the Registry of Toxic Effects of Chemical Substances and from the Toxicology Data Bank. Six reference books were consulted for information on each chemical (Doull et al., 1980; Hawley, 1977; Physicians Desk Reference, 1981; Sollmann, 1957; Weast and Astle, 1979; Windholz et al., 1976). In addition, other sources were searched for particular types of chemicals (U.S. Environmental Protection Agency, 1980; Estrin et al., 1982; Hayes, 1975; National Research Council, 1981). The nature and extent of the information in these references depend on the uses of a given substance and, hence, are correlated with the list(s) on which the substance appears.

NRC TOXICOLOGY INFORMATION CENTER (TIC)

Because of the need to gain access both to pre-1965 data and to very recently published data, both of which were not present on computer data bases, the card catalogs of the TIC were searched manually.

The 26-yr-old collection of toxicologic information in the TIC includes data produced before the collection began and acquired through retrospective literature searches, in addition to data generated from the establishment of the collection to the present. The TIC is also a repository of reprints gathered for reports of NRC committees, as well as of some private reprint collections donated by toxicologists. Although comprehensive in its collection of information on chemicals encountered environmentally and occupationally, the TIC generally excludes some types of substances, notably drugs and nutrients.

GOVERNMENT AGENCY FILES

Some research or regulatory agencies--e.g., the National Institute for Occupational Safety and Health (NIOSH), FDA, EPA, and the Department of Defense (DOD)--have repositories of data on toxicity, manufacturing processes, production volumes, and intended use (including the amounts of substances associated with each type of intended use). Because the committee's intent was not to report the production process and volume data, but rather to use the data as components of an exposure profile, it asked each agency for permission to gain access to confidential information pertaining to substances in the sample. All the regulatory agencies expressed concern about granting the committee access to trade-secret information provided by industries. In some cases, the agency responses were severely constrained by law, regulation, agency policies, and restrictions imposed by organizations that provided the data. These constraints varied markedly from agency to agency. No criticism of agency responses should be inferred from discussions in the following paragraphs.

EPA provided the toxicology portions of pesticide registration applications to NRC staff and to some committee members not involved in occupations associated with organizations that could benefit from company trade secrets. The information augmented the data in the published literature. Little specific information on extent of environmental contamination was obtained from EPA data files. Because those files are not centralized, it was difficult to gain the required information from them.

FDA was reluctant to open files; however, its Bureau of Foods provided access to the SCOGS (Select Committee on GRAS Substances) reports on substances in the sample. Most of the data in these reports were also identified by other means. The FDA Bureau of Drugs readily provided information on clinical use, route of administration, duration of use, and types of formulations for active drugs, as well as estimates of population exposure to those drugs and adverse reactions to them. Access to information contained in the Bureau of Drugs division reviews, new drug applications (NDAs), investigative new drug (IND) files, and abbreviated new drug applications (ANDAs) was more complex. Regardless of degree of confidentiality, these data were either inaccessible or were made accessible only with great difficulty, because of the excessive costs of manpower required to locate relevant information. There were long delays before the bureau responded to requests from the committee--not because of unwillingness to be responsive, but rather because of acute deficiencies in the bureau's data management. These deficiencies were most apparent after two major requests were made by the committee:

● The committee asked for a list of currently marketed prescription drugs, over-the-counter drugs, and excipients used in formulations of these drugs. The product was provided 6 months after the request was made, because manpower in the Bureau of Drugs was insufficient and the automated data management systems were incapable of generating the required list readily.

● The committee asked for toxicity information on chemicals that were sampled from the list eventually provided by the bureau. Within several weeks of the request, some NDAs, ANDAs, INDs, and division reviews were made accessible to cleared NRC staff members, who then had to locate the files and identify pertinent information in them. Most files consisted of many volumes that lacked indexing or content organization, except for chronologic entry of documents. Thus, the search for relevant information required manual scanning of every page in every volume. In many cases, the desired volumes were not available, because they were stored in a warehouse in a manner that made their retrieval extremely difficult or they were lost and could not be traced. Much of the information identified as lost had been submitted to FDA by industries that did not later publish the material.

Information collected by NIOSH on some of the substances was organized and readily accessible through that agency's health-hazard evaluations, criteria documents, and Current Intelligence Bulletins. However, these documents contained only a few of the substances of interest to the committee. Toxicity data on several chemicals were found in unclassified documents maintained by NIOSH (National Institute for Occupational Safety and Health, 1982).

MANUFACTURERS, COMMERCIAL USERS, AND TRADE ORGANIZATIONS

Manufacturers and trade associations are repositories of otherwise unobtainable information on potential occupational exposures, manufacturing processes, waste disposal practices, and production. Their assistance in obtaining information on the 100-substance subsample was requested through the Federal Register on March 16, 1982 (Public Health Service, 1982), and through correspondence with manufacturers of the 40 representative chemicals in commerce selected for the subsample from the TSCA Inventory. Approximately 600 companies were identified in the TSCA Inventory as being manufacturers of at least one of the 40 chemicals in commerce in the subsample. Each company was contacted by telephone, a brief explanation of the project was given, and the information to be sent to them was described. A followup letter--which included a complete description of the project, an alphabetized list of the 100 substances, a request for unpublished toxicity data, and a questionnaire--was sent to each company. The questionnaire contained the following questions:

● Is the material that is produced in your work environment regulated by FDA, OSHA [the Occupational Safety and Health Administration], EPA, or DOT [Department of Transportation] (and are there guidelines for limiting exposure)?

   ● If so, how do you control the chance of exposure?
      A. By engineering control?
      B. By personal protective clothing (e.g., the use of respirators)?
      C. By ventilation (e.g., roof fan, open exhaust)?
      D. Other (please specify)?

● Is the material measured in the air?

● Do you store the material? If so, for how long?

● Do you run a continuous or batch operation to produce the material?

● How old is the equipment used to manufacture the material?

● How frequently is maintenance required on the equipment?

● Can you indicate how much of this material you produce per year and at what site it is produced?

● What are the potential uses of this chemical in manufacturing, in commerce, and by consumers?

Some trade associations were contacted directly; others were forwarded the questionnaire by the manufacturers. The Research Institute for Fragrance Materials, the Cosmetic Ingredient Review, the Cosmetic, Toiletry, and Fragrance Association, the American Petroleum Institute, the Soap and Detergent Association, and the Flavor and Extract Manufacturers' Association were some that responded to inquiries. The committee also contacted persons whose professional expertise equipped them to provide additional information on potential human exposure. When information was particularly limited, authors of research papers were contacted in an attempt to obtain more information.

Responses to the inquiries were mixed, both in quantity and in usefulness. The few respondents to the Federal Register notice provided much information that was not in the open literature. Approximately 30% of the 600 manufacturers who were contacted responded. Approximately 15% of the 600 stated that they did not manufacture any of the substances. Responses from approximately 3% of the companies indicated that a general willingness to cooperate was frustrated or delayed by resource constraints, lack of expertise, etc. Approximately 5% of the companies supplied answers to the questionnaire, and an additional 5% provided documents and other information on the chemicals of interest. Only 1% of the companies that were contacted responded by indicating that they would not cooperate.

Reluctance or inability of industry to cooperate in studies like this was found to be only one factor inhibiting the collection of information on industrial practices and occupational exposure. It is very difficult--often impossible--to locate and contact all the current manufacturers of a given substance.

AVAILABILITY AND ROLE OF INFORMATION OTHER THAN TOXICITY DATA

Information on intended uses was readily available from the reference textbooks for most cosmetic ingredients, drugs and excipients in drug formulations, and food additives. Similar information on pesticides and inert ingredients of pesticide formulations was more often obtained from computer searches (including patent abstracts), government files, or other special sources. Information on the intended uses of chemicals in commerce (as listed in the TSCA Inventory) was obtained less often from either of these sources than from patent applications and primary literature. Chemical and physical properties were most readily available, although that information was not located for approximately 20% of the substances selected from the TSCA Inventory. Availability of information on manufacturing processes was variable. Even when information on synthesis could be located, several processes were often presented with no indication of which ones were used for bulk production. Limitations of resources did not allow for the case-by-case collection of these data from manufacturers or other sources beyond the collection already described.

Data on the 100 substances in the subsample (chemistry, production and consumption, intended uses, methods of waste and product disposal, and environmental persistence) were considered to be essential for evaluating the adequacy of the toxicity data base. Such data can be used to estimate the number of people potentially exposed, the magnitude and duration of exposure, and the routes of direct and indirect exposure. Without such data, only the most crude and untestable subjective estimates can be made.

From the outset, the committee recognized that the desired data on manufacturing processes and environmental persistence would be difficult to obtain, even for high-volume substances. A major limitation was the absence of known data bases in which such information was systematically stored. Furthermore, the data kept in industry files were largely inaccessible or were organized in ways that made it difficult to retrieve pertinent information. If such information were made available, the effort needed to assemble and process it would have added substantially to the workload, and that would have reduced the capacity to acquire and process the toxicity data. Although the task was, at the least, formidable or, more likely, infeasible, an earnest effort was made to address it. Letters sent to known producers requested information on production, occupational exposure, and waste disposal. The responses received contained little usable information. Likewise, requests to government agencies for data on occupational exposure and environmental contamination in general yielded very little specific information on the chemicals of interest. Nominal data had been accumulated by these agencies and not in accessible files.

Even the simplest of the relevant data (e.g., reliable annual production rates) could not be obtained in most cases. There were strong indications that three of the 10 subsample substances listed in the TSCA Inventory as having 1977 production of at least 1 million pounds (454 metric tons) were no longer in production or were produced in markedly smaller quantities. As a result, the committee's assessment of the adequacy of the toxicity data on many substances had to be based on sketchy data and subjective estimates of the numbers of persons exposed and the routes, durations, and intensities of their exposure. In most cases, the committee relied solely on information about the products' intended uses, their chemical and physical properties, and the general background knowledge of committee members.

The serious weakness of the exposure data base limits the committee's confidence in the adequacy of the toxicity-testing protocols for conducting health-hazard assessments. The lack of suitable exposure data places an even more severe limitation on the application of inferences drawn from analyses of the chemicals in the final sample to the larger "select universe" and on the development of quantitative dose-response models for chemicals in the environment. The assembled toxicity data on the 100 substances provide a good base for examining the predictive nature of toxicity-testing models; however, the absence of exposure data prevents a similar examination of exposure models.

The unavailability of reliable exposure data will be a continuing limitation for NTP in its planning of toxicity testing. NTP would benefit greatly if data of this kind were obtained in cooperation with other federal agencies--such as NIOSH, FDA, EPA, and the Consumer Product Safety Commission--that have an interest in and a need to collect similar information on exposure.

REFERENCES

Doull, J., C. D. Klaassen, and M. O. Amdur, Eds. 1980. Casarett and Doull's Toxicology: The Basic Science of Poisons. New York: Macmillan. 778 pp.

Estrin, N. F., P. A. Crosley, and C. R. Haynes. 1982. CTFA Cosmetic Ingredient Dictionary, 3rd ed. Washington, D.C.: The Cosmetic, Toiletry and Fragrance Association, Inc. 610 pp.

Hawley, G. G., Ed. 1977. The Condensed Chemical Dictionary, 9th ed. New York: Van Nostrand-Reinhold. 957 pp.

Hayes, W. J. 1975. Toxicology of Pesticides. Baltimore, Md.: Williams & Wilkins. 580 pp.

National Institute for Occupational Safety and Health, Chemical Systems Laboratory. 1982. Subfile excerpted from the NIOSH Registry of Toxic Effects of Chemical Substances. Rockville, Md.: National Institute for Occupational Safety and Health.

National Research Council. 1981. Food Chemicals Codex, 3rd ed. Washington, D.C.: National Accademy Press. 735 pp.

Physicians Desk Reference. 1981. 35th ed. Oradell, N.J.: Medical Economics Co. 2,047 pp.

Public Health Service. 1982. National Toxicology Program. Fed. Reg. 47:11321.

Sollmann, T. 1957. A Manual of Pharmacology and Its Applications to Therapeutics and Toxicology, 8th ed. Philadelphia, Pa: W. B. Saunders. 1,535 pp.

U.S. Environmental Protection Agency. 1980. Code of Federal Regulations 40. 1980. Subpart D. Exemptions from tolerances, pp. 277-297. In Tolerances and exemptions from tolerances for pesticide chemicals in or on raw agricultural commodities. Section 180.1001.

Weast, R. C., and M. J. Astle, eds. 1979. CRC Handbook of Chemistry and Physics, 59th ed. (1978-1979). West Palm Beach, Fla.: CRC Press.

Windholz, J., S. Budavari, L. Y. Stroumtsos, and M. N. Fertig, Eds. 1976. The Merck Index. An Encyclopedia of Chemicals. 9th ed. Rahway, N.J.: Merck & Co. 1,313+ pp.

PART 2

SETTING PRIORITIES FOR TOXICITY TESTING

CONTENTS

Tables

201

Figures

1

INTRODUCTION

The information available on the potential toxicity of most chemicals is scanty, and the resources available for toxicity testing do not suffice to test all chemicals for every possible health effect. Hence, a priority-setting system is needed for selecting substances to be tested and selecting tests with which to evaluate them. Because of limitations in available data and methods, the priority-setting system must be designed to operate in the presence of considerable uncertainty.

The Committee on Priority Mechanisms has sought to develop a priority-setting approach applicable to the large number of chemicals of potential concern with respect to human health. The approach, which is drawn from systems analysis and decision theory, is based on the thesis that the rationale of any priority-setting system should be explicit, open to inspection, and scientifically defensible. In view of the rapidity with which the art of toxicity testing is evolving, the committee does not propose a particular priority-setting scheme now, but rather suggests an approach for designing a system that can keep pace with advances in the field. The approach is presented in this report largely in conceptual form, although examples are given to illustrate how it could be applied to the selection of chemicals for carcinogenicity testing.

Testing priorities have traditionally been assigned on the basis of expert judgment, which is now supplemented with a variety of analytic, data-based techniques, such as scoring systems. The committee believes that this basic pattern should continue, with further improvement in techniques that allow expert judgment to be most effective. The committee recognized that no priority system, scheme, or procedure can be perfect, because the knowledge needed for unerring selection of the most important chemicals and tests is the same as the knowledge resulting from a complete and accurate testing program for all chemicals, which would of course make priority-setting unnecessary. The priority-setting system and the testing program form a continuum whose overall objective is to yield the most information about the overall hazards of chemicals.

Given a goal for the priority-setting system, the committee needed to decide whether improvements over current procedures for selecting chemicals for testing were possible. It concluded that improvements were possible--at least at the margin--by injecting additional systematic information-gathering and -processing procedures.

Because the total number of chemicals far exceeds the number that can be evaluated in depth at any one time, the committee's approach seeks to arrive at the combination of information-gathering procedures that, for a given investment of resources, will yield the most useful toxicity data

on the universe of chemicals to be considered. To enable the entire
select universe of chemicals to be scanned for compounds that may warrant
testing, a multistage scheme is presented for consideration; it begins
with automated processing of machine-retrievable data and proceeds
through successive stages to costlier procedures that depend on expert
judgment.

A model is presented to illustrate how the performance of this
four-stage priority-setting system can be optimized by selecting the most
effective set of decisions for a given investment of resources.

Lack of knowledge about the effectiveness of toxicity testing, the
extent and distribution of toxic properties among chemicals, or the
exposure to chemicals hinders the design of an optimal system. However,
the report shows not only how a priority-setting scheme might operate
with incomplete and uncertain information, but also how to determine what
information is most needed to improve the system itself.

The work of the committee is presented as a review of the elements of
systems analysis as applied to priority-setting for toxicity testing in
Chapter 2. These principles are applied to an illustrative system in
Chapter 4. An explanation of the operation of the illustrative system is
presented in Chapter 3, and suggestions for implementation are presented
in Chapter 5.

The committee believes that a fully developed version of the outlined
system not only is a plausible extension of current practice, but also
would provide at least marginal improvements over existing priority-
setting procedures toward the goal defined earlier. Obviously, it might
not provide improvements toward other goals, but it should not impede
them. Even at the margin, the improvements would probably easily justify
the costs of developing, implementing, and operating the system.
However, the implementation of these concepts in the illustrative system
or one of similar scope would require adjustments in the established
patterns of thinking about testing priorities. Specifically, full
application of the proposed analytic techniques will require that each
information-gathering procedure be described quantitatively with respect
to its ability to identify and to characterize potentially toxic
chemicals. This requirement is not readily fulfilled in our present
state of knowledge. Hence, efforts toward further quantification of the
performance characteristics of toxicologic methods would be essential to
full implementation of the priority-setting approach proposed herein.
For this reason, the approach can be pursued initially on a pilot scale,
with further implementation depending on the development and availability
of the necessary data. The committee believes that it should be possible
to institute changes in current procedures gradually, without
irreversibly committing resources to the novel features of its
suggestions.

2

DESIGN OF THE PRIORITY-SETTING SYSTEM

Developers of priority-setting systems must decide the following: the goal of the system; the chemicals to be considered; and the structure of the system, i.e., the information to be gathered in each step of the system, the type of analysis to be performed in each step, and the decision rules that determine which chemicals are to be tested and which are to be considered for other action (removed from consideration, placed on a holding list for future consideration, etc.). Those tasks can be accomplished by experts, using their skills and common sense, but some issues are not easily addressed unless the designers of the system make special provisions for addressing them. For example, what is the most effective sequence in which to gather different kinds of information? How effective is the system? What is the impact of changing some component of the system? To address the latter two issues, it is helpful to use elements of decision theory and systems analysis. Such an approach to toxicity testing requires that the system priority-setting criteria and toxicity tests be described in measurable terms and that goals be defined to enable the system's effectiveness to be determined in relation to its goals. This description requires the system's designers to be explicit about the assumptions on which the design is based. Components of the system must also be described explicitly in terms of the number of toxic chemicals among the chemicals considered, the effectiveness of each procedure or toxicity test, and the resources required for each such procedure or test.

The value-of-information concept is used to address the issue of which information is best to collect in each stage. This concept underlies a strategy in which decisions are based on the relationship between the importance of a piece of information and the degree of uncertainty about it (Raiffa, 1968). Each additional piece of information is sought to enable decisions about the potential health hazard of a chemical to be made with less risk of error. The value of the information is defined as the difference between the expected costs of error in classifying the chemical with and without the additional information.

The value-of-information concept is a simplification that may be difficult or even impossible to apply in complex decisions about testing priorities. To apply it rigorously, strong assumptions are needed to relate the multiple dimensions of chemical hazards to a single objective that is quantifiable. Although the committee recognizes that full implementation of a system based solely on these principles may in fact be impossible, it also holds that value of information is a useful and important concept. By designing an illustrative priority-setting system using the concept, the committee sought to analyze the factors that

207

are important for priority-setting and to explore how they interact, and to do so in a systematic and integrated manner. The discipline of a relatively formal analysis has produced insights that should be useful in a more flexible and operational priority-setting procedure.

GOAL OF PRIORITY-SETTING

When formulating a goal for use in designing a priority-setting system, one should consider that the goal of the system should reflect the mission of the user, provide guidance in designing the priority-setting system, and provide a mechanism to measure the performance of the system. The Committee on Priority Mechanisms considers that a goal to meet the need of NTP's priority-setting system could be the ability to assess accurately the potential public-health impacts of chemicals to which humans are exposed.

As demonstrated below, that goal influences all elements of the priority-setting process, including selection of the chemicals to be considered for priority-setting, measurement of the effectiveness of the system, and the means for judging the information gathered and how it is analyzed. It also leads to viewing the priority-setting process and testing program as related parts of a larger effort to assess the potential public-health impact of exposure to toxic substances. The goal thus calls for the system to produce the accurate assessment of all chemicals, not just the proper selection of chemicals for testing. The system should, in general, dispose of chemicals of low concern, rather than selecting them for testing, and should not suggest additional testing of a chemical if adequate information is already available for an accurate assessment of its potential public-health impact.

The above goal also implies that toxicity testing and the gathering of information on human exposure should be considered jointly in assessing the potential public-health impact of a chemical. Because concern about a substance depends not only on its intrinsic toxicity, but also on the extent of human exposure to it, the gathering of information on the numbers of people exposed to various concentrations of the chemical may be just as important in setting its priority for testing as is determining its potential for toxicity.

CHEMICALS CONSIDERED

Defining the universe of chemicals to be considered is an important design decision and is influenced by the goal chosen for the system. As indicated above, all chemicals to which there is potential human exposure should in principle be considered by the system. Although such a goal might cause the number of chemicals to be too large to manage or too difficult to estimate, several approaches can be used to solve the problem. One approach would attempt to define additional categories so that all chemicals to which there is potential human exposure would be included in a category and then estimate the number of chemicals in each category; another would remove some categories of chemicals. A third approach could be to redefine the goal.

The universe developed by the Committee on Sampling Strategies—which consists of food chemicals, pesticide chemicals, cosmetic ingredients, drug ingredients, and industrial chemicals—totals over 50,000 discrete (although not always well-characterized) substances. This universe could be expanded by adding categories—such as combustion products, pollutants, and some categories of naturally occurring substances—until most chemicals with a substantial potential for human exposure were included. Adding these categories could expand the universe of chemicals considered by several thousand chemicals. If a smaller universe were desired, some categories of chemicals could be deferred or removed. In either case, the designers could make an explicit decision concerning the system.

Testing priorities may be set for pure, well-defined compounds, commercial grades of such compounds, elements and all their compounds, categories of compounds (e.g., cyanides), mixtures of known or unknown composition, radicals, or other classes of chemical entities. The terms "substance" and "chemical" are used interchangeably in this report to include all these classes. When designing an exposure assessment, a toxicity assessment, and their interface, however, it is important to define as precisely as possible the identity of the substances being considered. The most commonly accepted (and usually unambiguous) identifier for a substance is its Chemical Abstracts System (CAS) Registry number.

Establishing the universe of substances usually involves some initial screening or the exclusion of some candidate chemicals. Some of the schemes reviewed by the Committee on Priority Mechanisms have been applied only to specific classes of chemicals (such as food additives or drugs); others have been applied only to chemicals on existing priority lists or to chemicals nominated by panels of experts (see Appendix A). In the latter case, the chemicals have already been screened through a process that involves scientific judgment, so chemicals on which there is little information are very likely to have been excluded without adequate review. Thus, the establishment of the initial universe in itself constitutes a major step in the priority-setting process.

In several of the schemes reviewed by the committee, the universe is immediately narrowed by the deletion of substances that are judged to be either irrelevant to the exercise or difficult to review. Classes of substances deleted in this way include the following:

● Chemicals already regulated, such as pesticides, drugs, and food additives.

● Substances not subject to regulation by a member agency of NTP, such as natural products.

● Chemicals only nominally subject to regulation and adequately tested under existing regulations.

● Substances without CAS numbers, including complex and ill-defined mixtures.

● Other substances difficult to characterize, including combustion products, pyrolysis products, and environmental breakdown products.

● Environmental mixtures, such as extracts of air pollutants and water pollutants.

Although the omission of such substances and mixtures can usually be understood on the grounds of convenience and practicality, it should be recognized that the classes of substances that are omitted include many that are both poorly characterized and potentially harmful.

STRUCTURE OF THE SYSTEM

The entire process of priority-setting and toxicity testing consists of a series of interconnected steps, each containing an information-gathering component and a decision-making component. An example of one such step, or stage, is diagramed in Figure 1. It begins with an information-gathering activity, which includes a search for and interpretation of specific data, or "data elements." Combined with information already on hand, the new information enables one to achieve a better understanding of the public-health concern of a chemical (middle box on right side). If the information is useful, the understanding will be more certain than it was in the initial state of knowledge (top box on right side). The figure represents concern as a position on a two-dimensional map of exposure and toxicity. In this example, the information-gathering activity is a toxicity test that narrows the uncertainty about toxicity without providing information on exposure. Other representations could include an estimated probability distribution for the concern (e.g., the probability that a given number of people would incur a specific effect during the next year) or a discrete probability distribution concerning the degree of hazard (e.g., the probability that the substance involves or does not involve a "significant" concern worthy of control activity).

After the new state of knowledge is determined, a decision must be made (lowest box on right) to determine what additional pieces of information would be most valuable or that no additional information is required to make a reliable classification of public-health concern. The latter decision is usually made when it appears that economic, health, or other costs of misclassification are unlikely to be reduced enough to justify additional information-gathering activities. Thus, exit from a step is followed either by no further consideration of the substance or by a new information-gathering step.

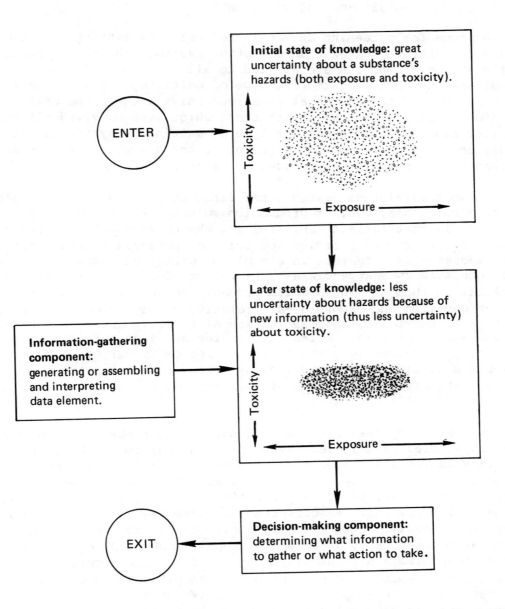

FIGURE 1 Example of one stage or building block in process of investigation and control of toxic substances. Information-gathering activity here is toxicity test that produces useful data.

Any priority-setting and testing scheme can be represented, at least conceptually, by a network of such information-gathering and decision-making steps. Traditional methods depend on evaluation by an expert committee that reviews a dossier on a substance (information-gathering component) and then makes recommendations about its disposition (decision-making component).

Systems for screening chemicals for priority-setting may have one stage or multiple stages. In one-stage systems, the same screening criteria and procedures are applied to all the chemicals under consideration. In the simplest type of multistage system, chemicals are screened out of the system at each successive stage; the only chemicals considered in the last stage are those which have survived all the earlier stages. A more complex type of multistage system is the decision tree, in which the screening criteria applied at each stage depend on the outcome of the previous stage.

In most multistage systems, the first stage is a simple screen based on chemical class, use, or production volume; the second stage is based on criteria that reflect exposure; and the third and later stages are based on criteria that reflect toxicity or potential risks. Although this sequence is a feature in six of the different systems reviewed by the committee (see Appendix A), the reasons for its choice were not made explicit in them; it was adopted probably because crude indexes of use--production and exposure--are relatively easy to obtain for large numbers of chemicals, whereas indexes of toxicity are more difficult to obtain and require more scientific review and judgment. In more elaborate systems--e.g., the decision tree of Cramer et al. and the six-stage linear screen described by Nisbet--the late stages require fairly extensive compilations of toxicity and exposure data and fairly detailed scientific review.

Multistage systems are advantageous, in that the screening process can use simple, readily retrieved data, so many chemicals of low priority can be eliminated from consideration quickly and scientific attention can be focused on the chemicals of greatest interest. Systems of this kind appear to offer a practical way to deal with very large numbers of chemicals. An offsetting disadvantage is that the criteria used in the early stages are necessarily crude, so some chemicals may be eliminated erroneously in an early stage. Another disadvantage is that exposure information is usually considered in less detail than toxicity information; hence, chemicals with unusual pathways of exposure may not be identified. These problems may be alleviated by providing for reconsideration of chemicals eliminated in early stages or reconsideration of exposure in later stages. These features are included in the Interagency Testing Committee's system.

Designing a multistage screening system, under the constraint of an overall budget, involves balancing of the costs of generating information on a large number of chemicals in early stages against the costs of generating more detailed information on fewer chemicals in late stages. The efficiency of such a system depends on the number of stages, the amount of information considered in each stage, and the number of chemicals correctly classified in each stage. In the systems reviewed by the committee (Appendix A), these characteristics appear to have been chosen rather subjectively, and it is not clear that maximal efficiency was achieved.

Decision-tree systems are in principle more flexible than linear multistage systems, because they can use more appropriate criteria for screening at various stages. However, the systems that have been proposed to date require rather precise information and would be difficult to apply to several classes of chemicals, especially those with little or no toxicity testing.

ANALYSIS OF INFORMATION

Data on a chemical must be analyzed so that decisions can be made about its disposition. Most priority-setting systems assess the public-health concern about chemicals with respect to exposure, toxicity, and social considerations.

The different degrees of concern can be thought of as ranges on a meter, with high readings generally warranting serious social concern and low ones suggesting that little or no concern is called for. The meter would register higher readings to the extent that there was an increase in any of the following:

- Number of people exposed.

- Per capita exposure in that population.

- Frequency of exposure.

- Probability of a toxic response at that exposure.

- Severity of a toxic response at that exposure.

- Cost to society or to an individual related to the manifestation and treatment of health effects.

On such a hazard scale, cigarette smoke would rate high, because:

- Millions of people smoke cigarettes, and additional millions are exposed to others' smoke.

- Exposures are high: grams of material are inhaled for every pack smoked.

● Frequencies of exposure are high—one or more packs per day for many smokers.

● Cigarette smoke is known to increase significantly the probabilities of lung cancer, other cancers, heart disease, and other adverse health effects.

● Lung cancer, heart disease, and some of the other effects are frequently fatal or incapacitating for long periods.

● Costs of treatment and the implicit cost of pain, suffering, and early death are great.

Other chemicals would rate low, for various reasons. In between would be yet other chemical hazards. For example, one assessment estimated that vinyl chloride might cause about 10 deaths per year (Kuzmack and McGaughy, 1975), compared with 1.3×10^5 cancer deaths per year from cigarette-smoking (U.S. Department of Health and Human Services, 1981). This example also demonstrates another feature of such assessments: their uncertainty. The figure cited for deaths associated with cigarette smoke is probably an accurate statement of the "true" value, to within a factor of 3; for vinyl chloride, the uncertainty is at least a factor of 10, and probably greater. Although there might be other reasons to test cigarette smoke further, an analysis based on the value of information would imply that the classification of high public-health concern was good enough, and that further testing would not markedly improve it. Another study of the carcinogenicity of vinyl chloride at very low exposures might or might not be warranted, but it would probably have a lower priority than a lifetime bioassay on a chemical that was similar to vinyl chloride in structure, uses, and exposures and that was mutagenic in vitro, but had never been tested in a rodent bioassay.

These examples are based on the argument that hazard depends on both the human exposure to a chemical and toxic responses in humans. Exposure and responses interact in a complex way. For example, some carcinogens may be presumed to exhibit a nearly linear relationship between per capita exposure and the probability of causing cancer. In this case, the hazard would be similar, regardless of whether 100 people were each exposed to 1 g/yr or 100,000 people were each exposed to 1 mg/yr. But, if the chemical exhibited a threshold for effects or had a distinctly nonlinear dose-response relationship, the hazard would depend markedly on the distribution of doses; 1 g/yr might be fatal to all of 100 people exposed, but 1 mg/yr might have little effect on 100,000 people. Or the outcome could depend on the fractionation of the dose over the year, on the route of exposure (oral, respiratory, or dermal), on synergistic exposures, or on the age, sex, race, and health status of the people exposed.

An ideal priority-setting system would be able to capture the importance of all information regarding these subtleties of exposure and toxicity and to modify accordingly the probabilities with which a chemical should be assigned to the various degrees of hazard. However, such information is generally scarce and expensive to retrieve and interpret, and the system would become excessively complicated if it attempted to treat the information in such detail. Consequently, the committee assumes that hazard is defined by the combination of exposure and toxicity, each capable of being defined and estimated independently of the other. For example, if the exposure range and the toxicity range were each divided into three segments (high, medium, and low), there would be nine categories of hazard, each corresponding to a different pairing of exposure and toxicity. (Some might turn out to be equivalent to one another, e.g., high toxicity and low exposure, low toxicity and high exposure, and medium toxicity and medium exposure.)

Data and tests are selected to help decrease the uncertainty in estimating the probabilities with which exposure or toxicity will truly fall into defined categories. For example, production volume is not a direct measure of human exposure (as are number of people, per capita exposure, and frequency of exposure); but, in the absence of more direct information on exposure, knowing the production volume does reasonably influence our perception of the probability that exposure is high. The results of an Ames <u>Salmonella</u>/microsome test do not prove or disprove toxicity in humans, but a positive result would increase, rather than decrease, our concern that the substance might be carcinogenic.

The following subsections contain more detailed discussions of the concepts of exposure and toxicity.

<u>ESTIMATING EXPOSURE</u>

In an ideal exposure assessment, the investigator would attempt to answer each aspect of each of the following questions:

- Who is exposed?

    - age
    - sex
    - race
    - health status

- How many are exposed?

- How are they exposed?

    - occupationally
    - in the community
    - as consumers
    - environmentally

- By what route?

  - oral
  - dermal
  - inhalation
  - single route of exposure
  - multiple routes of exposure (e.g., orally and dermally, as workers and consumers)

- In what pattern?

  - g/yr
  - g/incident
  - mg/m^3 or mg/L

- How frequently?

  - continuously
  - regularly, periodically (e.g., 8 h/d, 4 pills/d, once every month)
  - irregularly, but repeatedly
  - in single incidents

- Through what chain of events?

  - production or extraction
  - manufacturing process
  - transportation
  - storage
  - use
  - environmental transformation
  - bioaccumulation
  - industrial discharge
  - waste disposal
  - environmental transport

If all such information could be acquired, one could imagine dividing each dimension of exposure (route, amount, frequency, etc.) into ranges and then classifying all exposed people into groups defined by every combination of these ranges. Clearly, however, it would not be technically feasible to collect such details on most chemicals. Moreover, it would not be economically possible for a priority-setting process to acquire and use this much information, nor would it be easily understood. The committee therefore proposes to use available exposure-related information (in a given stage of the priority-setting process) merely to help estimate the probability that human exposure to the chemical will be relatively high, intermediate, or low. Thus, if all other factors were equal, exposure would be considered higher than average if the data element suggested that:

- Large numbers of people are or may be exposed.

- Concentrations in air, water, soil, food, or other environmental sources are higher than average.

- Exposure is more frequent or lasts longer than average.

- Exposure occurs by a route suspected of eliciting the most toxic response.

- Exposure occurs among groups that are especially susceptible because of age, race, sex, health status, or other conditions.

When a surrogate for exposure data is used, the objective is to choose a variable that reflects exposure as accurately as possible, in view of the extent of the toxicity information available at that stage of the priority-setting process. If, for example, reliable information shows an abrupt rise in the dose-response curve (e.g., from essentially zero effect to 100% occurrence of a serious effect), then the definitions of high, low, and medium exposure would clearly correspond to the per capita doses above, below, and near the point where the response curve rises rapidly. At the other extreme, when there is no toxicity information whatsoever, it is appropriate to assume a linear dose-response relationship of unknown slope, in which the extent of hazard varies continuously and proportionally with the extent of exposure. Here, the definitions of high, medium, and low exposure are largely arbitrary.

Of the entire universe of chemicals, probably only a few substances are usually considered to be in the high exposure region, more are in the medium region, and most are in the low region. This distribution conforms with common perceptions about the number of toxic chemicals to which humans are exposed at high concentrations. That is, if toxicity is rarely produced at a dose less than 1 mg/kg of body weight for single acute doses, then exposure would be considered high for a 50-kg adult only if typical doses were higher than approximately 50 mg in a single exposure incident.

In the absence of a quantitative model of the entire exposure-toxicity-hazard phenomenon, it will ordinarily be necessary to work with information that is much more ambiguous than the above guidelines suggest. For example, recourse only to some crude production and use information is unlikely to produce a convincing prediction of the number of people who will breathe a substance in concentrations greater than 30 mg/m^3, even if it is assumed that all other properties of the substance are "typical." In such cases, therefore, classification of exposure could not be based on the concentration at which humans are actually exposed, but must instead be expressed in terms of the pieces of information (data elements) that are available.

To continue the example, exposure is likely to be considered high if production is greater than 10^6 kg/yr and the substance is used predominantly in consumer products with high potential for exposure, e.g., foods, drugs, and cosmetics. This logic is not extended to estimate that exposure is greater than 10^6 kg/$(225 \times 10^6$ people) per year, i.e., 4 g/yr per person on the average; that figure might not be at all relevant to the assessment of toxicity.

ESTIMATING TOXIC EFFECTS

The process of predicting or estimating the biologic activity and toxic potential of a chemical is extremely complex. It requires the consideration of several types of information, coordination of a number of scientific disciplines, judgment and intuition, and an appropriate schedule of testing, if warranted. There are no firm rules for toxicologic prediction, nor are there guidelines that will ensure reliable answers. Attention is generally focused on human health effects for which there are attributable chemical causes. Surrogates (i.e., laboratory animals) are commonly used in testing for human health effects. In the best circumstances, toxicologists may be able to predict a particular activity of a chemical, ascribe a potential effect, identify a useful end point in an established animal model, elucidate the mechanism of the effect, and arrive at the conclusion that the chemical either will have no toxic effect in humans or will cause a measurable effect.

The toxic response to a chemical is multidimensional because of the wide variety of possible effects on human health, many of which are dose-dependent. In some cases, one human health effect may be clearly more serious (and thus have a higher priority for testing resources) than is another. For example, many more testing resources are devoted to cancer bioassay than to skin irritation tests, and severity is a major reason for the difference. In contrast, the relative severity of other pairs of effects may be difficult to evaluate and may differ according to individual perceptions. Nevertheless, choices between tests for different effects bring with them value judgments about relative severity, and a systematic priority-setting procedure will reflect some set of judgments. For the illustrative system in Chapters 3 and 4, the committee evaluated only one health effect--cancer--and avoided comparing the severity of effects. A complete and general system would need to consider explicitly the values that society assigns to different effects.

Chemically induced health effects in humans can be cataloged and

218

organized in a number of ways. One of the most convenient is to classify them by target organ. The range of such effects and their corresponding organs or tissues can be seen in the list of examples presented in Table 1.

Toxicity data derived from human exposure are necessarily limited—and are nonexistent for most substances. Therefore, it is customary to depend on data from toxicity tests in animals, from which extrapolations are made to predict health effects in humans. The most useful data are those derived from an accepted animal model for a given chemically induced condition in humans, but adverse effects in animals not related to a known disease in humans are also of value. Results of toxicity tests in animals have been recorded in machine-readable files for approximately 15,000 chemicals, whereas chemicals to which there is potential human exposure number several tens of thousands. Therefore, information on many chemicals is limited merely to molecular structure and physical constants.

Some idea of the relative importance of toxic effects can be gained from examining their consequences to the injured person and to society. For example, is an impairment structural or functional? Is it reversible or irreversible? Is it progressive? These questions are implicitly considered in most priority-setting schemes, including those that depend exclusively on human judgment. They imply value judgments about relative severity and involve not only judgments about biologic consequences, but also individual perceptions of harm. To examine consequences for the individual, a number of factors should be considered:

● An impairment may be caused by structural or functional damage or both. Irritation and corrosion may be inconvenient and painful, but are primarily structural. Lead-induced behavioral changes may be regarded as functional. Hexane-induced azonal neuropathy is both structural and functional.

● An impairment will generally be considered to be of greater concern if it is irreversible than if it is reversible. Many biologic effects are expressed only in the presence of the causative chemical and cease with or soon after its disappearance. Others, such as death, are obviously irreversible. Irritation, depression, and methemoglobinemia are examples of reversible effects; sensitization, retinal damage, pulmonary fibrosis, and carcinogenesis may be regarded as irreversible effects.

● An impairment may result immediately after acute exposure, may be delayed until some time after acute exposure, or may require repeated and prolonged exposure to become manifest. Corrosion is the most obvious immediately apparent impairment. Delayed hypersensitivity frequently results from an initially innocuous exposure. Alcohol-related cirrhosis of the liver is an example of an impairment manifested only after repeated and prolonged insult.

TABLE 1 Target Organs and Chemically Induced Effects in Humans

Target Organ or Tissue	Effects	Target Organ or Tissue	Effects
Skin	Altered appearance Irritation Sensitization Corrosion	Musculoskeletal system	Osteoporosis Corrosion
Eye	Irritation Corneal opacity Retinal damage Corrosion	Liver	Mixed-function oxidase induction Cholestasis Neoplasia Adenoma, carcinoma Cirrhosis Necrosis
Mucous membrane	Irritation Corrosion	Kidney	Aminoaciduria, proteinuria Uremia Lithiasis
Lung	Irritation Sensitization Fume fever Pneumoconiosis Fibrosis Neoplasia Adenoma, carcinoma Asphyxiation	Reproductive system	Germ cell mutation Embryotoxicity Teratogenesis Infertility
		Blood and hematopoietic system	Methemoglobinemia Bone marrow depression Aplastic anemia Leukemia
Nervous system	Behavioral changes Peripheral neuropathy Central nervous system depression Cholinesterase inhibition Locomotor ataxia Narcosis Convulsions Respiratory paralysis	Immune system	Immune suppression
		Fetus	Abortion Malformation Neonatal death

● Some impairments may be regarded as threats to life. Although many structural and functional impairments may be undesirable, painful, and debilitating, they are nevertheless clearly less objectionable than life-threatening impairments.

These factors exemplify the criteria that may be used for assessing the relative importance of health effects, such as are presented in Table 1. They are obviously incomplete in a number of respects. For example, no account is taken of acute lethality, individual susceptibility, dose-response relationships, and age and sex relationships.

ASSESSMENT AND DECISION-MAKING

Priority-setting systems include some procedure to reach a judgment about chemicals with the information collected. On the basis of this judgment, chemicals are considered further, recommended for testing, or removed from consideration.

Several assessment procedures may be used: scoring, modeling, sorting, ordinal ranking, and expert judgment. The choice of procedures depends on the number of chemicals considered, information available, cost, and type of personnel required. In addition, designers could consider the likelihood that the procedure might introduce bias, the ability to function even though some data are missing, and the ability to respond to uncertain or inaccurate data.

● In scoring systems, the data used as ranking criteria are assigned numerical scores (usually integers), and the scores are combined by a rule (often a weighted addition) to yield a single score that represents relative toxicity, relative exposure, or relative overall hazard. Scoring systems have the advantages of being easy to use and providing consistent treatment of all chemicals considered. However, it is difficult to combine scores in a valid way. Missing data often require substitution of default values. Uncertainty in the data is not reflected in the scores.

● Modeling-based systems use the data elements directly (kilograms of chemical produced, LD_{50} in milligrams per kilogram, etc.) and combine them into an index that represents the degree of human exposure, the degree of toxicity, or the overall health hazard. Modeling-based systems require more analysis than other procedures and require moderately skilled personnel. Uncertainty in the data can be dealt with more easily in models than in other procedures.

● Sorting (or screening) procedures answer questions regarding aspects of exposure and toxicity and sort chemicals into categories in accordance with the answers. The categories and sorted chemicals in each category are then ranked according to judgments as to which ones represent greater or more important hazards. Sorting procedures are easy to use and may be applied to large numbers of chemicals.

● In ordinal ranking, chemicals are ranked on each of various elements of exposure and toxicity, and the ranks are combined, according to a rule, to derive an overall ranking. Ranking procedures are easy to use and may be computerized. When rankings are combined, it is difficult to understand the meaning of the combined ranking. Missing data require substitution of default values. Uncertainty in the data is not indicated in the ranking.

● Expert judgment requires one or more highly skilled persons to review data, make an assessment, and recommend further action, such as testing. Methods of improving the elicitation of expert judgment are described in Appendix C. Although the use of experts is probably the most accurate assessment method, it is also the most expensive and requires highly skilled personnel.

DESIGN FACTORS

DESCRIBING THE PRIORITY-SETTING SYSTEM

While choosing the elements of the priority-setting system, the designer should consider that the system must respond to the following external parameters: the nature of the chosen universe of chemicals, accuracy of selection stages or other components and accuracy of the toxicity test(s) for which the system is selecting chemicals, and costs of the system components and toxicity tests. Each of these factors may be described in as much detail as desired by the designers of the system or as needed for the analytic technique being applied. The minimal description is a word description. Use of a mathematical model designed to optimize system performance requires a quantitative description.

Even a minimal description may be useful, because it requires system designers to address issues explicitly. For example, if one attempts to estimate the proportion of chemicals that cause a specific type of toxicity, several elements must be defined (e.g., what types of toxicity and which health effects are specified?). And one must specify the population of chemicals to which the estimate applies.

ACCURACY OF STAGES OR TESTS

Priority-setting systems may be regarded as classification procedures. Each stage may assess the degree of public-health concern separated into categories of toxicity and exposure for different types of health effects. Degree of public-health concern may be based on any number of categories of exposure or toxicity. The main limitation in defining categories of exposure and toxicity is the amount of information that is available. It does little good to define a large number of narrow categories if almost no information is available for assigning chemicals to them. Because the information is far from perfect, there will be errors. If there were only two categories of toxicity (or exposure)--such as toxic and nontoxic--there would be four possible classifications: true-positive, false-positive, true-negative, and false-negative. If the exposure and toxicity categories were divided into high, medium, and low toxicity and high, medium, and low exposure, there would be nine possible combinations, each with some degree of error (see Table 2).

Figure 2 illustrates misclassification in a complex situation. Public-health concern is a function of exposure and toxicity. Categories of concern are labeled S_1 to S_6, in order of increasing severity. For example, S_1 might refer to a situation of low exposure and low toxicity, and S_6 to one of high exposure and high toxicity. Each cell in the diagram stands for a classification. For example, if a chemical is truly in category S_2, but is classified in category S_5, it would be in cell $C_{5,2}$. Chemicals that are correctly classified lie along the diagonal--$C_{1,1}$, $C_{2,2}$, etc. These diagonal cells represent zero misclassification. As one moves from the diagonal to either the upper right or lower left corner, the extent of misclassification becomes greater. The cell in the top right corner ($C_{1,6}$) represents the most serious false-negative classification--a chemical with high public-health concern (high exposure and high toxicity) classified as one with low public-health concern (low exposure and low toxicity). The bottom left corner ($C_{6,1}$) represents the most serious false-positive classification.

MEASURING PERFORMANCE

In a perfect priority-setting and testing program, all chemicals would be classified correctly, so they would appear in cells along the diagonal ($C_{1,1}$ to $C_{6,6}$) in Figure 2. Even with unlimited funds, this would not be possible in practice, because even the best test batteries may sometimes yield false-positive and false-negative results. For any priority-setting system, there will be a spread of chemicals away from the diagonal. A testing and priority-setting program should make this spread as narrow as possible; more precisely, it should minimize the misclassifications for a given investment of resources.

In designing a priority-setting process, value judgments are unavoidable. Two such judgments are especially important. First, it is necessary to make some judgment about the relative importance of a false-negative and a false-positive for a given effect. This judgment is built into the design by the relative weights attached to the upper right and lower left corners of Figure 2. An illustrative method of deriving the relative weights is given in Appendix E. Second, it is necessary to make some judgment about the relative importance of health effects. This is built into the design by relating the cost of a false-positive for one effect to the cost of a false-positive for another effect (or comparing the relative costs of false-negatives). In principle, one type of true classification could be treated as more important than another type of true classification. In general, however, available information appears insufficient to make such a judgment (see Appendix B).

TABLE 2 Possible Results of Test Having Three Results: High, Medium, and Low

Result of Test	True Toxicity		
	Low	Medium	High
Low	Correct	False-negative	Greatest false-negative
Medium	False-positive	Correct	Greater false-negative
High	Greatest false-positive	Greater false-positive	Correct

TRUE CATEGORIES

	S_1	S_2	S_3	S_4	S_5	S_6
S_1	$C_{1,1}$					$C_{1,6}$
S_2		$C_{2,2}$				
S_3			$C_{3,3}$			
S_4				$C_{4,4}$		
S_5		$C_{5,2}$			$C_{5,5}$	
S_6	$C_{6,1}$					$C_{6,6}$

ASSIGNED CATEGORIES

FIGURE 2   Illustration of misclassification. Concern (S) is based on exposure and toxicity and ranges from S_1 (lowest) to S_6 (highest). C and its subscripts indicate cell into which chemical is categorized and false and true classifications. Cells $C_{1,6}$, $C_{5,2}$, and $C_{6,1}$ are misclassifications. Cells on diagonal ($C_{1,1}$, $C_{2,2}$, etc.) are correct classifications.

Once the extent of misclassification is at least conceptually defined, it is possible to compare alternative designs of priority-setting processes and to select the one with the lowest cost of misclassification within a given budget for priority-setting and testing.

COST

The costs of components of priority-setting systems and toxicity tests (in dollars per substance) are used in designing selection processes to ensure the selection of the most productive information-gathering activities per dollar spent. Cost data for toxicity tests are described in Stage 4 in Chapter 4.

3

BRIEFLY FOLLOWING A CHEMICAL
THROUGH AN ILLUSTRATIVE SYSTEM

This chapter presents a simplified overview of the operation of an illustrative priority-setting system. It assumes that the system has been fully designed and implemented and is ready for operation. Although it uses the decision rules from an optimization model, it does not attempt to show how those rules were derived (see Appendix B).

As currently envisioned, the principal operations of the system occur in three stages (Stages 1, 2, and 3), with successively increasing costs per chemical processed and decreasing number of chemicals needing scrutiny. This chapter also discusses the formation of the universe of chemicals of concern (Stage 0) and the actual testing of chemicals (Stage 4). Figures 3 through 5 illustrate the steps in each stage. The reader is encouraged to refer to the figures while reading the text.

The system can treat a chemical only if it is in the universe of chemicals of interest (or is nominated for Stage 3 by an agency or individual). The universe must be assembled as a list containing unique identifiers for each chemical, such as its number in the CAS Registry. The universe is made up of several lists, the largest and most accessible of which is the select universe compiled by the Committee on Sampling Strategies. Other sources are the environment (which could be characterized in the sampling programs of EPA); food ingredients, including natural constituents, contaminants, and conversion products (which might be available from lists maintained by FDA); and the growing group of chemicals on which premanufacturing notices have been filed with EPA. The additional lists must be acquired or constructed, and identifiers and associated data added; then these new lists must be merged with the existing lists by eliminating duplication and sorting the chemicals into the order in which they will be maintained on the master list.

To illustrate the operation of the remaining stages, it is useful to have an example chemical, rather than dealing in the abstract with an unlimited range of possibilities. We have chosen bisphenol A as our example chemical, although at some places in our analysis we have speculated about its properties to create the best illustration.

STAGE 1

Bisphenol A is identified by its CAS Registry number, 80057. This number links the priority-setting system with the various automated data bases on which it depends. In Stage 1, only a few fragments of information can feasibly be retrieved and processed. Among them are the following:

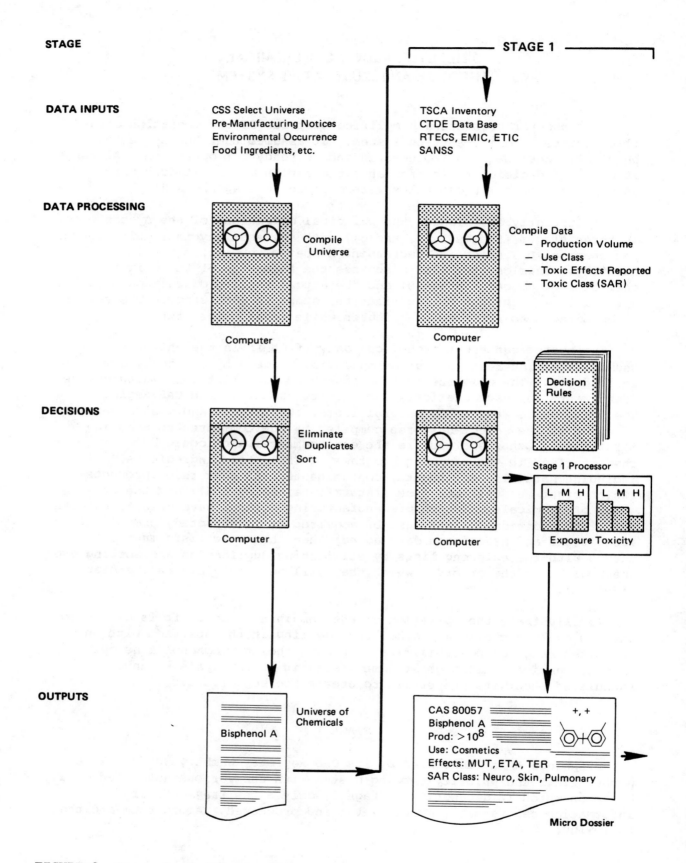

FIGURE 3 Stage 1 of illustrative system.

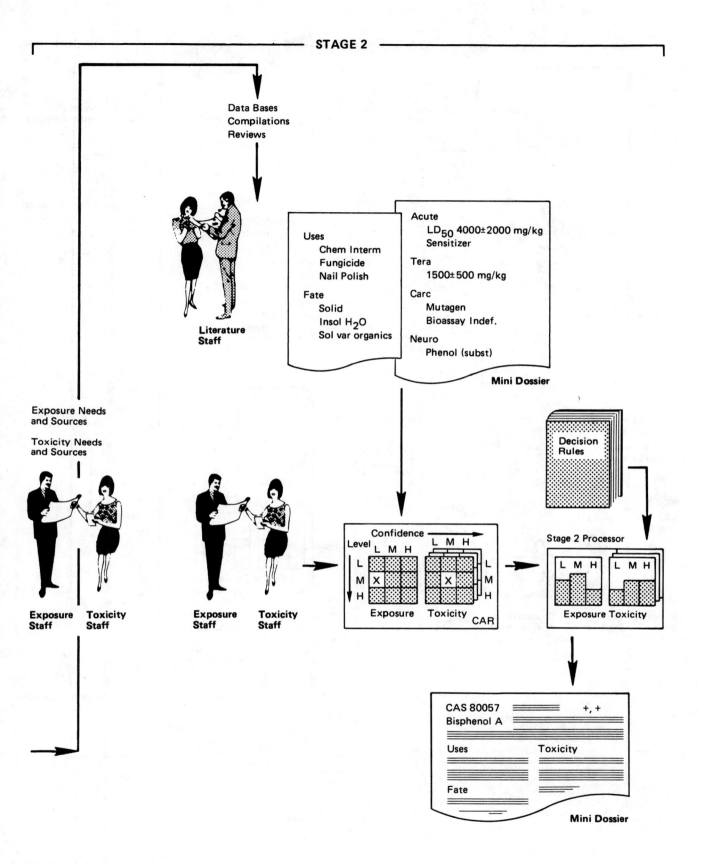

STAGE 2

Data Bases
Compilations
Reviews

Literature
Staff

Uses
    Chem Interm
    Fungicide
    Nail Polish

Fate
    Solid
    Insol H_2O
    Sol var organics

Acute
    LD_{50} 4000±2000 mg/kg
    Sensitizer
Tera
    1500±500 mg/kg
Carc
    Mutagen
    Bioassay Indef.
Neuro
    Phenol (subst)

Mini Dossier

Exposure Needs
and Sources

Toxicity Needs
and Sources

Exposure   Toxicity
Staff   Staff

Exposure   Toxicity
Staff   Staff

Decision
Rules

Confidence
Level L M H   L M H
L   L
M   X   X   M
H   H
Exposure   Toxicity   CAR

Stage 2 Processor

L M H   L M H

Exposure Toxicity

CAS 80057 ======== +, +
Bisphenol A

Uses   Toxicity

Fate

Mini Dossier

FIGURE 4   Stage 2 of illustrative system.

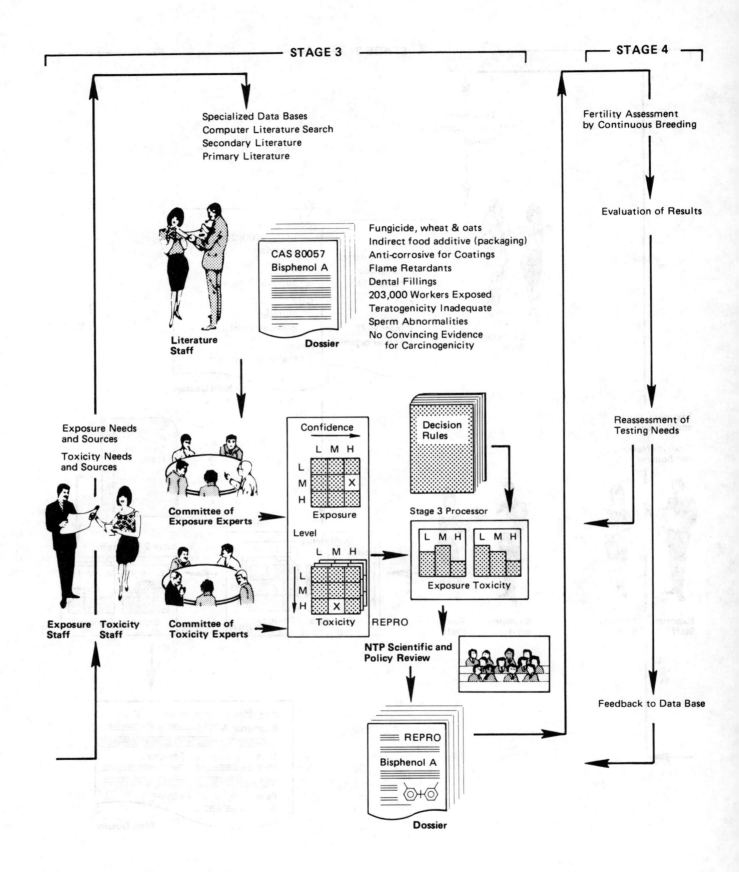

FIGURE 5   Stages 3 and 4 of illustrative system.

● Information on the magnitude of production of the chemical from the TSCA Inventory (U.S. Environmental Protection Agency, 1979, 1980), the International Trade Commission's synthetic organic chemicals data set, the Toxicity Data Bank (TDB), or any other accessible automated data banks. In our example, bisphenol A is found in the TDB, where it is shown as being produced at more than 10^{11} g/yr, or more than 2×10^8 lb/yr.

● Intended-use class in the Committee on Sampling Strategies data base for the select universe, or an equivalent use code from another data base. Bisphenol A is classified as a cosmetic ingredient in the committee's data base, although we shall see that it could well have been listed in another category.

● Chemical structure, if it has a defined structure and is in an accessible data base, such as the CAS connection tables or the Chemical Information System data base. Bisphenol A proves to have two phenol rings joined by a propyl group at the central carbon, as shown in the "microdossier" in Figure 3.

● Registry of Toxic Effects of Chemical Substances (RTECS) data on observed toxic effects. RTECS includes several listings for bisphenol A in the categories of mutagenic effects (MUT), equivocal tumorigenic effects (ETA), and teratogenic effects (TER) and has several entries for acute toxicity.

● Chemical fragments found in the structure by the Structure and Nomenclature Search System (SANSS) and the corresponding suspected effects from Table 5. Our current table would show bisphenol A only in the "phenol (substituted)" category and would show some suspicion of neurotoxicity, skin sensitization, and pulmonary irritation.

When the data on a given chemical have been collected from the various automated data bases, the computer implements the decison rules derived from the optimization model as the "Stage 1 processor." These computer-generated rules (described in Appendix B) effectively compute a pair of probability distributions of exposure and toxicity, as indicated by the two small bar charts in Figure 3. Because the production of bisphenol A is greater than 10^8 lb/yr and it is used in cosmetics, the processor decides that exposure to it is more likely to be high than is exposure to a random chemical, but still shows that the most probable exposure is low. Similarly, the occurrence of so many RTECS entries and the suspicion of even more toxic effects from the structure-activity-relations (SAR) class suggest with more than usual confidence that the toxicity is likely to be at least moderate, so moderate toxicity is shown as having highest probability.

On the basis of these distributions, the computer decides not only that bisphenol A should be passed to Stage 2, but that it deserves the higher level of investigation for both exposure and toxicity, as indicated on Figure 4 by "+, +," meaning that a moderate amount may be spent on gathering and assessing data both on exposure and on toxicity. In practice, the computer need not generate the explicit probability distributions in making the decision, because the rules used to select chemicals for further consideration in Stage 2 are always the same, given the same initial data ranges. However, the distributions can be useful inputs to Stage 2 and can be included in the microdossier that is the output of Stage 1. Figure 6 shows the selection rules when carcinogenicity is the only toxic effect considered. There would be other rules for suspicion of teratogenicity, neurotoxicity, and so on. The possibility of all such effects would be mentioned in the microdossier for consideration in Stage 2.

STAGE 2

The microdossier (if available) is received by NTP staff and examined by two persons, one expert in toxicity assessment and one in exposure assessment. The exposure assessor may or may not deal with the toxicity part of the microdossier, and the toxicity assessor may or may not deal with the exposure part. Knowledge of the suspected toxic effects can help guide the exposure search, but can also bias it.

The first task is to check the status of the chemical in the testing program to ensure that the effects of concern are not currently under examination and not yet reported in RTECS. Let us assume that no tests of bisphenol A are in progress.

Each expert scans the microdossier and develops a strategy for searching the appropriate data bases, compilations, and reviews, noting what sorts of information are probably most important to retrieve. In the case of bisphenol A, the exposure assessor calls for a routine printout from the Toxicity Data Bank and checks the Cosmetic Ingredient Dictionary (Estrin et al., 1982), the Merck Index (Windholz et al., 1976), and the Directory of Chemical Producers for further information on production, uses, and physical and chemical properties. This strategy is carried out by the literature staff of NTP. In our example, they find that at least 90% of the bisphenol A produced is used as an intermediate in the production of other chemicals, especially resins. In cosmetics, it turns out to be a component of nail polish, but there are other dispersive uses, e.g., as a fungicide. The melting point implies that bisphenol A is ordinarily a solid; and it is relatively insoluble in water, but soluble in many other solvents. The exposure assessor integrates this information subjectively and decides that the best guess is still moderate exposure; but confidence in that guess is still low, because the amounts used in nail polish and fungicides are so uncertain.

Meanwhile, the toxicity assessor requests a full printout of the RTECS listings and examines the TDB printout. The literature staff

STAGE 1: EXPOSURE DATA ELEMENT

                                                     Go to Stage 1
                                                     Toxicity Data
Intended Use and Production   Elements Part

Food chemical, unclassified or pesticide with
unknown production   4

STAGE 1: TOXICITY DATA ELEMENTS

Part 4

                                                Go To Stages 2-4
                                                via Branch

    Member of a chemical group associated
    with cancer with "low suspicion
         not in RTECS.................................1
         in RTECS, no mention of MUT or CAR...........2
         in RTECS, mention of MUT, but not CAR........6
         in RTECS, mention of CAR.....................6

STAGES 2-4

Branch 2

Stage 2: Perform search for exposure data and limited search for
          toxicity data

Assessments Based on Data Gathered in Stage 2		Stage 3: Data Gathering	Assessments Based on Data Gathered in Stage 3	Stage 4: Recommended Testing[a]
Exposure	Toxicity			
Medium	Medium or high	Dossier	High toxicity and high exposure	LT
			Otherwise	ST

_____
[a]ST = Short-term test; LT = Long-term test

          FIGURE 6 Decision rules used in processing example
          chemical bisphenol A.

supplies copies of the data found in such handbooks as <u>Dangerous Properties of Industrial Materials</u> by Sax (1979) and the <u>Merck Index</u>, which shed little additional light on the toxicity picture. However, it becomes evident that bisphenol A is not very acutely toxic (LD_{50}, around 4,000 mg/kg) and that it has shown some teratogenicity (also at high doses) and has been judged indefinite in a cancer bioassay. The cancer hazard is still judged as moderate, but with only moderate confidence, and the reproductive hazards are considered high, also with only moderate confidence. No support arises for any of the SAR suspicions, and they are eliminated as concerns.

Both assessors have filled out the matrix of concern, as shown in Figure 4 immediately under the minidossier. Using the decision rules (summarized in Figure 6 for the carcinogenicity data), they find that a dossier covering both exposure and toxicity searching is justified in Stage 3. They can also generate the exposure and toxicity probability distributions (Figure 4, "Stage 2 Processor") either automatically from the matrix assignment or on the basis of their own judgment. These estimated probability distributions are passed to Stage 3 with the augmented minidossier containing all information gathered to date.

STAGE 3

Exposure and toxicity experts (not necessarily the same ones) now examine the minidossier and determine the Stage 3 search strategy (Figure 5). Clearly, it is desirable to acquire more information about consumer exposures to bisphenol A in both cosmetics and fungicides and about occupational exposures to bisphenol A in its large use as an intermediate. Searches of the <u>Chemical Economics Handbook</u> (SRI International), the NIOSH National Occupational Hazard Survey (U.S. Department of Health, Education, and Welfare, 1974, 1977, 1978), EPA contractor reports, and so on are ordered. The literature staff finds additional uses of bisphenol A as an indirect food additive (because of residua in plastic food containers), as a fire retardant, as an anticorrosive in coatings, in dental fillings, and in various other minor applications. The fungicidal application turns out to be for use on wheat and oats. In the manufacture of bisphenol A and its derivatives, over 200,000 workers may be exposed.

Considerable effort goes into computer-assisted literature searches of the secondary and primary literature for citations on the toxicity of bisphenol A. The teratology experiments are found to be rated as inadequate, but there has been a finding of sperm abnormalities in rats after intraperitoneal injection. The "indefinite" cancer bioassay is now assessed as showing "no convincing evidence of carcinogenicity."

Once the toxicity and exposure parts of the dossier are compiled, they are submitted to expert committees on toxicity and exposure, respectively. Again, there are pro's and con's as to whether the committees should see both dossiers or only the parts pertinent to their own expertise. The dossiers are studied by the committee members both individually and collectively, but the full toxicity committee is asked to evaluate the evidence and reach a consensus on the toxicity for each human effect, and the full exposure committee likewise for exposure. Each committee is also asked to state its members' confidence in the rating. If there is much disagreement over the rating, there is good reason to lower the stated degree of confidence. The ratings are reported on a matrix as in Stage 2, or the committee may prefer to generate a probability distribution directly, instead of using the standard one generated by the Stage 3 processor. The important difference from current procedure, however, is that the committees are not asked to make testing recommendations, but only to assess the available data in their field of expertise.

Whether the committees generate their own probability distributions or simply use matrix rankings, the Stage 3 processor uses the previously developed decison rules to decide what tests should be recommended to NTP. In the case of bisphenol A, the committee on exposure has rated the exposure as still only moderate, but now with high confidence. The toxicity committee now believes that there is comparatively little evidence of carcinogenicity, but is highly confident of moderate teratogenicity. It now believes that other reproductive effects may be of high concern, but has only moderate confidence in that judgment. The carcinogenicity rule, reproduced in Figure 6, suggests that no repeat of the bioassay is advisable now, but that an expansion of the battery of short-term tests for mutagenic potential may be worth while. We assume that the rule for reproductive effects would call for a test, as shown in the box on the final version of the dossier in Figure 5. Before this recommendation is accepted, however, the dossier and recommendations are reviewed by the NTP panels for scientific merit--especially to see whether the recommendation is unrealistic for some reason not perceived in the formal process--and for such policy considerations as public concerns, costs and feasibility of controls, or regulatory responsibility. Recommendations may be overridden on the basis of results of either type of review.

Assuming that the NTP panels accept the processor's recommendations, the dossier and test directions are sent on to the testing program in Stage 4.

STAGE 4

Finally, the substance goes into the specific type of reproductive study warranted by the particular combination of concerns about exposure and toxicity and the degree of confidence in each. In the case of bisphenol A, fertility assessment by continuous breeding of rodents is the preferred test. When the test is complete, its results are evaluated with respect to public policy and their relevance to further refinement of the selection process (Figure 5, "Feedback to Data Base"). The test results can also be fed back into the specific data base on bisphenol A to determine whether further tests are justified. Although it is not shown in Figure 5, the argument for a new bioassay would be much more compelling if the short-term mutagenicity assays had proved consistently positive.

4
DETAILED DESCRIPTION OF THE
OPERATION OF AN ILLUSTRATIVE SYSTEM

INTRODUCTION

An illustrative priority-setting system has been developed to demonstrate the application of principles described in Chapter 2 and to demonstrate the feasibility of the approach to designing priority-setting systems.

The system responds to the approach taken in Part 1, in that it uses the same classes of intended use and incorporates two of the toxicity tests used by the Committee on Sampling Strategies and the Committee on Toxicity Data Elements. The approaches differ in that the illustrative system is structured to be specific for a human health effect, whereas those two committees take a more general approach.

ELEMENTS OF DESIGN

Authors of priority-setting systems have several choices to make: goal of the system, universe of chemicals considered, structure of the system, and assessment.

GOAL

The goal chosen for this illustrative priority-setting and testing system is to assess accurately the public-health concern about all chemicals to which there is human exposure. This goal was chosen to reflect the wide range of chemicals of interest to NTP. A goal might be chosen to be oriented more toward regulation, such as minimizing the impact on human health caused by chemicals to which there is exposure. The Committee on Priority Mechanisms did not choose this goal, because NTP does not have regulatory authority and because NTP does testing for purposes other than regulation.

UNIVERSE

The universe of chemicals to be considered is defined to include all chemicals to which there is potential human exposure. The universe defined by the Committee on Sampling Strategies has five categories of intended use: food chemicals, drug ingredients, pesticide chemicals, cosmetic ingredients, and general industrial chemicals (TSCA Inventory). There are two additional categories: "other," such as pollutants; and "unknown."

ASSESSMENT

The degree of public-health concern is a combination of degree of human exposure and degree of toxicity to humans. For purposes of designing this illustrative system, toxicity is limited to carcinogenesis. This health effect was chosen because of its severity and the high degree of public interest in it. Also, more effort has been devoted to discovering carcinogens; therefore, more data are available for designing the system.

Both toxicity and exposure are expressed as high, medium, or low, because available information seems inadequate to define more than three levels. Toxicity, in this case carcinogenic potency, is defined in terms of TD_{50} (Ames et al., 1982): high is a TD_{50} less than 10^2 µg/kg-d, medium between 10^2 and 10^6 µg/kg-d, and low greater than 10^6 µg/kg-d. This definition assumes a linear dose-response curve. The category of "low" toxicity also includes noncarcinogens. For exposure, high is greater than 10^7 person-grams per year (p-g/yr), medium is between 10^7 and 10^5 p-g/yr, and low is less than 10^5 p-g/yr.

Note that, in this population-exposure formulation, large exposure of a small number of people could be equivalent to small exposure of a large number of people. Which situation is more important is a matter of social judgment.

STRUCTURE

A multistage structure was chosen, so that the system could handle many thousands of chemicals and fit the current institutional structure of the NTP selection system. A multistage structure provides for using small amounts of data to assess large numbers of chemicals in the early stages and examining fewer chemicals in depth in later stages.

RULES FOR SELECTION

The result of each stage is an assessment of the toxicity and exposure of each chemical considered in that stage and an estimate of confidence in each assessment. Having these assessments from each stage, it is still necessary to decide whether a chemical should receive further consideration or be removed from consideration. A mathematical model chooses rules for selection such that the errors in assessment are minimized, for a chosen amount of resources for selection and testing (Appendix B).

That optimization model not only guided the Committee in designing the illustrative priority-setting scheme, but also allowed it to examine the values of and interactions among various indicators of exposure and toxicity. Even the simplified model used for choosing the rules of the illustrative system validated some intuitive judgments about testing priorities and provided valuable additional insights; further development of the model may be justified to sharpen the rules for chemical selection in an operational priority-setting system and to modify them as additional information is gained by operating the system and conducting the tests it selects.

DESIGN PARAMETERS

Several characteristics of the universe were estimated and serve as data for the model used to develop the rules for the illustrative system. These estimates were based on fragmentary data and are intended to be replaced with better estimates based on data obtained from operating the system. The universe was estimated to consist of 70,000 discrete chemicals. The Committee on Priority Mechanisms intends that the definition of the universe be considered flexible and subject to change as needed. The estimate of 70,000 is not intended to become either a maximal or a minimal number of chemicals to be considered. On the basis of the above definitions of degree of exposure, the overall distribution of low, medium, and high exposure is 0.7, 0.2 and 0.1. On the basis of the definition of carcinogenic potencies, the estimated distribution of potency is 0.95 low, 0.04 medium, and 0.01 high. These estimates are only for purposes of designing the illustrative system and are highly speculative. Estimates of the proportion of carcinogens in a group of chemicals may vary widely (U.S. Congress, 1981), because of the definition of "carcinogen" and the group of chemicals considered.

STAGE 1

The purpose of Stage 1 is to provide a mechanism for scanning the entire population of chemicals to be considered by the system. It is intended that this scanning could be performed in many ways: by category or combination of categories of intended use or by chemical structure. The scanning need not be performed in one year, but might be performed over a period of a few years.

Stage 1 is limited to data in machine-readable data bases, so that human intervention is minimized. The Committee on Priority Mechanisms chose to use the data bases and searching capability of the Chemical Information System (CIS) developed by the National Library of Medicine (NLM) and EPA. This system was chosen because many chemicals of interest are already included in the data bases. In addition, the Structure and Nomenclature Search System of CIS provides a searching capability for chemical structures and thus makes possible a crude analysis of structure-activity relationships.

EXPOSURE DATA ELEMENTS

The purpose of exposure data elements in all stages is to contribute to an attempt at estimating exposure on the basis of surrogate data--data that are related to exposure only incompletely or inaccurately. "Exposure" is a concept that embraces the quantities of material that reach people, the number of people exposed, the rates and patterns of exposure, the characteristics of the exposed population, and other factors. A fully developed system for priority-setting not only would take into account as many of these dimensions as feasible, but also would ensure that they interacted properly with the dimensions of toxicity. However, complexity may not be justified for screening decisions at early stages of the system, and it also obscures the logic of the illustrative system. Therefore, the committee conducted its analysis with a single exposure variable to which degree of hazard would be proportional for a constant degree of toxicity. If a linear dose-response relationship is assumed for toxicity as a first approximation, then exposure can be measured by the total mass of a substance that is ingested, inhaled, or otherwise taken in by all members of the population, in units of person-grams per year. This quantity can be estimated by summing the products of the per capita annual intakes and the number of persons for all exposed groups with different intakes.

Two surrogates of exposure--class of intended use and production volume--are used to define the possible classifications of a Stage 1 exposure data element (Table 3); each classification consists of a pair of subelement classes--use and production volume. The first is derived from the definitions of intended use devised for the select universe, and the second from the TSCA Inventory or other automated sources.

We assume that, on the average, a higher fraction of food-chemical and drug production than of cosmetic, general TSCA chemical, or other (as yet undefined) chemical production eventually results in human exposure. New classes would need to be examined to test this assumption. Prime candidates for new classes are general environmental chemicals in air and water (including degradation products and "natural" pollutants) and food constituents and their products. If one accepts the exposure assumption, then the production volume required for a high probability of exposure should be lower for food chemicals and drugs than for chemicals in the other use classes. Table 3 lists an illustrative exposure classification, in which there is a decreasing probability that exposure is high; where the probabilities are identical, the list is in decreasing order of confidence in the estimates. The ability of the classifications to identify high-exposure chemicals cannot be determined without estimating the numbers of chemicals in the universe of concern that fit these classifications. Table 4 is a hypothetical classification and would need to be revised as data and experience accumulate.

TABLE 3 Illustrative Stage 1 Estimates of Probability of Exposure
in Relation to Use and Production

		Probability That Exposure Is:		
Use[a]	Production, lb/yr	Low	Medium	High
F,D	$> 10^4$	0.40	0.40	0.20
P,C	$> 10^5$	0.40	0.40	0.20
G,O	$> 10^8$	0.40	0.40	0.20
U	$> 10^6$	0.40	0.40	0.20
G,O	10^6–10^8	0.50	0.35	0.15
P,C	10^4–10^5	0.50	0.35	0.15
U	10^4–10^6	0.68	0.20	0.12
F,D	U	0.68	0.20	0.12
F,D,U	$< 10^4$	0.73	0.18	0.09
P,C,G,O,U	U	0.73	0.18	0.09
P,C	$< 10^4$	0.75	0.17	0.08
G,O	10^4–10^6	0.75	0.17	0.08
G,O	$< 10^4$	0.76	0.17	0.07

a F = food chemical.
   D = drug.
   P = pesticide.
   C = cosmetic.
   G = general commerce (TSCA).
   O = other (known, not previously classified).
   U = unclassified.

TABLE 4 Illustrative Estimates of Distribution of Production Volumes in Relation to Use Categories

Use Class	Fraction of All Chemicals		Fraction by Production Volume, lb/yr				
		$>10^8$	$10^6 - 10^8$	$10^5 - 10^6$	$10^4 - 10^5$	$<10^4$	U[a]
Food	0.12	b --	0.001	0.003	0.003	0.003	0.11
		c 0.001	0.009	0.020	0.03	0.06	--
Drug	0.10	b --	--	0.001	0.002	0.004	0.093
		c 0.001	0.004	0.015	0.025	0.055	--
Pesticide	0.05	b --	--	0.001	0.002	0.002	0.045
		c 0.0002	0.006	0.005	0.013	0.026	--
Cosmetic	0.05	b --	0.001	0.001	0.002	0.002	0.044
		c 0.001	0.002	0.003	0.014	0.03	--
General	0.066	0.02	0.06	0.06	0.09	0.28	0.3
Other	0.01[d]	0	0	0	0	0	0.01
Unknown	0.01[d]	0	0	0	0	0	0.01

a   In inventory, but without stated production volume.
b   Estimated distribution of production volumes listed in CIS.
c   Estimated distribution of production volumes if all were known.
d   At present--these categories will grow.

TOXICITY DATA ELEMENTS

Two surrogates of toxicity, RTECS status and chemical class, are used in Stage 1. Each is used to estimate the likelihood that a substance may have toxic effects. These surrogates indicate that the likelihood that a given chemical or class of chemicals will have health effects is different from the likelihood that any substance in the universe of chemicals will have those health effects.

Given the present state of knowledge, it is only slightly possible to assess the potential for an adverse response on the basis of chemical structure. But it would be desirable to use knowledge of structure-activity relationships (SARs) to identify potentially toxic substances by analyzing substructures that have been associated with adverse effects in humans or animals. Such an assessment is ordinarily based on expert knowledge, experience, and intuition and is used in considering the type of testing that may be required.

A number of systems have been created to place SAR analysis on a more formal footing. They range from simple classifications of key types of substances to sophisticated statistical treatments that involve weighting of subgroups and from detailed treatments of specific health effects to general considerations of toxicity. These approaches have merit as research efforts, but none has evolved enough to provide a practical or accurate method for identifying potentially toxic chemicals. A more detailed review of the possible contribution of SAR analysis is given in Appendix D.

To be usable in Stage 1, any data base used for SAR analysis must contain a large proportion of the universe of chemicals being considered. The data must also be susceptible to a search for chemical subgroups that can be associated with specific types of toxicity. The SANSS data base may provide a potentially useful data base. It can be searched for any of 271 functional groups containing at least two nonhydrogen atoms, any of 137 specific cyclic nuclei, and a number of hydrocarbon radicals. Each structural group or feature is described by a code that can be modified to allow for other structural features, such as the attachment of phenyl nuclei and aromaticity in ring structures. At the simplest level, the system provides a machine-readable way of identifying all the compounds in the system's collections that contain a specified structural group. The system may also be programed to identify structures with specified subgroups; thus, the code for phenols can be modified to produce separately monocyclic, dicyclic, and tricyclic phenols. And the system allows identification of compounds that have more than one specified functional group; this permits the identification, in lists of compounds that contain a given structure, of substances in which that structure is accompanied by another structure that might modify its biologic activity.

243

About 80 of the SANSS specific functional-group codes are associated with one or more human health effects, as shown in Table 5. The groups are assumed to produce the ascribed effect either by direct action or after metabolic activation. The table is illustrative and should not be considered comprehensive or definitive. The estimated degree of association between a given chemical structure and a specific form of toxicity is indicated in the table as low (L), medium (M), or high (H). An association between a chemical structure and an effect and the degree of that association imply that chemicals that contain the structure are more likely to cause the particular health effect in question than are randomly selected chemicals in the universe being considered.

RTECS, published by the National Institute for Occupational Safety and Health, is the most comprehensive (but not the only) summary of information on the toxicity of chemicals that is available in easily accessible, computerized form. In the 1980 edition, positive results of toxicity testing of some 45,000 chemicals are reported. In general, negative results of toxicity testing are not reported, so RTECS is biased toward toxicity. But RTECS does report negative results of lifetime bioassays for carcinogenicity by NTP (and earlier by the National Cancer Institute).

Consequently, mere listing in RTECS slightly raises the probability of high or moderate toxicity at the expense of the probability of low toxicity. RTECS also reports specific positive results, such as carcinogenicity, mutagenicity, teratogenicity, reproductive effects, and a variety of acute and other chronic toxicities. Nevertheless, in practice, even listing as a carcinogen carries with it a false-positive rate, because of the imperfect correlation in cancer hazards among different species, experimental uncertainties, faulty experimental procedures, or mere error. The absence of a CAR code implies that the substance may have proved negative in a bioassay, but the false-negative rate for such an assumption is high. Similar problems beset other toxicity codes.

In a fully developed priority-setting system, each toxic effect of concern would be related to at least one RTECS code. Matrices of probability for a given effect would be constructed to show how the underlying prevalence rates are modified for appearance or nonappearance of particular codes, which are retrievable by computer search. In this report, we discuss only carcinogenicity, both because time was too limited to develop other "performance characteristics" and because the theory and data base for carcinogenicity are better developed than those for other kinds of toxicity.

Six codes are thought to be related to carcinogenicity: NTP POS,* NTP NEG, CAR, NEO, ETA, and MUT (MTDS in the latest RTECS). NTP POS

---

*Many RTECS codes refer to NCI-positive or -negative results, rather than to quantitative results of tests conducted according to NTP-approved protocols.

TABLE 5 Chemical Groups Associated with Human Health Effects Either Directly or After Metabolic Activation

Chemical Group	Molecular Entity	SANSS Code	In TSCA	Carcinogenesis	Mutagenesis	Developmental Effect	Neuropathy	Pulmonary Disorder	Cardiovascular Disorder	Immune System Disorder	Hepatic Renal Disorder	Blood and Bone Marrow Disorder	Skeletal Effects	Skin or Mucous Membrane Disorder
Acetals, aliphatic	(structure)	Fg 99	253	L	L	L	L							H
Acridine	(structure)	Fg 93, SCN 119, SCN 120	16	M	M									L
Alcohols: Aliphatic, saturated C$_6$	--C-O	Fg 80	764	L		M	M				M			L
Aliphatic, unsaturated C$_6$	---C-O	Fg 80		L										M
Aldehydes	--C=O	Fg 85		L	L		L							M
	~C=O	Fg 88	1711	L	L		L							M
	(structure)	Fg 130		L	L			L						M
	---(R)C=O	Fg 130, Fg 85R		L	L									M
Allyl	C=C-C	HR 12E	458	L	L									M

[a] Health Effect

TABLE 5 (continued)

Chemical Group	Molecular Entity	SANSS Code	in TSCA	Carcinogenesis	Mutagenesis	Developmental Effect	Neuropathy	Pulmonary Disorder	Cardiovascular Disorder	Immune System Disorder	Hepatic Renal Disorder	Blood and Bone Marrow Disorder	Skeletal Effects	Skin or Mucous Membrane Disorder
Amines: aliphatic C_6	-N	Fg 143		L	L	L	M							M
aromatic	-N-	Fg 144	678	L	L	L	M			L				M
	-N-	Fg 145		L	H	L	M				M	M		M
	RN-	Fg 143R	409	H	H	M	L							L
Anhydrides, carboxylic	$\overset{O}{\underset{}{C}}-O-\overset{O}{\underset{}{C}}$	Fg 133	1					M						H
Anthracene	(anthracene ring structure)	SCN 124	27	M	L									L
			622	L	L									
Arsenic and derivatives	[As]	Fg 3–Fg 17	33	M								M		
Aziridines	(aziridine ring, N)	SCN 1	65	H	H	M		L						L
Azo aliphatics	-N=N-	Fg 174	23	L	L									
Azo aromatics	-N=N-R	Fg 172R	2462	H	H							L		L
Azoxy compounds	$\overset{O}{\underset{}{-N=N}}$	Fg 174	23	L	L									

246

Chemical Group	Molecular Entity	SANSS Code	In TSCA	Carcinogenesis	Mutagenesis	Developmental Effect	Neuropathy	Pulmonary Disorder	Cardiovascular Disorder	Immune System Disorder	Hepatic Renal Disorder	Blood and Bone Marrow Disorder	Skeletal Effects	Skin or Mucous Membrane Disorder
Benzyl derivatives	El-C-R	HR2ER	2045	L	L						L			L
Borates	-O-B-O- O-	Fg 19-22	58				L							L
Cadmium compounds[b]														
Carbamic acids: esters	O=C N-C-O	Fg 51, 52	248	L		M	M	L	L		L			
halides	O=C N-C-X	Fg 48 & R	6	H	H	L	M	M		L				M
Carbazoles	(structure)	SCN 111	22	L	L									L
Carbodiimides	-N=C=N-	Fg 62	4											L
Carboxylic acids	O=C-O	Fg 94 (R)												M

247

TABLE 5 (continued)

Chemical Group	Molecular Entity	SANSS Code	In TSCA	Health Effect										
				Carcinogenesis	Mutagenesis	Developmental Effect	Neuropathy	Pulmonary Disorder	Cardiovascular Disorder	Immune System Disorder	Hepatic Renal Disorder	Blood and Bone Marrow Disorder	Skeletal Effects	Skin or Mucous Membrane Disorder
Chromium compounds[b]				L				L			L			L
Cyanohydrins		Fg 123	11					H						
1,4-Dioxanes		SCN 43	7								M			
Disulfides	-S-S-	Fg 251	110											L
Ethylene oxides		SCN 2	833	H	H	H		H			L			L
Flourene		SCN 112	15	M	M									L

Health Effect

Chemical Group	Molecular Entity	SANSS Code	In TSCA	Carcinogenesis	Mutagenesis	Developmental Effect	Neuropathy	Pulmonary Disorder	Cardiovascular Disorder	Immune System Disorder	Hepatic Renal Disorder	Blood and Bone Marrow Disorder	Skeletal Effects	Skin or Mucous Membrane Disorder
Glycolethers	$-OCH_2CH_2O-$	Fg 96 & Fg 98 R	30			M	L				L			
Halides, aliphatic, saturated	$---C-X$	Fg 112	1190	L	L		L	L		L	L			L
Halides, aliphatic, unsaturated	CX	Fg 112	630	L	L		L			L	L			L
Halides, aromatic	RCX	Fg 112R	2874	L	L	L	L			L	L			
Haloaliphatic amines	$X(CH_2)_x-N$	Fg 13	54	M	M			M		M				M
Haloalkyl sulfides	XCH_2-S-	Fg 250	8	L	L			M		M				M
Haloethers	XCH_2-O-	Fg 112	88	H	H			L						L
Haloformic esters	$XC-O--$	Fg 101	40	L				M						L
Hexyls	$C_6H_{13}El \sim$	HR 31E	167				M							

249

TABLE 5 (continued)

Chemical Group	Molecular Entity	SANSS Code	In TSCA	Carcinogenesis	Mutagenesis	Developmental Effect	Neuropathy	Pulmonary Disorder	Cardiovascular Disorder	Immune System Disorder	Hepatic Renal Disorder	Blood and Bone Marrow Disorder	Skeletal Effects	Skin or Mucous Membrane Disorder
Hydrazines: aliphatic	-N-N(-)	Fg 167	31	M			M	M			M	M		M
aromatic	RN-N-	Fg 167R	61	M			M	M	M		M	M		M
	RN-N-	Fg 168R												
Hydrocarbons, polyaromatic			565	M	M									
Hydroxylamines	N-O-	Fg 169	47	M	M			M				M		M
Isocyanic acids and esters	-N=C=O	Fg 40	847					L						L
Isocyanides, aryl	R-N=C=O	Fg 40R	627					M		M				M
Ketenes, aliphatic	C=C=O	Fg 129	2					M						M
Ketones, aliphatic	-C-C=O	Fg 86	211				M	L						L

Health Effect

Health Effect

Chemical Group	Molecular Entity	SANSS Code	In TSCA	Carcinogenesis	Mutagenesis	Developmental Effect	Neuropathy	Pulmonary Disorder	Cardiovascular Disorder	Immune System Disorder	Hepatic Renal Disorder	Blood and Bone Marrow Disorder	Skeletal Effects	Skin or Mucous Membrane Disorder
Lead (organolead compounds)	-Pb	Fg 244	10				H				H	L	L	L
Mercury compounds	-Hg	Fg 139	25			M	H				H			
Nickel compounds[b]				L	L		L	L						M
Nitric acid esters	O_2N-O-	Fg 161	27						L					L
Nitro compounds, aliphatic	$O=N-O-$	Fg 154	46	L	L		L		L					L
Nitro compounds, aromatic	$RN(=O)=O$	Fg 154R	1781	L	L							M		L
N-Nitroso compounds, aliphatic	$-N=O$	Fg 151		H	H						M			L
N-Nitroso compounds, aromatic	$RN=O$	Fg 151	24	L	L									L

251

TABLE 5 (continued)

| | | | | Health Effect | | | | | | | | | | | |
Chemical Group	Molecular Entity	SANSS Code	In TSCA	Car-cino-gene-sis	Mut-agen-esis	De-velop-mental Effect	Neu-rop-athy	Pul-monary Dis-order	Car-dio-vascu-lar Dis-order	Immune Sys-tem Dis-order	Hep-atic Renal Dis-order	Blood and Bone Marrow Dis-order	Skel-etal Ef-fects	Skin or Mu-cous Mem-brane Dis-order
Phenanthrenes		SCN 121	0	L	L									
Phenanthridines		SCN 121	0	L	L									
Phenols		Fg 83	4179				M	L						L
Phosphine		Fg 236–Fg 238	22				L	L						
Phosphoric acid derivatives		Fg 239	26				M							L
Phosphoric esters		Fg 231 Fg 204	493					M						L

252

Health Effect

Chemical Group	Molecular Entity	SANSS Code	in TSCA	Carcinogenesis	Mutagenesis	Developmental Effect	Neuropathy	Pulmonary Disorder	Cardiovascular Disorder	Immune System Disorder	Hepatic Renal Disorder	Blood and Bone Marrow Disorder	Skeletal Effects	Skin or Mucous Membrane Disorder
Pyridines		SCN 44	497				L							L
Quinolines		SCN 99	221	L	L									
Selenides	–Se–°°°	Fg 254	7	L										
Steroids		SCN 127 –SCN 134	1820	L		L					L			
Sulfonamides		Fg 159 Fg 159R Fg 160 Fg 160R	722								L	L		
Sulfonates, alkyl		Fg 223	396	L	L									
Sulfonyl halides		Fg 209	124					M						M

TABLE 5 (continued)

				Health Effect										
Chemical Group	Molecular Entity	SANSS Code	In TSCA	Car-cino-gene-sis	Mut-agen-esis	De-velop-mental Effect	Neu-rop-athy	Pul-monary Dis-order	Car-dio-vascu-lar Dis-order	Immune Sys-tem Dis-order	Hep-atic Renal Dis-order	Blood and Bone Marrow Dis-order	Skel-etal Ef-fects	Skin or Mu-cous Mem-brane Dis-order
Sulfuric acid esters	(structure)	Fg 232	633	L	L			M						M
Tellurides	-Te-	Fg 265	1				L				L			
Thiophosphoric esters		Fg 205 Fg 220 Fg 221	60				M							L
Thiourea		Fg 42	51	L	L	L								
Tin (organo derivatives)	-Sn	Fg 264	225				M							
Vinyl: aliphatic	C=C El	HR 4E	773	L	L	L								
aromatic	C=C-R	HR 4R	885	L	L	L								L

a Estimated degree of association between chemical structure and form of toxicity: L, low; M, medium; H, high.

b Searched by name.

254

means that an NTP test with an approved protocol has been conducted and that results were positive in at least one species, whereas NTP NEG means that the test results were negative in at least one species. If both NTP POS and NTP NEG are listed, we assume equivalence to CAR, which means that acceptable, but not necessarily NTP-approved, bioassays have been assessed as positive. NEO means that a bioassay of careful design produced tumors, but not necessarily malignant ones. ETA (equivocal tumorigenic agent) means that some test was reported as positive for carcinogenesis, but failed to meet RTECS criteria for CAR or NEO. MUT means that a substance was found mutagenic in at least one of several recognized assays. It is generally assumed that mutagenesis is a better-than-random predictor of human carcinogenesis. The order in which these findings suggest carcinogenic potential is shown in Table 6. For our study, we chose to consider NTP POS and CAR together, to consider MUT and NEO together, and to ignore ETA and NTP NEG.

The estimated probabilities for the four remaining combinations are summarized in Table 7, which also shows the committee's arbitrary estimates of the proportions of chemicals in the universe with various carcinogenic potencies: 95% low, 4% medium, and 1% high. The estimated probabilities are based on reported false-positive and false-negative rates for the correlation between mutagenesis and animal carcinogenesis and on the committee's subjective assessments of animal-to-human extrapolation and of the certainty of RTECS criteria.

Only about 10% of the TSCA ("general") chemicals appear in RTECS; we have assumed that chemicals in other categories appear more often (Table 8). The 23% (100% - 77%) assumes that over half the food additives, drugs, cosmetics, and pesticides have had some reported toxicity information, even if only LD_{50}s, reported to RTECS. (Note that, if a substance showed both MUT and CAR, it would be assigned to CAR.)

Use of the RTECS data element is simple. The list of chemicals constituting the inventory is compared with the RTECS data base, using CAS registry numbers as the matching field. The RTECS text is scanned for the chosen terms; for the present illustration, only NTP POS, CAR, MUT, and NEO would be sought. Depending on the results of this test, the chemical class, and the exposure data elements, a substance in question is either removed from current consideration or passed to Stage 2 with a recommendation for a specific minidossier on toxicity and exposure. The rules for the choice are summarized below.

SELECTION RULES

If a computer is to assist in making the decision regarding which of the substances screened in Stage 1 to pass on to Stage 2--and to which level of Stage 2--the criteria for the decision must be expressed as an algorithm selection rule that can be implemented by a computer. That is,

TABLE 6 Significance of Ranking of RTECS Codes for Carcinogenic Potential

Code or Category	Meaning
NTP POS	High probability positive
CAR	Moderately high probability positive
MUT (MTDS)	Moderate probability positive
NEO	Moderate probability positive
ETA	Moderately low probability positive
In RTECS, with no cancer code	Slight probability negative
Not in RTECS	Moderate probability negative
NTP NEG	High probability negative

TABLE 7 Estimated Proportions of Chemicals with Given Carcinogenic Potency, by RTECS Code

Code or Category	Proportion with Carcinogenic Potency, %		
	Low	Medium	High
Chemicals in universe	95.0	4.0	1.0
Not in RTECS	97.4	2.3	0.3
In RTECS, with no cancer code	97.0	2.0	1.0
MUT (MTDS) or NEO	48.0	40.3	11.7
CAR or NTP POS	55.0	33.0	12.0

TABLE 8 Estimated Proportions of Chemicals with RTECS Codes Related to Cancer

Code or Category	Proportion of Chemicals in Universe, %
Not in RTECS	77
In RTECS, with no cancer code	18
MUT (MTDS), not CAR	2
CAR	3

the computer must be given precise criteria for causing a chemical to be removed from consideration, considered later, or moved forward. The committee has shown how such rules can be developed from a mathematical model of the system and has made more subjective evaluations of the meanings of the various data elements.

The rules are presented in Appendix B in the form of a hierarchic directory that is similar to a taxonomic field guide. By selecting an answer to a question on exposure, the reader (or, in Stage 1, the computer) is directed to a page on which other questions and their possible answers appear. When all Stage 1 questions have been answered, a decision may be made regarding the next step: do nothing (place the chemical on a list of chemicals to be considered later); go to Stage 2 and prepare some combination of high- or low-exposure and high- or low-toxicity minidossiers; or go to Stage 3 or 4 immediately.

The model in Appendix B shows all chemicals as advancing to some level of Stage 2. However, a large class (Branch 1) is shown to require an inexpensive exposure minidossier and no toxicity information or testing whatsoever. The model predicts that a small amount of additional information would be useful in reducing the probability of misclassification of such chemicals by category of exposure. However, such information might be virtually meaningless to the NTP testing program; in effect, chemicals classified as low exposure and low toxicity would be on a dormant list, as far as NTP were concerned.

STAGE 2

Data used in Stage 2 will include the data gathered in Stage 1, data submitted with nominations, or additional data that may be easily located in handbooks or machine-readable data bases. Two levels of possible data-gathering effort are envisaged for both toxicity and exposure. These data are compiled into a dossier for toxicity and a dossier for exposure.

Toxicity and exposure are estimated separately.

EXPOSURE DATA ELEMENTS

The data gathered in Stage 1 consist of:

- Annual production volume, if known.
- Use class.
- Reported types of toxicity, if any.
- Types of toxicity suspected from structure.
- Exposure and toxicity distributions.
- Recommended Stage 2 dossier budget.

The budget may or may not be split between exposure and toxicity. In any case, the exposure assessor for Stage 2 must decide the most cost-effective way to apply the budget on the basis of the fragmentary input data.

The types of toxic effects of concern with regard to a chemical will have some influence on the development of the exposure dossier. If the effect is likely to have a very low threshold or no threshold, as might be assumed for carcinogens, then both low exposures of large numbers of people and high exposures of smaller numbers of people may be of interest. If a high threshold is likely, as with some of the reversible effects that are clinically significant only when severe (methemoglobinemia, perhaps), then only high exposures are important. Furthermore, if the effect is likely to be related principally to one route of exposure (for example, inhalation in the case of lung cancer), then a corresponding exposure data set can be emphasized.

An initial check of regulatory and testing status should be made for every chemical. If a chemical is already being tested because of suspected toxicity, there is little benefit in gathering further information at this point.

If production volume is reported as unknown, some attempt should be made to estimate it from such data bases as the Chemical Information System or SRI International's Directory of Chemical Producers. These sources may also supply information on uses, numbers of producers, and so on. If production volume is high, it may be important to estimate it more accurately, because for TSCA chemicals it may have a range of a factor of 100.

Intended use is the principal determinant of the exposure search strategy. The broad Stage 1 use class should be both refined and augmented. For example, the type of pest (insect, fungus, nematode, weed, rodent, etc.) should be identified as a minimal refinement for pesticides. And such standard sources as the Kirk-Othmer Encyclopedia of Chemical Technology (1978) or the Merck Index (1976) should be consulted. These sources are far from complete or current, but they can aid further investigation. The finding that uses extend beyond the "intended" use defined by the Committee on Sampling Strategies data base will be the rule, rather than the exception, for chemicals of continuing interest.

After further details as to use have been established, the dossier strategy may become fairly obvious. Each use class tends to have its own "standard" reference works or data base; Table 9 lists a few of these. Furthermore, some pieces of data become more or less relevant, depending on the use of the chemical. Table 10 indicates the relative importance of various kinds of data for the use class in question; an exposure assessor should find it easy to make the choices in a specific case. Budget constraints might imply that even a highly rated piece of

TABLE 9 Selected Sources of Exposure Information by Use Class

Food Chemicals

Carroll, M. D. 1983. Dietary Intake Source Data: United States,
    1976-80. Hyattsville, Md.: U.S. Department of Health and Human
    Services, National Center for Health Statistics. 486 pp. (DHHS
    Publ. No. (PHS) 83-1681; Vital and Health Statistics Series 11, No.
    231.

Federation of American Societies for Experimental Biology. 1980.
    Reviews of GRAS Substances for Food and Drug Administration. For a
    list see FASEB, "Evaluation of GRAS Monographs (Scientific Literature
    Reviews)." (Available from NTIS, Springfield, Va., as PB-80-203789.)

Furia, T. E., Ed. 1973 (v. 1), 1980 (v. 2). Handbook of Food Additives,
    2nd ed. Boca Raton, Fla.: CRC Press. [1,430 pp.]

Furia, T. E., and N. Bellanca, Eds. 1975. Fenaroli's Handbook of Flavor
    Ingredients. Boca Raton, Fla.: CRC Press. 1,504 pp.

National Academy of Sciences/National Research Council. 1965. Chemicals
    Used in Food Processing. Washington, D.C.: National Academy of
    Sciences. 294 pp. (NAS Publication 1274)

U.S. Food and Drug Administration, Bureau of Foods. 1982.
    Toxicological Principles for the Safety Assessment of Direct Food
    Additives and Color Additives Used in Food. [245 pp.]

Drugs

Gilman, A. G., L. S. Goodman, and A. Gilman, Eds. 1980. Goodman and
    Gilman's The Pharmacologic Basis of Therapeutics, 6th ed. New York:
    Macmillan. 1,843 pp.

IMS America, Rockville, Md. Computer data bases: U.S. Drugstore,
    U.S. Hospital, and National Prescription Audit.

Opdyke, D. L. J. 1979. Monographs on Fragrance Raw Materials.
    Oxford: Pergamon Press. 732 pp.

Physicians' Desk Reference, 36th ed. 1982. Oradell, N.J.: Medical
    Economics. 3,060 pp. [updated annually]

TABLE 9 (continued)

Cosmetics

Estrin, N. F., P. A. Crosley, and C. R. Haynes, Eds. 1982.
CTFA Cosmetic Ingredient Dictionary, 3rd ed. Washington, D.C.:
The Cosmetic, Toiletry and Fragrance Association, Inc. 610 pp.

U.S. Food and Drug Administration, Division of Cosmetics Technology,
Washington, D.C. Frequency of the Use List gives nonquantitative
information on presence of substances in cosmetics.

Pesticides

Gosselin, R. E., H. C. Hodge, R. P. Smith, and M. N. Gleason. 1976.
Clinical Toxicology of Commercial Products, 4th ed. Baltimore:
Williams & Wilkins. [1,791 pp.]

Johnson, R. D., D. D. Manske, D. H. New, and D. S. Podrebarac. 1981.
Pesticide, heavy metal, and other chemical residues in infant and
toddler total diet samples (II): August 1975–July 1976. Pestic.
Monit. J. 15:39-50.

Johnson, R. D., D. D. Manske, and D. S. Podrebarac. 1981. Pesticide,
metal, and other chemical residues in adult total diet samples (XII):
August 1975–July 1976. Pestic. Monit. J. 15:54-69.

Miller, S. A. 1982. Compliance Program Report of Findings: FY 79
Total Diet Studies--Adult. Washington, D.C.: U.S. Food and Drug
Administration, Bureau of Foods. 48 pp. (Available from NTIS,
Springfield, Va., as PB83-112722.)

Commercial Chemicals

Kirk-Othmer Encyclopedia of Chemical Technology, 3rd ed. (25 volumes
projected) New York: John Wiley & Sons, v. 1, 1978 - .

Lawler, G. M., Ed. 1977. Chemical Origins and Markets, 5th ed.
Menlo Park, Calif.: SRI International. 118 pp.

Sax, N. I. 1979. Dangerous Properties of Industrial Materials,
5th Ed. New York: Van Nostrand Reinhold. 1,118 pp.

SRI International. Chemical Economics Handbook Program. SRI
International, Menlo Park, Calif. [a continuously updated program]

Synthetic Organic Chemicals: United States Production and Sales, 1980.
1981. Washington, D.C.: U.S. International Trade Commission.
(USITC Publication 1183 [annual])

TABLE 9 (continued)

Consumer Products

Gosselin, R. E., H. C. Hodge, R. P. Smith, and M. N. Gleason. 1976.
   Clinical Toxicology of Commercial Products, 4th ed. Baltimore:
   Williams & Wilkins. [1,791 pp.]

Environmental Pollutants

Anderson, D. Emission Factors for Trace Substances. 1973. Research
   Triangle Park, N.C.: U.S. Environmental Protection Agency, Office of
   Air Quality Planning and Standards. 80 pp. (Report no.
   EPA-450/2-73-001) (Available from NTIS, Springfield, Va., as
   PB-230894.)

Leo, A., C. Hansch, and D. Elkins. 1971. Partition coefficients and
   their uses. Chem. Rev. 71:525-616.

STORET Water Quality Data Base. Produced by U.S. Environmental
   Protection Agency, Office of Water Regulations and Standards,
   Washington, D.C. Available through Freedom of Information Act
   request or through NTIS, Springfield, Va.

Verschueren, K. 1977. Handbook of Environmental Data on Organic
   Chemicals. New York: Van Nostrand Reinhold.

Occupational Exposures

Clayton, G. D., and F. E. Clayton, Eds. 1978-1982. Patty's Industrial
   Hygiene and Toxicology, 3rd rev. ed. 3 vols. New York: Wiley.
   [5,864 pp.]

Documentation of the Threshold Limit Values, 4th ed. 1980. Cincinnati,
   Ohio: American Conference of Governmental Industrial Hygienists, Inc.
   [various paging] (supplements issued each year)

U.S. Department of Health, Education, and Welfare. 1974. National
   Occupational Hazard Survey, Vol. 1: Survey Manual. NIOSH Publ. No.
   74-127. Public Health Service, Center for Disease Control, National
   Institute for Occupational Safety and Health, Cincinnati, Ohio. 202
   pp.

U.S. Department of Health, Education, and Welfare. 1977. National
   Occupational Hazard Survey, Vol. 2: Data Editing and Data Base
   Development. NIOSH Publ. No. 77-213. Public Health Service, Center
   for Disease Control, National Institute for Occupational Safety and
   Health, Cincinnati, Ohio. 154 pp.

TABLE 9 (Continued)

U.S. Department of Health, Education, and Welfare. 1978. National
    Occupational Hazard Survey, Vol. 3: Survey Analysis and Supplemental
    Tables. NIOSH Publ. No. 78-114. Public Health Service, Center for
    Disease Control, National Institute for Occupational Safety and
    Health, Cincinnati, Ohio. 799 pp.

Solvents

Scheflan, L., and M. B. Jacobs. 1983. Handbook of Solvents. New York:
    Van Nostrand.

General

IARC Monographs on the Evaluation of the Carcinogenic Risk of Chemicals
    to Humans. Lyon: International Agency for Research on Cancer, v. 17,
    1978 - . (Continues IARC Monographs on the Evaluation of the
    Carcinogenic Risk of Chemicals to Man, v. 1, 1972 - v. 16, 1978.

Minerals Yearbook 1981. 1982. Washington, D.C.: U.S. Department of the
    Interior, Bureau of Mines. (revised annually)

SOCMA Handbook: Commercial Organic Chemical Names. 1966. Washington,
    D.C.: American Chemical Society.

Toxicology Data Bank (TDB). Computer data base available from National
    Library of Medicine, Specialized Information Services, Toxicology
    Information Program, Bethesda, Md.

Windholz, M., S. Budavari, L. Y. Stroumtsos, and M. N. Fertig, Eds. 1976.
    The Merck Index. An Encyclopedia of Chemicals and Drugs, 9th ed.
    Rahway, N.J.: Merck & Co. 1,313 pp.

TABLE 10 Data Needs for Use Classes[a]

Data	Food Chemicals	Drugs	Cosmetics	Pesticides	Consumer Products	Industrial Chemicals	Environmental Pollutants	Constituents of Food
Production volume	2	3	2	3	2	2	1	--
Production locations	1	1	1	2	1	2	2	--
Fraction used as intermediate	2	1	2	1	2	3	1	--
Detailed uses	2	1	3	2	3	2	1	--
Volume by use	2	2	3	2	3	2	1	--
Measured concentrations	2	1	2	2	2	2	3	3
Molecular weight	2	1	2	1	2	2	2	2
Structural diagram	2	2	2	2	2	2	2	2
Solubility in water	2	2	2	3	1	2	3	2
Partition coefficient	1	1	2	3	1	1	3	1
Melting point	1	1	1	1	2	2	2	1
Boiling point	1	1	1	1	1	1	1	1
Vapor presure	1	1	1	3	1	2	3	2
Reactivity	1	1	1	2	2	3	1	--
National Occupational Hazard Survey	2	2	2	3	2	2	3	--
Regulatory status	2	2	2	3	2	2	3	--

[a] 3 = highly desirable; 2 = desirable; 1 = useful; -- = irrelevant.

information may never be found--or even sought. Nevertheless, the assessor should record whether the search was made and, if so, what was found; and extra, easily available data should be retained in the dossier if little extra expense is involved.

Once the dossier is assembled--by the assessor or by an information-retrieval specialist--the assessor should review its contents (usually between 1 and 10 pages). The assessor should then mentally integrate the information and make a judgment about degree of exposure, using toxicologic relevance for the effects of concern, as well as a mental model of the substance's progress from production to eventual human exposure.

When the material has been assimilated, the assessor is asked to make two judgments: degree of concern about exposure and degree of confidence in or reliability of that judgment. The checklist in Figure 7 could be used to record those judgments. The assessor could record an explicit probability distribution if he or she does not choose to use the "standard" distribution corresponding to the choice of concern and confidence made earlier. The distribution of degrees of exposure estimated by the committee is shown for comparison; the assessor should make his or her own estimates.

If the assessor has not entered an explicit distribution, the codes for concern and confidence will be entered into the information system, and a corresponding exposure distribution will be generated. Hypothetical distributions (L-M-H) are shown in Table 11. If the estimated concern about exposure is low and confidence in the estimate is low, then the probability distribution is 75-17-8--where 75 is the probability (in percent) that the true exposure is low, 17 the probability that it is medium, and 8 the probability that it is high; and so on.

TABLE 11 Hypothetical Estimates of Probability Distributions of Degree of Exposure Generated in Stage 2 in Relation to Estimated Degrees of Concern and Confidence

Estimated Degree of Exposure[a]	Degree of Confidence in Estimate (Estimated Probability Distribution of True Exposure), %		
	Low	Medium	High
Low	75-17-8	80-15-5	85-12-3
Medium	50-35-15	45-40-15	40-45-15
High	30-40-30	25-40-35	20-40-40

[a]See text for explanation.

<u>Stage-2 Exposure Assessment</u>

CAS or other identifier _____

Substance name _____

<u>Degree of Concern</u> (Circle one)

    L - No evidence of special features suggesting significant exposure

    M - Evidence suggesting that exposures may be greater than for typical
        substances

    H - Evidence suggests that exposures may be significantly greater than
        for typical substances

Explanation: _____

_____(point out reason for considering exposure to be____

_____medium or high)_____

_____

<u>Degree of Confidence</u> (Circle one)

    L - Evidence spotty and uncertain; poor understanding of exposure
        process

    M - Evidence moderately available, but still often missing, uncertain,
        or difficult to interpres

    H - Evidence missing only for a few of the less important kinds of
        information; implications of information generally clear;
        quantitative exposure estimates may be possible

Explanation: _____

_____

_____explain reason for medium or high confidence_____

_____

_____

<u>Probability Distribution</u>

    Indicate on the histogram below (for comparison with the "standard"
distribution) your estimates of the probability that exposure to the
chemical is low, medium, or high.

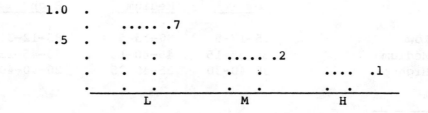

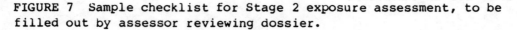

FIGURE 7 Sample checklist for Stage 2 exposure assessment, to be
filled out by assessor reviewing dossier.

266

The estimates of degree of exposure and of confidence in those estimates have similar implications, whether they are made using a dossier resulting from a minimal or a moderate effort to search for data. However, the number of estimates scored as deserving low confidence should be lower for the dossier resulting from the moderate data search, and the number of estimates deserving high confidence should be higher, as indicated in Table 12.

TABLE 12 Hypothetical Percentages of Chemicals as Estimated by Assessor to Have Given Degrees of Concern and Confidence in Estimate

Effort to Search for Data	Degree of Concern about Exposure, %			Degree of Confidence, %		
	Low	Med.	High	Low	Med.	High
Minimal	50	35	15	80	15	5
Moderate	50	35	15	65	25	10

The distribution of degrees of confidence reflects the presumed enrichment between Stages 1 and 2, at least from the perspective of the assessor.

TOXICITY DATA ELEMENTS

Toxicity data elements are analogous to exposure data elements, except for the content of the information. The Stage 1 information is reviewed for relevance to Stage 2 data-gathering decisions. The Stage 1 exposure inputs influence the choice of toxicity information to seek: use in food implies that toxicity through oral exposure would be most relevant; industrial use implies more concern for inhalation and for higher exposures of fewer people (that makes effects with thresholds more significant than they might otherwise be); and cosmetic use implies repeated dermal exposure.

As with exposure, the toxicity output of Stage 1 will suggest the effects of greatest interest, but again--as with uses--additional effects should be recorded as they are discovered.

All easily accessible data bases should be searched for reported toxicity-test results, whether positive or negative: RTECS, Chemical Information System (CIS), Toxicology Data Bank (TDB), Environmental Mutagen Information Center (EMIC), Environmental Teratogen Information Center (ETIC), and so on. Major compendia should also be consulted, such as those in Tables 9 and 13. Chemical structure should be reviewed to see whether

TABLE 13 Selected Sources of Toxicity Information by Effect

Mutagenicity

Environmental Mutagen Information Center (EMIC). Computer data base
available from EMIC, Oak Ridge National Laboratory, Tenn., and as subfile
of TOXLINE.

Carcinogenicity

Survey of Compounds Which Have Been Tested for Carcinogenic Activity
2nd ed. 1951. Bethesda, Md.: U.S. National Cancer Institute. (Public
Health Service Publication No. 149) (Additional volumes cover later
literature: Supplement 1; Supplement 2; 1961-67 volume, sections I and II;
1968-69 volume; 1970-71 volume; 1972-73 volume; 1978 volume)

IARC Monographs on the Evaluation of Carcinogenic Risk of Chemicals to
Humans. 1978-present. Lyon: International Agency for Research on
Cancer. v. 17. (Continues IARC Monographs on the Evaluation of
Carcinogenic Risk of Chemicals to Man, v. 1, 1972- v. 16, 1978.)

Teratogenicity

Shepard, T. H. 1980. Catalog of Teratogenic Agents, 3rd ed. Baltimore:
Johns Hopkins University Press.

Environmental Teratogen Information Center (ETIC). Computer data
base available from ETIC, Research Triangle Park, N.C., and as a subfile of
TOXLINE.

Neurotoxicity

Spencer, P. S., and H. H. Schaumburg, Eds. 1980. Experimental and
Clinical Neurotoxicology. Baltimore: Williams & Wilkins.

Skin effects

Maibach, H. I., and G. A. Gellin, Eds. 1982. Occupational and Industrial
Dermatology. Chicago: Year Book Medical Publishers, Inc.

Marzulli, F. N., and H. I. Maibach, Eds. 1983. Dermatotoxicology, 2nd ed.
Washington, D.C.: Hemisphere Publishing Corp.

Eye effects

Grant, W. M. 1974. Toxicology of the Eye, 2nd ed. Springfield, Ill.: Charles C Thomas.

Liver effects

Zimmerman, H. J. 1978. Hepatotoxicity: The Adverse Effects of Drugs and other Chemicals on the Liver. New York: Appleton-Century-Crofts.

Other effects

Documentation of the Threshold Limit Values, 4th ed. 1980. Cincinnati, Ohio: American Conference of Governmental Industrial Hygienists, Inc. (Supplements issued each year)

Registry of Toxic Effects of Chemical Substances, 1980 edition. 1982. Cincinnati, Ohio: U.S. National Institute for Occupational Safety and Health. (More recent updates available as microfiche from NIOSH or as computer data base from National Library of Medicine, Bethesda, Md., or from Chemical Information Systems, Inc., Baltimore, Md.)

predicted Stage 1 effects still appear consistent with structure and whether any other possibilities are suggested. If the budget for the toxicity dossier permits, a beginning search of bibliographic information, such as the number of entries in TOXLINE, may be conducted. Table 13 shows some standard sources by category of effect.

The search for toxicity data and preparation of the toxicity dossier may require more effort than the corresponding work for exposure, but the toxicity dossier may well include little information other than that pertaining to chemical structure. It is reviewed by the toxicity assessor, and a record almost identical with Figure 7 is completed (except that the word "toxicity" is substituted for "exposure" everywhere). A separate record is completed for each major human health effect of continuing concern. For each health effect, the appropriate table of estimated probability distributions should be consulted. The hypothetical distributions for carcinogenicity shown in Table 14 are consistent with the assumption that 90% of the chemicals considered in Stage 2 are noncarcinogens, 7% are weak carcinogens, and 3% are strong carcinogens.

Finally, the distribution of chemicals by degrees of concern and confidence, as assigned by the assessor in Stage 2 for toxicity (carcinogenicity), might look as follows:

Effort to Search for Data	Degree of Concern, %			Degree of Confidence, %		
	Low	Med.	High	Low	Med.	High
Minimal	80	15	5	80	15	5
Moderate	80	15	5	70	20	10

SELECTION RULES

The Stage 2 rules (Appendix B) specify what action is to be taken in Stage 3 on the basis of the degrees of concern about exposure and toxicity estimated in Stage 2. Depending on the relative values for exposure and toxicity, the Stage 2 rules specify either that a dossier should be prepared in Stage 3 or that it should not be (i.e., consideration of the chemical should be deferred). More detailed rules could be developed to cover degree. of confidence, as well as degree of concern, but these are not illustrated in Appendix B.

STAGE 3

The input to Stage 3 from Stage 2 will consist of the following:

- Stage 2 exposure dossier, if written.
- Stage 2 toxicity dossier, if written.

● Distributions for high, medium, and low exposure and toxicity (by effect) derived from the subjective judgments of the Stage 2 assessors.
  ● Recommendations regarding need for a Stage 3 exposure dossier, toxicity dossier, or both.

<u>or</u>

  ● Agency nomination of substances for specific tests, with supporting documentation that may exceed the information content of the best Stage 2 dossier.

TABLE 14 Hypothetical Estimates of Probability Distributions of Carcinogenic Potency as Generated in Stage 2, in Relation to Assigned Degree of Concern about Carcinogenicity and Degree of Confidence in Assignment

Degree of Concern about Carcino-genicity	Degree of Confidence in Estimate (Estimated Probability Distribution of True Potency), %		
	Low	Medium	High
Low	95- 3- 2	96-3- 1	97-2- 1
Medium	80-15- 5	65-30- 5	50-45- 5
High	50-35-15	40-40-20	30-40-30

For illustrative purposes, the Committee on Priority Mechanisms has assumed that the standard budget for preparation and review of Stage 3 dossiers would be $1,000 for exposure and $1,000 for toxicity. We also assumed that the Stage 3 dossier managers would adjust budgets for specific substances according to the perceived availability of and need for information; thus, some dossiers would require only a few hundred dollars, whereas others might require twice the standard budget. And we assumed that dossiers will automatically be prepared for both toxicity and exposure for all agency-nominated substance, but that the supporting documentation might supply much of the needed information.

EXPOSURE DATA ELEMENTS

The procedure for processing exposure information in Stage 3 is nearly identical with that in Stage 2. First, the dossier manager determines what types of information need to be sought for what types of substances. (The bases for selection of the information to be sought include use, structure, effects of concern, physical and chemical properties, and other information in the Stage 2 dossier.) A list of possible information categories appears in Table 9. Next, the dossier information is sought and assembled under the direction of the manager. The dossiers then go in batches of, say, 20-50 to an expert committee* for review. (The committee would presumably assign primary reviewers for each substance, but it should be encouraged to form its own procedural rules.) The committee attempts to reach a consensus regarding the degree of concern and confidence about exposure, as in Stage 2. Difficulty in reaching a consenus about degree of concern is a good reason to assign a low degree of confidence. However, we assume that the criteria for high and medium confidence will be more stringent than those for the same terms in Stage 2. (Again, if the committee felt comfortable about doing so, it might generate a probability distribution for exposure directly.) The decision should be documented on forms similar to those shown for Stage 2 in Figure 3, but with more room for explanations.

If the committee does not generate an explicit probability distribution for exposure, one can be selected from Table 15, using the degrees of concern and confidence shown.

TABLE 15 Hypothetical Estimates of Probability Distributions of Degree of Exposure Generated in Stage 3, in Relation to Assigned Degree of Confidence in Assignment

Degree of Exposure	Degree of Confidence in Estimate (Estimated Probability Distribution of True Exposure), %		
	Low	Medium	High
Low	80-15-5	85-12-3	90-9-1
Medium	45-40-15	40-50-10	35-60-5
High	20-40-35	15-40-45	10-35-55

---

*Separate exposure and toxicity committees may be useful, but a single committee could serve if instructed to keep the assessments separate.

It is difficult to predict the distribution of ratings generated by the Stage 3 process, because of uncertainties as to how the committee will assess the dossiers. Hypothetical initial assumptions of the distribution of the assignments are as follows:

	Degree of Concern, %			Degree of Confidence, %		
	Low	Medium	High	Low	Medium	High
Stage 3 exposure dossier	25	50	25	50	30	20

TOXICITY DATA ELEMENTS

As in Stage 2, the toxicity dossier should be oriented toward the effects suggested as being of greatest concern, but should present any other information on observed or suspected effects that turns up in the course of the search. As with exposure, the Stage 3 toxicity dossier strategy should be designed to yield the information of most use to the expert committee that will be reviewing the dossiers. Information on obscure details of biochemistry and on effects whose clinical significance has not been interpreted probably is not useful. However, it is possible in Stage 3 to include information that would be difficult to process in earlier stages, such as known or suspected synergism with other chemicals; in such cases, it may prove useful to search for information on joint exposures to the chemical of concern and those with which it acts synergistically.

The expert committee on toxicity will assess the dossiers and make judgments about degree of concern and degree of confidence for each effect of interest. Although such judgments will normally be limited to the effects nominated by agencies or indicated on Stage 2 dossiers, any other effect of serious concern that becomes evident may also be assessed. If the committee does not generate an explicit probability distribution for toxicity, one can be selected from Table 16, in which an illustrative set of distributions for carcinogenic potency is presented.

TABLE 16 Hypothetical Estimates of Probability Distributions of
Carcinogenic Potency as Generated in Stage 3, in Relation to Assigned
Degree of Concern about Carcinogenicity and Degree of Confidence in
Assignment

Degree of Toxicity	Degree of Confidence in Estimate (Estimated Probability Distribution of True Toxicity), %		
	Low	Medium	High
Low	96-3-1	97-2-1	98-1 1
Medium	75-21-4	57-40-3	40-57-3
High	40-40-20	30-40-30	25-35-40

The corresponding distribution of judgments by the expert committee
might be as follows:

	Degree of Concern, %			Degree of Confidence, %		
	Low	Med.	High	Low	Med.	High
Stage 3 toxicity dossier	60	30	10	60	25	15

These distributions are consistent with an assumpton that 81% of the
chemicals moving from Stage 2 to Stage 3 are noncarcinogens, 15% are weak
carcinogens, and 4% are strong carcinogens.

Note that the expert committee is not asked to make any judgment
about the value of proposed tests. We propose that systematic and
largely predetermined rules be used to determine the value of a
mutagenicity assay or a lifetime bioassay in clarifying the concern about
carcinogenicity for a specific chemical that has already been assessed by
an expert committee.

The recommendations for testing would then be reviewed by the board
of scientific counselors, or a similar body, to ensure that they are
reasonable, given the status of existing information and the detailed
properties of the substance in question. Unfortunately, this whole
approach to testing decisions is not as well suited to evaluating
reconnaissance testing (e.g., 90-d feeding studies), in which an effect
of concern may be specified only generally, if at all. The performance
characteristics for such tests would have to be either created from the
aggregate of several effects that tests could detect or defined for a
generalized toxicity measure, which would be sought only because exposure
is fairly high and fairly certain.

SELECTION RULES

As in Stage 2, the Stage 3 rules (Appendix B) operate on the finding of high, medium, or low toxicity and high, medium, or low exposure; later, they should be revised to consider degree of confidence. The choices for Stage 4 are a short-term test, a long-term cancer bioassay, and no action.

STAGE 4

Because the objective of priority-setting must be the minimization of misclassification of the potential public-health concern about chemicals, the system cannot be complete without an analysis of the predictive power of the tests to which it leads. We discuss here some problems related to the selection of toxicity tests and the relations between operating characteristics of tests and selection of chemicals for toxicity testing.

At this stage of the priority-setting process, the goal is to select, for the chemicals to be evaluated, tests that provide the most useful toxicity information for the resources invested. Because testing resources are limited, selection of an optimal test or combination of tests involves a tradeoff between the total number of chemicals evaluated and the completeness with which any one chemical can be characterized.

To illustrate how this tradeoff can be made, we constructed a mathematical model for the allocation of resources for one kind of toxicity, carcinogenicity (Appendix B). This choice was made both because methods for carcinogenicity testing are better than methods for most other toxicity testing and because carcinogenicity testing consumes a larger fraction of NTP resources than does testing for any other toxic effect.

Tests of other effects, like tests to assess carcinogenicity, entail consideration of both cost and accuracy of testing. These tests range in cost from a few hundred dollars for some short-term tests with bacteria to more than a half-million dollars for a standard long-term bioassay in rats and mice (Table 17). The cost of a given test may vary among laboratories, depending on the substance being tested, the conditions of testing, institutional overhead expenses, and other factors (Lave et al., 1982).

Tests vary in predictive accuracy. None is considered accurate enough to predict with complete reliability the carcinogenicity of a substance for humans; that can be established only by epidemiologic evidence of carcinogenic effects in humans. Nevertheless, the demonstration that a substance is carcinogenic in appropriately conducted animal tests is generally considered an adequate basis for classifying it a presumptive human carcinogen (IRLG, 1979; U.S. Congress, 1981; IARC, 1982). Interpretation of animal tests is uncertain, however, and requires expert judgment. Extrapolation from animal test results to quantitative estimation of human risks entails large uncertainties in

TABLE 17 Estimates of Costs of Some Carcinogenicity Tests[a]

Tests[b]	Type of Test		Estimated Average Cost, $	No. Responses
	Bacterial cell:			
1-4,6-13	Salmonella	his-	1,200	10
19	Escherichia coli	WP2	400	2
21	Bacillus subtilis	rec-	800	1
22-24	Escherichia coli	rec-	1,500	1
29	Degranulation		2,500	1
37	Yeast cell:			
	Saccharomyces cerevisiae	D7	1,400	1
	Mammalian cell:			
40-42	Unscheduled DNA synthesis-- human fibroblasts, HeLa cells		5,200	3
43,45	Sister-chromatid exchange-- CHO cells		3,000	1
44	Chromosomal aberrations-- CHO cells, rat liver cells		7,500	3
c	Transformation--CHO cells		1,400	1
c	Transformation--C3H-10T 1/2		5,400	4
48	TK +/- L5178Y mouse lymphoblasts		4,900	1
50,51	HGPRT-CHO cells, V79 cells		6,500	4
	Whole Animal:			
56-58	Sex-linked recessive lethal-- Drosophila melanogaster (injection)		10,000	1
59	Sister-chromatid exchange--mouse		3,000	1
60-62	Micronucleus--mouse		3,400	2
63	Sperm morphology--mouse		11,400	1
	Whole-animal two-species rodent bioassay		500,000[d]	1

[a] Modified from Lave et al., 1982.
[b] See Table 19 for list of tests.
[c] Test was not conducted in International Collaborative Program (de Serres and Ashby, 1981) in either CHO or C3H cells, but in BHK-21 cells.
[d] From Weinstein, 1983.

connection with mechanisms of carcinogenesis and species differences in carcinogen metabolism (IRLG, 1979; Calkins et al., 1980; IARC, 1982). The difficulties in interpreting tests for carcinogenicity have led the International Agency for Research on Cancer (1982) to publish guidelines for interpretation.

Long-term bioassays in rats and mice are so expensive (Table 17) and time-consuming (requiring up to 5 yr for completion) that such tests are not feasible for more than a small percentage of the many thousands of chemicals to be tested. Faster and less expensive assays are needed. To this end, more than 100 short-term tests have been introduced (Hollstein et al., 1979). Thus far, however, such tests have been used only for screening purposes, and not for definitive predictions of carcinogenicity, pending further standardization and validation of their accuracy. The need for standardization of such tests is indicated by the variability in results of the most widely used short-term screening test--the Salmonella typhimurium/microsome plate mutagenicity assay. Table 18 shows the degree of agreement found by the International Collaborative Program, which had 12 laboratories apply the test to a series of 19 carcinogens and noncarcinogens (de Serres and Ashby, 1981). The discrepancies are attributable at least in part to the use of differing procedures.

Despite the variations, the Salmonella mutation assay yielded preponderantly positive results with chemicals known to be carcinogenic in rodents. Nevertheless, the observed correlation between mutagenicity and carcinogenicity has varied from one class of chemicals to another. For some carcinogens (e.g., asbestos and halogenated hydrocarbons, such as DDT) and for most tumor-promoting agents (such as phorbol esters and some naturally occurring hormones), the test has given negative results (Rinkus and Legator, 1979; Ames and McCann, 1981; Purchase, 1982). For chemicals of all classes tested to date, its overall accuracy is 60-80% (see Table 19) (Ames, 1979; Purchase, 1982; Upton et al., in press). Other short-term tests have received less systematic evaluation than has the Salmonella test; only limited data are available on their comparative results (see Table 19). One naturally successful combination of the two tests is the Salmonella test with the in vitro cell-transformation assay.

Combinations of various short-term tests, in batteries or in tiers, have been observed to yield greater accuracy than any one of the tests alone (Bridges, 1976; Weisburger and Williams, 1981; Lave et al., 1982). Six combinations of tests have been calculated to have predictive accuracies of 81.6-89.7% (Table 20) for a limited number of animal carcinogens and noncarcinogens (Table 21). None of the combinations appears capable of avoiding a substantial percentage of false-positives and false-negatives. However, where a false-positive is observed, it may be suspected that the carcinogenicity of the chemical under consideration escaped detection because of deficiencies in the animal test used as a criterion.

277

TABLE 18 Correlation among Results of _Salmonella_/microsome Tests Performed by 12 Investigators[a]

Investi-gators	1	2	3	4	6	7	8	9	10	11	12	13
1	1.00	0.81	0.53	0.47	0.47	0.34	1.00	0.62	0.81	0.47	0.34	0.22
2		1.00	0.65	0.62	0.62	0.53	0.81	0.81	1.00	0.65	0.53	0.15
3			1.00	0.65	0.65	0.34	0.53	0.53	0.65	0.42	0.34	0.19
4				1.00	0.33	0.53	0.47	0.47	0.62	0.65	0.26	0.15
6					1.00	0.53	0.47	0.47	0.62	0.35	0.26	0.03
7						1.00	0.34	0.34	0.53	0.46	0.30	0.05
8							1.00	0.62	0.81	0.47	0.34	0.22
9								1.00	0.81	0.47	0.34	0.07
10									1.00	0.65	0.53	0.15
11										1.00	0.46	0.02
12											1.00	0.21
13												1.00

[a] From Lave et al., 1982. Entries are squared correlation coefficients computed from test results on 19 chemicals reported to International Collaborative Program by 12 investigators. 1.00 indicates complete agreement and 0 indicates no agreement between investigators.

TABLE 19 Predictive Accuracy of Various Short-Term Tests
for Carcinogenicity[a]

Test Code Number	Type of Test	Accuracy of Test Results[b]	
		N	Accuracy
Bacterial Mutation Assays			
1	S. typhimurium/plate	37	0.70
2	S. typhimurium/plate	37	0.68
3	S. typhimurium/plate	36	0.59
4	S. typhimurium/plate	38	0.71
5	S. typhimurium/fluctuation	31	0.71
6	S. typhimurium/plate	38	0.63
7	S. typhimurium/plate	33	0.64
8	S. typhimurium/plate	38	0.68
9	S. typhimurium/plate	37	0.57
10	S. typhimurium/plate	37	0.65
11	S. typhimurium/plate	37	0.73
12	S. typhimurium/plate	38	0.68
13	S. typhimurium/plate	38	0.71
14	S. typhimurium/plate norharman	28	0.57
15	S. typhimurium/ fluctuation	33	0.55
16	S. typhimurium/ azaguanine res	38	0.66
17	S. typhimurium/E. coli W2/fluctuation hepatocytes	36	0.67
18	S. typhimurium/E. coli WP2/plate	36	0.78
19	E. coli WP/2/plate	34	0.65
20	E. coli 343	18	0.72
Bacterial Repair, Phage Induction, Degranulation, and Nuclear Enlargement Assays			
21	B. subtilis, M45 rec-	38	0.79
22	E. coli, RecA/PolA	34	0.59
23	E. coli, RecA-	37	0.62
24	E. coli, RecA-/PolA	36	0.58
25	E. coli, Pol	36	0.58
26	λ-induction (gal+)	20	0.65
27	λ-induction (lysis)	36	0.56

TABLE 19 (continued)

Test Code Number	Type of Test	Accuracy of Test Results[b]	
		N	Accuracy
28	Degranulation of RER	33	0.39
29	Degranulation of RER ribonuclease post-treatment	30	0.67
30	Nuclear enlargement, HeLa cells	22	0.32
31	Nuclear enlargement, human fibroblasts	22	0.64
Yeast Assays			
32	S. cerevisiae XV185-14C	30	0.70
33	S. pombe P1	29	0.66
34	S. cerevisiae PG 148 PG-154 PG-155 PG-166	35	0.43
35	S. cerevisiae D4	31	0.42
36	S. cerevisiae D6	37	0.65
37	S. cerevisiae D7	35	0.66
38	S. cerevisiae JD1	32	0.72
39	S. cerevisiae rad	32	0.72
In Vitro Mammalian Test Systems			
40	Unscheduled DNA synthesis, human fibroblasts	18	0.39
41	Unscheduled DNA synthesis, human fibroblasts	21	0.62
42	Unscheduled DNA synthesis, human fibroblasts	38	0.68
43	Sister chromatid exchange, CHO cells	19	0.53
44	Sister chromatid exchange, CHO cells	18	0.67
45	Sister chromatid exchange, CHO cells	33	0.52
46	Cytogenetic analysis, micronucleus test	21	0.57
47	Cytogenetic analysis, micronucleus test	18	0.72
48	Forward-mutation assay, mouse lymphoma cells L518Y	19	0.53

TABLE 19 (continued)

Test Code Number	Type of Test	Accuracy of Test Results[b]	
		N	Accuracy
49	Gene-mutation assay, CHO cells, HGPRT gene	9	0.44
50	Gene-mutation assay, CHO cells, HGPRT gene	3	0.67
51	Gene-mutation assay, V79 hamster cells	5	0.60
53	Cell transformation	34	0.59
54	Cell transformation, BKH-21 cells	38	0.82

In Vivo Assays

56	Sex-linked recessive lethal	9	0.44

Droposphila

57	Sex-linked recessive lethal	15	0.53
58	Sex-linked recessive lethal	9	0.44
59	Sister chromatid exchange, mouse	16	0.63
60	Micronucleus assay, mouse	29	0.66
61	Micronucleus assay, mouse	17	0.47
62	Micronucleus assay, mouse	33	0.45
63	Sperm morphology, mouse	15	0.47

[a]From Lave et al., 1982.

[b]Figures tabulated indicate accuracy (number of chemicals correctly identified divided by number of chemicals tested) for tests 1 through 63. Data subset includes all 42 chemicals listed in Table 21 except presumptive noncarcinogens 2, 16, 20, 22, and 27, which had high frequencies of positive results in international study (de Serres and Ashby, 1981). Diphenylnitrosamine (22) was found to be carcinogenic.

TABLE 20 Predictive Accuracy of Several Short-Term Tier Testing Regimens[a]

Tier Number	Tests Included in Tier[b]	No. of Chemicals on Which Results are Based	Predictive Accuracy,
1	S. typhimurium/plate (4) Differential killing, B. subtilis M45 Rec⁻ (21) Cell transformation, BHK-21 cells (54)	38	84.2
2	S. typhimurium/plate (4) Cell transformation, BHK-21 cells (54) Forward mutation, S. pombe (33)	29	86.2
3	S. typhimurium/plate (4) Forward mutation, S. pombe (33) Cell transformation, BHK-21 cells (54) Unscheduled DNA synthesis, HeLa cells (42)	29	89.7
4	S. typhimurium/plate (4) Unscheduled DNA synthesis, HeLa cells (42) Cell transformation, BHK-21 cells (54)	38	81.6
5	S. typhimurium/plate (4) Cell transformation, BHK-21 cells (54) Rabin's test (degranulation) rat liver cells (29)	30	83.3
6	S. typhimurium/plate (4) Unscheduled DNA synthesis, HeLa cells (42) Cell transformation, BHK-21 cells (54)	38	81.6

[a] From Lave et al., 1982.
[b] Numbers in parentheses are code numbers of tests (Table 19).

TABLE 21   Chemicals Analyzed by Short-Term Tests (International Collaborative Study)[a]

	Chemical	Carcinogenicity Classification[b]
1	4-Nitroquinoline-N-oxide	+
2	3-Methyl-4-nitroquinoline-N-oxide	−
3	Benzidine	+
4	3,3',5,5'-Tetramethylbenzidine	−
5	4-Dimethylaminoazobenzene(butter yellow)	+
6	4-Dimethylaminoazobenzene-4-sulfonic acid, Na salt	−
7	Benzo[a]pyrene	+
8	Pyrene	−
9	ß-Propiolactone	+
10	γ-Butyrolactone	−
11	9,10-Dimethylanthracene	+
12	Anthracene	−
13	Chloroform	+
14	1,1,1-Trichloroethane	−
15	2-Acetylaminofluorene	+
16	4-Acetylaminofluorene	−
17	Dimethylcarbamoyl chloride	+
18	Dimethylformamide	−
19	2-Naphthylamine	+
20	1-Naphthylamine	−
21	N-Nitrosomorpholine	+
22	Diphenylnitrosamine	−
23	Dinitrosopentamethylene tetramine	−
24	Urethane	+
25	Isopropyl N-(3-chlorophenyl)carbamate	+
26	Methylazoxymethanol acetate	−
27	Azoxybenzene	+
28	DL-Ethionine	−
29	Methionine	+
30	Hydrazine sulfate	+
31	Hexamethylphosphoramide (HMPA)	+
32	Ethylenethiourea	+
33	Diethylstilbestrol	+
34	Safrole	+
35	Cyclophosphamide	+
36	Epichlorhydrin	+
37	3-Aminotriazole	+
38	4,4'-Methylenebis(2-chloroaniline) (MOCA)	+
39	o-Toluidine hydrochloride	+
40	Auramine (technical grade)	+
41	Sugar (sucrose)	−
42	Ascorbic acid	−

[a] From de Serres and Ashby, 1981.
[b] Based on effects in human populations or experimental animals.

Other approaches being emphasized in current research efforts include those for identifying nongenotoxic carcinogens, including cocarcinogens and tumor-promoting agents (Sivak, 1982; Upton, in press), and the exploitation of in vivo short-term bioassays (Ashby, in press). It may be envisioned that the test systems of the future will include various assays in addition to those represented in Table 19. It may also be expected that the particular combinations of tests used and the sequences in which they are used will depend on the nature of the chemicals being tested and their patterns of use. Whether a tier regimen culminates in a long-term bioassay in rats and mice will depend on the results of the antecedent short-term tests and their apparent predictive accuracy.

The principles involved in setting priorities for carcinogenicity testing apply also to the setting of priorities for testing chemicals suspected of other forms of toxicity. It must be recognized that the process of predicting the toxic potential of a chemical is extremely complex. There are no firm rules, or even guidelines, for obtaining reliable answers. Attention is generally focused on devising animal models or tests for human health effects with attributable chemical causes. Thus, in the best circumstances, it may be possible to predict the biologic activity of a given chemical, identify a useful end point with an established animal model or test, and draw some tentative conclusion regarding the chemical's likely toxicity in humans.

Results of toxicity tests on animals have been recorded in RTECS, a machine-readable file of approximately 15,000 chemicals, although the universe of chemicals defined by the Committee on Sampling Strategies contains over 70,000 chemicals. Even information on well-characterized chemicals is often limited to data on molecular structure and physical characteristics. A chemical may have several effects on human health that are individually dose-dependent. To set testing priorities among different types of toxic effects, two tasks must be accomplished: a catalog of human health effects must be established, and the various effects must be ranked according to relative severity.

The relative importance of different types of health effects depends on their consequences to the affected persons and to society. Ranking of effects according to severity is implicit in most priority-setting schemes. It involves not only technical judgments of toxicity, but also individual perceptions of harm. An approach to determining the relative severity of toxic effects is described in Chapter 2.

When a chemical is suspected of causing more than one toxic effect, which is often the case, the combined impacts of all its potential effects must be taken into account in setting priority for its testing. Because of the multiplicity, diversity, and complexity of the health effects of different chemicals, no attempt is made here to develop a detailed or comprehensive system for this purpose.

5
FUTURE DEVELOPMENT, IMPLEMENTATION, AND REFINEMENT OF THE SYSTEM

The priority-setting approach presented in this report would require further elaboration and further development of methods and data bases before it could be implemented fully. This chapter describes the developments needed to make the system operational and then discusses strategies for implementation and refinement.

DEVELOPMENT

The system as illustrated is potentially capable of screening tens of thousands of substances and determining--after specified information-gathering steps--for which of them it would be warranted to apply a battery of short-term tests or a long-term carcinogenicity bioassay. To a lesser extent, it could screen the same universe of chemicals to select those which should be tested for other health effects. Before this system could be implemented, however, the following steps would need to be taken:

● Refinement of the estimated frequencies of outcomes of various information-gathering activities. For example, the numbers of chemicals from the universe that would fall into the chemical classes described in Table 5 should be better determined.

● Refinement of estimates of the accuracy of the data elements in all stages of the system. For example, we need to consider further the power of an RTECS entry as a data element to reflect carcinogenic hazard. Perhaps more important, what should be the impact of a subjective determination of "high" exposure by an NTP staff assessor? Outside verification that the data elements and their operating characteristics are reasonable is needed.

● Refinement of the choice of penalties for incorrect assessments of public-health concern. More attention needs to be given to the relative importance of false-positive and false-negative findings and their implications for decisions of whether to consider chemicals further. There is a need to integrate the intuitive decisions about selecting chemicals with the decisions derived from the mathematical model of the system--and it is probably necessary to depart from both to find a satisfactory synthesis.

● Expansion of the model to incorporate more data elements and outcomes. For example, both ratings of degree of concern (exposure and toxicity) and confidence in the ratings need to be included in the model as outcomes of Stages 2 and 3. The confidence information has not yet been included.

● Testing of the model with a wider range of data and, by exploring the sensitivity of the model to those data, establishment of a set of design rules that either are intuitively satisfying or can be explained as stemming logically from the data, instead of from limitations in the model itself.

● Expansion of the guidance to system operators in designing Stage 2 minidossiers and Stage 3 dossiers and strategies to search for information.

● Development of further guidance for the evaluation of Stage 2 minidossiers with respect to exposure and toxicity ratings and their corresponding degrees of confidence.

● Determination of which health effects are worthy of treatment by the techniques presented here for carcinogenicity. This will require both a value judgment as to the severity of an effect and a scientific judgment as to the availability of tests, the number of toxic substances in the universe, accuracy of the tests, and the importance of classifying chemicals correctly with respect to such effects.

● Development of data elements, estimation of their accuracy, estimation of the number of chemicals that cause a health effect, determination of misclassification costs, and development of corresponding decision rules for the additional health effects selected.

● Expansion of the system to include all toxicity tests for which chemicals are being given priorities.

IMPLEMENTATION

A priority-setting system based on value-of-information analysis may be implemented by several possible approaches, and these vary considerably in the time and resources required. At the most modest level, the approach to implementation is qualitative. A greater emphasis on estimated probabilities (Appendix D) may help in the evaluation of the uncertainties that play such a fundamental role in priority-setting, design of tier testing, and characterization of risk of chemicals. The discussion of possible data elements (Chapter 4) may suggest indicators to help select chemicals in the initial stages of a priority-setting system, where information is especially fragmentary. The walkthrough with the example chemical bisphenol A (Chapter 3) may suggest ways for chemical managers to pick from among the many possible sources of information, especially for Stages 2 and 3. Without describing the priority-setting system quantitatively, much of the value-of-information analysis can be implemented with little extra expenditure of time and resources. In its most modest form, the computer-assisted Stage 1 could be foregone in favor of a smaller select universe of chemicals.

At the most elaborate level, a complete implementation could be attempted from the start. The program for the model could be rewritten for a mainframe computer to accommodate several effects, additional Stage 4 candidate tests, and more Stage 2 and Stage 3 data elements and possible outcomes. The resulting large number of factors could be estimated and the software developed for reading the files for Stage 1. This effort might require a couple of years and considerable resources.

With a gradual approach, implementation would start at a modest level and increase. Files and software for Stage 1 would be developed in steps, perhaps without addressing the entire universe of 70,000 chemicals immediately. It would probably be useful to rewrite the computer program for the optimization model, adapt it to available computers, and familiarize the technical staff with the algorithm. Carcinogenesis and perhaps another health effect could be included initially. A small number of data elements could be evaluated and their accuracy estimated during the first year. In later years, these estimates could be refined on the basis of new data (some from the tests), and other data elements and toxic effects could be added. With this more gradual approach, the system could be implemented in the first year at a low cost.

In the process of setting priorities, information is collected on toxicity and exposure; some of it is recorded in the dossiers (including minidossiers and microdossiers), and some is embodied in the estimates of accuracy (e.g., the estimated false-positive and false-negative rates). Additional information is developed by the testing program itself. Putting this information into a usable format is an important potential contribution of NTP. For example, it is very helpful to use false-negative rates for estimating rates of detection, especially for clinical tests with negative outcomes.

The following tasks would be required, to implement the demonstration system:

● Training of staff in probabilistic thinking and value-of-information analysis.

● Training of staff in the special skills needed for dossier design and evaluation.

● Definition of a universe of substances and preparation or augmentation of data bases with environmental chemicals, food constituents, pyrolysis products, etc.

● Development of software for building computerized files for Stage 1 operation, including provisions for maintaining status lists ("dormant," "on test," "Stage 2 minidossiers complete," etc.).

● Design of forms and procedures for moving substances through Stage 3. For example, how often does the expert evaluation committee need to meet, if the required frequency is different from the current NTP schedule?

● Design of a feedback and control system for continuous updating of the system.

● Provision of an effective interface between the established agency nomination process and the proposed "long-list" process.

● Expansion of the role of the expert committee to include estimation of degrees of exposure and toxicity, in addition to test recommendations.

FURTHER REFINEMENT

Once the system becomes operational, a number of refinements can be comtemplated, as experience with its use accumulates. The whole design is built on a series of estimates, such as the estimates of the number of carcinogens among the chemicals considered and the accuracy of the selection stages. Although by the time of implementation these estimates will have been subjected to a great deal of examination, they still will be quite uncertain. Thus, the system must be tested and refined as new information becomes available. These are some of the questions to be addressed:

● How many* chemicals are unstructurable?

● How many have no RTECS listing?

● How many have no listed production?

● How many reach a dormant list in Stage 1?

● How much does it cost to construct a minidossier?

● How often does an assessor score a chemical as having "high" exposure or "low" confidence in the toxicity rating?

● Which is the most useful information for the Stage 3 committees? How much does it cost to retrieve? For how many chemicals does one find useful information?

● How many chemicals are retained for short- and long-term tests?

● How many of those prove positive?

_____

*That is, what fraction of the universe?

● What is the distribution of test recommendations among various types of tests?

After the system has operated for a while, the answers to these and other questions can be assessed for relevance to estimates used to design the system. Changes can be made as deemed necessary, and the model used to calculate an improved set of rules to decide whether to consider chemicals further.

Another activity to be considered is a continuing sensitivity analysis of the system. What would happen if a decision rule related to an RTECS code were changed? Would more chemicals reach Stage 2? Could one spend less per minidossier and still get a reasonably effective assessment from the minidossiers? These kinds of questions, which could be addressed without committing any of the testing budget, might suggest new ways of allocating resources in the next cycle of operating the priority-setting system.

One can anticipate continuing development of possible data elements. For example, various empirical structure-activity systems--such as those described by Craig and Enslein (1980), Hodes (1981), Tinker (1981), and Klopman (1983)--might be applied in Stage 2, or perhaps in Stage 1. The accuracy and costs of such data elements need to be explored. The new data elements and corresponding decisions could be added to the priority-setting system, if it were worth while.

EVALUATION

A distinctive feature of the proposed model is that each of the estimates used in the model may be verified by results of the priority-setting system. For example, the estimate of toxic chemicals among the chemicals considered by the system predicts the percentage of toxic chemicals that would be discovered if absolutely definitive tests were conducted on all chemicals in a particular class. As tests are conducted, by NTP and others, it will be possible to validate these predictions by comparing them with test outcomes. Doing so allows refinement of the estimates used in the model, hence improving its efficiency. It also helps to identify better sources of information about these parameters. If estimates of accuracy prove to be poor predictors of performance, it may be because the users of the tests are overestimating or underestimating what the tests can provide.

Evaluation is essential to good science and good priority-setting. Not only will an effort to measure performance of the system generate unique information, but the credibility of any scientific or policy-making process is increased if it invites responsible evaluation. Designing an evaluation scheme will not be easy. One must, for example, consider what constitutes "truth" for the purposes of evaluation when there is uncertainty about the interpretation of the results of tests in terms of health effects in humans.

Potential users of the recommended system should take care to forestall misinterpretations of its output. For example, it is clear that priority-setting for testing is quite different from priority-setting for other kinds of action. Failure to consider a chemical for further testing can have several meanings, including its having been exonerated and its being indistinguishable from many other chemicals about which little is known. Although exposure and toxicity are grounds for further testing, both are needed for health effects; and concern about one is not in itself a cause for regulatory action.

POSSIBLE NEW DATA ELEMENTS

A central feature of the approach to priority-setting recommended here is its flexibility. It can, in principle, accommodate any set of values, any substances, and any kinds of tests. The last-named capability is illustrated by the inclusion of a wide variety of data elements in the demonstration scheme. Once the components of the priority-system have been described quantitatively, evaluation of a proposed data element requires only an estimate of its costs and accuracy to determine its feasibility and to emphasize that the approach is not limited to the data elements of the demonstration scheme. This section describes some additional sources of information as possibilities, rather than proposals. More detailed investigation may show that some are not worthy of incorporation, whereas others may require only time to become widely accepted.

THE USE OF SURVEYS AND EPIDEMIOLOGIC INFORMATION

There are a variety of methods for obtaining toxicologic information other than laboratory tests. These include analysis of existing health records and questioning of people in particular jobs or neighborhoods. The costs and value of information collected by these techniques vary greatly; however, their utility may be examined in terms of accuracy and cost, just as more traditional types of information have been examined elsewhere in this report.

Populations that might be surveyed include members of the general public, persons at risk because of their occupation or lifestyle, and medical or health personnel. Each group could be asked questions to ascertain its perceptions on exposure or health effects.

Using techniques from epidemiology (Buffler and Sanderson, 1981), psychology (Barker, 1963), safety research (Rentos and Kamin, 1975), and organization analysis (Mintzberg, 1975), investigators could study the daily routine of respondents to identify potential exposures. Such techniques could serve as a basis for analyzing the accuracy of surveys in which responses to survey questions are not supplemented by direct confirming observations. Considerable care would be needed to design questions that were scientifically meaningful yet comprehensible to respondents (National Research Council, 1981; Buffler, 1982).

The benefits of exposure surveys might include the identification of aspects not covered by traditional approaches, where existing data are often proprietary, dated, or mute about chemicals that are created in the home or are encountered only as intermediates in an industrial process. Positive reports might influence priority-setting by increasing estimates of potential exposure. They might also prompt further studies to clarify degrees of exposure or might increase the perceived need for toxicity testing.

Because toxicity testing is intended to serve the public good, it could be guided by what the public considers to be important. Two kinds of judgments are particularly significant for the priority-setting process. One concerns perceptions of the relative harmfulness of various health effects; in general, the amount of resources devoted to the study of a health effect varies as a function of the degree of public concern about it. The second issue concerns the relative costs of errors in misclassifying chemicals. Although Appendix E presents arguments from economic theory, the process of choosing chemicals to test necessarily involves value judgments as to the relative costs of misclassifying a chemical as safe (false-negative) and misclassifying it as harmful (false-positive). Surveys afford one way of assessing the views of a representative sample of the public on such matters and comparing them with the views of technical experts. It need not be assumed that all disagreements between the public and experts about the nature of risk reflect poorly on either group. Research has shown that what appear to be disagreements about the magnitude of risk can often be traced to other explanations, such as differences in the definition of key terms (e.g., risk) or political values (Fischhoff et al., in press).

DEVELOPING VERY-LOW-COST (VLC) TESTS

Of the universe of about 70,000 chemicals considered by the illustrative system, it appears that there is no characterizing information on tens of thousands except chemical name and (usually) structure. The Chemical Abstracts Service list contains about 6 million chemicals, on perhaps 1 million of which there is little or no information. If 1% of the noncharacterized chemicals are highly likely to have important health effects, then it may be assumed that some 6,000-10,000 of them are of sufficiently high concern to be candidates for further testing.

These large numbers preclude intensive testing of any substantial fraction of the noncharacterized chemicals. We believe that it would be useful to consider development of very-low-cost (VLC) data elements (or tests) that can provide at least rudimentary screening of large numbers of chemicals. Because data (sometimes even structural information) are lacking on many chemicals, this screening would involve testing the chemicals (albeit briefly). Taking a "quick look" at chemicals would yield some new knowledge, in contrast with the use of computerized data scoring, which capitalizes on old knowledge. Indeed, comparing the results of VLC tests with the results of computerized screening would be a way of validating both methods and uncovering potentially interesting discrepancies.

The results of VLC tests, like the results of low-cost screening, would permit a limited characterization of each chemical. The precision of VLC tests would necessarily be low. They would be used solely for screening, rather than for refining one's degree of concern. The ideal test in this regard is one that is most sensitive to the most toxic chemicals. Because chemicals about which this test generated no concern would not be considered further, the cost of false-negatives would be high, but lower than if no "testing" of this population had been done at all. Because the next stage for a chemical that generated concern would be consideration for low-cost testing, the cost of false-positives would be relatively low. These costs of misclassification make it possible to choose between candidate VLC procedures once one has an idea of their accuracy.

One direction that seems promising, but that requires considerable additional analysis before it can be evaluated for development, is to use simplified versions of existing tests and thus save money by avoiding the methodologic refinements needed for definitive reliable results. As unsatisfying as such tests might be--in contrast with traditional, more sophisticated forms--they can serve a purpose in priority-setting. One requirement is that their results be able legitimately to alter expert opinion regarding degree of concern about a chemical. Whether the potential for such alteration justifies even their low costs can be addressed systematically by value-of-information analysis in the framework described in this report. Indeed, one of the singular features of this framework is that it allows a defensible way to address such questions.

Most oncogenes appear to exhibit base-pair substitution; hence, a test like the Salmonella/microsome assay (Salmonella with S-9, but with duplicate plating at only a single dose of chemical) would be a VLC test. With a similar TA-98 strain, both base-pair and frameshift mutagens might be identified in a VLC test. This example is only illustrative, not definitive; the idea is to use a simplified, and hence less expensive, test carefully conceived to screen and identify likely toxic chemicals of high potency. Examples of other possible screening tests are unreplicated skin-exposure tests and unreplicated acute-toxicity tests. VLC tests might be used in any decision stage to improve the decision-making process.

Such a direct approach may be the only means of ensuring that materials of unknown structure and mixtures are given some attention. Once the chemicals of high concern are identified and carried forward, the low-cost tests may be less effective, because the remaining chemicals presumably will have lower toxicity or exposure potential and hence be more difficult to detect, in view of test variability.

RANDOM SAMPLING

The most disturbing circumstances facing those who must set priorities are the magnitude of the problem and the deficiencies in information. When ignorance about all members of a set of chemicals is equal (and total), then one has no choice but to treat them equally from a priority-setting perspective. All should be advanced to the next stage, or all should be left behind. The decision will depend much more heavily on the resources available for testing than on the concern engendered by having a large set of poorly understood chemicals.

One response to this situation is to delegate responsibility to the scientific community as a whole, on the assumption that the diversity of interests in that community will ensure that troublesome chemicals quickly come to attention. Once some disturbing information gives cause for concern about a chemical, it can be treated more specifically. It might be argued that this leaves too much to chance, in view of the rate at which unanticipated health hazards have emerged in the past.

An alternative approach is to admit to ignorance and to sample at random from the universe of chemicals. The expected value of such sampling can be calculated roughly by assuming that evidence of toxicity will be found at the estimated rate of prevalence of toxicity in the population. The sampling would, of course, provide a basis for reviewing and refining that estimate. Ideally, the resources so invested would justify increasing concern about some chemicals that would not otherwise be considered and justify decreasing concern about many more. In addition, they would aid in refining the decision rules of the priority-setting system and thereby improve its functioning. There is a small probability that sampling would produce major scientific surprises. A careful analysis could reveal the proportion of the overall testing budget that could be usefully devoted to this kind of development activity. The analysis would also indicate which sampling strategy is most efficient (e.g., what tests in what order and whether to use stratified sampling). Testing a randomly chosen sample might also reveal biases in the use of a universe restricted to chemicals on which there is already some information.

6

CONCLUSIONS AND RECOMMENDATIONS

Far more chemicals are in the human environment than can be evaluated for potential toxicity with available methods and resources. Therefore, some chemicals have to be selected for testing for their potential impact on public health, and that requires a priority-setting process. The selection of chemicals for testing is made difficult by the existence of many possible health effects to test for and many possible tests and combinations of tests to choose from.

Much of the information needed to set priorities for testing is fragmentary or lacking; little if any toxicity information is available on most chemicals, and the information on human exposure to or potential toxicity of only a few is more than minimal. Moreover, many of the data that do exist are not easily retrievable or are not in a form that makes them readily usable or verifiable. An essential function of any priority-setting system must be to serve as a guide to strategies for improving the available data through the most rapid and effective combinations and sequences of information-gathering procedures.

On the basis of its review of existing priority-setting systems, the Committee on Priority Mechanisms has concluded that such systems must be logically and scientifically defensible, open to peer review, practical, and explicit about their underlying assumptions. Such analytic techniques as systems analysis and decision theory appear to permit design approaches with those characteristics. Application of these analytic techniques requires, however, that each information-gathering or testing procedure be described quantitatively with respect to its ability to identify potentially toxic chemicals; and the current art of toxicity testing and priority-setting does not readily permit such quantitative descriptions with great precision. Efforts toward further quantification of performance characteristics of toxicologic methods are essential to the development of an optimal priority-setting system.

In addition to information on the accuracy of testing procedures, the design of a priority-setting system involves the number and types of chemicals to be considered by the system, the accuracy of selection procedures, the availability of testing resources, penalties for misclassifying chemicals, costs of selection and testing, and adequacy of information on exposure to and the potential health effects of the chemicals to be considered.

Priority-setting systems must also be flexible, so that they can be applied by users with different missions. No system can anticipate and address all possible contingencies, and a given system should be adjustable to suit specific needs. Testing may be undertaken for a variety of reasons--e.g., to provide special scientific insights or to improve knowledge of testing procedures. At times, policy considerations may dictate testing of substances that would otherwise receive lower priorities. A priority-setting system should be able to respond to changes that result from advances in science and changes in the perceived relative importance of various toxicity and exposure problems.

There is no evidence that any priority-setting system can remedy the basic problem of insufficient data. In the light of that impediment, the following may be considered as characteristics of a system that would advance the state of the art of priority-setting:

● Use of explicit, detailed, and formal decision-making procedures in selecting and ranking chemicals for testing, such as are illustrated in the approach described in this report, which is based on principles of systems analysis and decision theory.

● Validation and quantification of the performance characteristics of all tests and other information-gathering procedures used in the testing program.

● Development of relevant data bases, including those related to the production and use of chemicals, to the potential for human exposure, to structure-activity relationships, and to toxic activity.

● Characterization of the universe of chemicals to be tested.

● Estimation of the prevalence of substances that can cause different types of toxicity.

● Development of systematic procedures to ensure that the priority-setting process corrects and refines itself on the basis of experience.

● Inclusion of a procedure for systematic screening of the entire select universe of chemicals as defined in this report, as a basis for nominating substances to be tested.

REFERENCES

Ames, B. N. 1979. Identifying environmental chemicals causing mutations and cancer. Science 204:587-593.

Ames, B. N., L. S. Gold, C. B. Sawyer, and W. Havender. 1982. Definition of carcinogenicity, pp. 663-670. In D. T. Sugimura et al., Eds. Carcinogenic Potency of Environmental Mutagens and Carcinogens. New York: Allen R. Liss, Inc.

Ames, B. N., and J. McCann. 1981. Validation of the Salmonella test: A reply to Rinkus and Legator. Cancer Res. 41:9192-4201.

Ashby, J. The future role of rodents in the detection of possible human carcinogens and mutagens. Mutat. Res. (in press)

Barker, R. G. 1963. The Stream of Behavior. New York: Appleton-Century-Crofts. 352 pp.

Bridges, B. A. 1976. Use of a three-tier protocol for evaluation of long-term hazards, particularly mutagenicity and carcinogenicity. pp. 549-568. In R. Montesano, H. Bartsch, and L. Tomatis. Screening Tests in Chemical Carcinogenesis. IARC Scientific Publications No. 12. Lyon, France: International Agency for Research on Cancer.

Buffler, P. A. 1982. Epidemiological approaches to defining sensitive employees. Ann. Amer. Conf. Gov. Ind. Hyg. 3:11-26.

Buffler, P. A., and L. M. Sanderson. 1981, pp. 55-107. In C. R. Shaw, Ed. Prevention of Occupational Cancer. Boca Raton, Fla.: CRC Press.

Calkins, D. R., R. L. Dixon, C. R. Gerbert, D. Zarin, and G. S. Omenn. 1980. Identification, characterization, and control of potential human carcinogens: A framework of federal decision-making. J. Natl. Cancer Inst. 64:169-76.

Craig, P. N., and K. Enslein. 1980. Application of structure-activity studies to develop models for estimation of toxicty, pp. 411-419. In D. B. Walters, Ed. Safe Handling of Chemical Carcinogens, Mutagens, Teratogens and Highly Toxic Substances. Vol. 2. Ann Arbor, Mich.: Ann Arbor Science Publishers, Inc.

de Serres, F. J., and J. Ashby, Eds. 1981. Evaluation of Short-Term Tests for Carcinogens. Report of the International Collaborative Program. Progress in Mutation Research. Vol. 1. New York: Elsevier North Holland. 828 pp.

Estrin, N. F., P. A. Crosley, and C. R. Haynes, Eds. 1982. CTFA Cosmetic Ingredient Dictionary, 3rd ed. Washington, D.C.: The Cosmetic, Toiletry and Fragrance Association, Inc. 610 pp.

Fischhoff, B., P. Slovic, and S. Lichtenstein. 1977. Knowing with certainty: The appropriateness of extreme confidence. J. Exp. Psychol. Hum. Percept. Perform. 3:552-564.

Fischhoff, B., P. Slovic, and S. Lichtenstein. The "public" vs. the "experts": Perceived vs. actual disagreements about the risks of nuclear power. In V. Covello, G. Flamm, J. Rodricks, and R. Tardiff, Eds. Analysis of Actual vs. Perceived Risks. New York: Plenum Press. (in press)

Hodes, C., G. F. Hazard, R. I. Geran, and S. Richman. 1977. A statistical-heuristic method for automated selection of drugs for screening. J. Med. Chem. 20:469-475.

Hollstein, M., J. McCann, F. A. Angelosanto, and W. W. Nichols. 1979. Short-term tests for carcinogens and mutagens. Mutat. Res. 65:133-226.

Interagency Regulatory Liaison Group (IRLG). 1979. Scientific bases for identification of potential carcinogens and estimation of risk. J. Nat. Cancer Inst. 63:241-268.

International Agency for Research on Cancer (IARC). 1982. IARC Monographs on the Evaluation of the Carcinogenic Risk of Chemicals to Humans. Chemicals, Industrial Processes, and Industries Associated with Cancer in Humans. IARC Monographs Supplement 4. Lyon, France: International Agency for Research on Cancer.

Kirk-Othmer Encyclopedia of Chemical Technology. 1978. Vol. I, 3rd ed. New York: John Wiley & Sons.

Klopman, G. 1983. Computer automated structure evaluation. A new method for evaluating biological activity of diverse molecules. Paper presented at the 33rd semi-annual meeting of the Chemical Manufacturers Association, November 1983, Washington, D.C.

Kuzmack, A. M., and R. E. McGaughy. 1975. Quantitative Risk Assessment for Community Exposure to Vinyl Chloride. Washington, D.C.: U.S. Environmental Protection Agency, Office of Research and Development. [61] pp.

Lave, L. B., G. S. Omenn, K. D. Heffernan, and G. Dranoff. 1982. Analysis of the Cost-Effectiveness of Tier-Testing for Potential Carcinogens. Paper presented at the First World Conference of American Association of Toxicologists, May 29, 1982.

Mendelsohn, M. L. The identification of chemical carcinogens: Performance using IARC data; strategies; optimization; interpretation. (in press)

Mintzberg, H. 1975. The manager's job. Folklore and fact. Harv. Bus. Rev. 53:49-61.

Moore, J. A., J. E. Huff, L. Hart, and D. B. Walters. 1981. Overview of the National Toxicology Program, pp. 555-574. In J. D. McKinney, Ed. Environmental Health Chemistry. The Chemistry of Environmental Agents as Potential Human Hazards. Ann Arbor, Mich.: Ann Arbor Science Publishers, Inc.

National Research Council, Committee on National Statistics. 1981. Surveys of Subjective Phenomena: Summary Report. C. F. Turner and E. Martin, Eds. Washington, D.C.: National Academy Press. 97 pp.

Purchase, I. F. H. 1982. An appraisal of predictive tests for carcinogenicity. Mutat. Res. 99:53-71.

Raiffa, H. 1968. Decision Analysis: Introductory Lectures on Choices under Uncertainty. Reading, Mass.: Addison-Wesley. 309 pp.

Rinkus, S. J., and M. S. Legator. 1979. Chemical characterization of 465 known or suspected carcinogens and their correlation with mutagenic activity in the Salmonella typhimurium system. Cancer Res. 39:3289-3318.

Sax, N. I. 1979. Dangerous Properties of Industrial Materials, 5th ed. New York: Van Nostrand Reinhold Co. 1,118 pp.

Sivak, A. 1982. An evaluation of assay procedures for detection of tumour promoters. Mutat. Res. 98:377-378.

SRI International. Chemical Economics Handbook Program (a continuously updated program). Menlo Park, Calif.: SRI International.

Upton, A. C., D. G. Clayson, J. D. Jansen, H. Rosenkranz, and G. Williams. Report of ICPEM Task Group on the Differentiation between Genotoxic and Non-Genotoxic Carcinogens. Mutat. Res. (in press)

U.S. Congress, Office of Technology Assessment. 1981. Assessment of Technologies for Determining Cancer Risks from the Environment, pp. 135-136. Washington, D.C.: U.S. Congress, Office of Technology Assessment.

U.S. Department of Health and Human Services. 1981. The Health Consequences of Smoking. The Changing Cigarette. A Report of the Surgeon General. DHHS(PHS) 81-50156. Rockville, Md.: U.S. Department of Health and Human Services, Public Health Service, Office on Smoking and Health. 252 pp.

U.S. Department of Health, Education, and Welfare, Public Health Service, Center for Disease Control, National Institute for Occupational Safety and Health. 1974. National Occupational Hazard Survey. Vol. 1: Survey Manual. NIOSH Publ. No. 74-127. Cincinnati, Ohio: U.S. Department of Health, Education, and Welfare, National Institute for Occupational Safety and Health. 202 pp.

U.S. Department of Health, Education, and Welfare, Public Health Service, Center for Disease Control, National Institute for Occupational Safety and Health. 1977. National Occupational Hazard Survey. Vol. 2: Data Editing and Data Base Development. NIOSH Publ. No. 77-213. Cincinnati, Ohio: U.S. Department of Health, Education, and Welfare, National Institute for Occupational Safety and Health. 154 pp.

U.S. Department of Health, Education, and Welfare, Public Health Service, Center for Disease Control, National Institute for Occupational Safety and Health. 1978. National Occupational Hazard Survey. Vol. 3: Survey Analysis and Supplemental Tables. NIOSH Publ. No. 78-114. Cincinnati, Ohio: U.S. Department of Health, Education, and Welfare, National Institute for Occupational Safety and Health. 799 pp.

U.S. Environmental Protection Agency. 1979. Toxic Substances Control Act (TSCA) Chemical Substance Inventory. Vol. 1. Initial Inventory. Washington, D.C.: U.S. Environmental Protection Agency.

U.S. Environmental Protection Agency. 1980. Toxic Substances Control Act (TSCA) Chemical Substance Inventory. Cumulative Supplement to the Inventory. Washington, D.C.: U.S. Environmental Protection Agency.

Weinstein, M. C. 1983. Cost-effective priorities for cancer prevention. Science 221:17-23.

Weisburger, J. H., and G. M. Williams. 1981. Carcinogen testing: Current problems and new approaches. Science 214:401-407.

Windholz, M., S. Budavari, L. Y. Stroumtsos, and M. N. Fertig, Eds. 1976. The Merck Index. An Encyclopedia of Chemicals and Drugs, 9th ed. Rahway, N.J.: Merck & Co. 1,313 pp.

APPENDIX A

REVIEW OF SELECTED PRIORITY-SETTING SYSTEMS

Initial examination of selected priority-setting schemes revealed that the multiplicity of approaches was more apparent than real. The appearance of dissimilarity arises more from differences in emphasis, or scope, than from differences in basic logic or strategy.

Selected for detailed description here are schemes that were thought to make important contributions to the developing science or art of priority-setting. The choices in some cases were related to uniqueness in the treatment of exposure, of toxicity, or of the interaction between the two. A comprehensive list of schemes has been compiled (U.S. Environmental Protection Agency, 1980). It has been recommended that federal agencies adopt priority-setting systems (Administrative Conference of the United States, 1982).

The Toxic Substances Control Act-Interagency Testing Committee (TSCA-ITC) scheme (Nisbet, 1979) is of particular interest, because it deals with a large part of the universe with which the NTP is concerned. Equally important, it has had to face the test of continued use over several years, and it has been systematically reviewed (Nisbet, 1979).

The schemes of Kornreich et al. (1979) and Ross and Lu (1980) are based on a systematic review of a substantial portion of the literature on priority-setting. The Food and Drug Administration (FDA) scheme (U.S. Department of Health and Human Services, 1982) is limited to one route of exposure, but otherwise is comprehensive in its approach. The scheme of Wilhelm (1981) is in large measure a response to what were perceived as deficiencies in the TSCA-ITC system. That of Astill et al. (1981) is designed to function with a sequential testing and feedback strategy. The ranking algorithm of Brown et al. (1980) is based on a simple mathematical model and is designed for multinational application. The proposed cyclic review procedure for FDA (U.S. Department of Health and Human Services, 1982) uses structure-activity considerations to establish initial "levels of concern," which are also found in the decision-tree approach of Cramer, Ford, and Hall (1978). Gori's scheme (1977) provides a ranking index based on exposure that is complementary to a second scheme that uses structure-activity analysis for assessing possible carcinogenic activity (Dehn and Helmes, 1974).

NATIONAL TOXICOLOGY PROGRAM

The NTP chemical nomination and selection process, as described in the NTP Fiscal 1983 Annual Plan (U.S. Department of Health and Human Services, 1983), is summarized below.

301

CHEMICAL NOMINATION

Member agencies of NTP and other sources (other federal agencies, state agencies, the public, labor, and industry) submit to NTP nominations of chemicals for various types of toxicity testing. Each nomination is expected to include the name of the chemical, the particular toxicity test(s) desired, the rationale for the nomination, and the available background data on production, use, exposure, environmental occurrence, and extent of toxicologic characterization. All nominations are considered, however, regardless of the depth of information submitted.

Nominations are referred to the NTP chemical selection coordinator for review, to determine which chemicals have been tested, are already on test or scheduled for test, or have been previously considered and rejected for testing by NPT or its predecessors. This may involve preliminary searches of on-line data bases and reference books. The nominations and background information are then forwarded to the chemical-review staff at the National Center for Toxicological Research (NCTR), who examine the available literature, assess the relevant data, and prepare draft executive summaries of the information. (Executive summaries are not prepared for chemicals nominated solely for mutagenicity testing.) Included in each draft executive summary are chemical identification, surveillance index (production, use, occurrence, and analysis), information on toxic effects, and a statement of the source of and reason for nomination.

EVALUATION OF NOMINATED CHEMICALS

The chemical-review staff sends the draft executive summaries to the Chemical Evaluation Committee (CEC), which is composed of representatives of the Consumer Product Safety Commission (CPSC), Environmental Protection Agency (EPA), FDA, Occupational Safety and Health Administration (OSHA), National Cancer Institute (NCI), National Institute of Environmental Health Sciences (NIEHS), National Institute for Occupational Safety and Health (NIOSH), NCTR, and NTP. Members are requested to search data bases peculiar to their agencies for further information on the nominated chemicals (and structurally related compounds), to improve the evaluation process. The CEC evaluates the summaries and recommends the types of testing to be performed. Primary and secondary reviewers are also assigned to each chemical, after consideration of the nature of exposure, so that appropriate regulatory concerns will be addressed.

At the CEC meetings, the primary reviewer for each chemical summarizes the data on that chemical and makes recommendations for testing. The secondary reviewer presents additional information, where available, and also discusses the testing of the compound. After discussion, the CEC votes on the recommended types of testing and assigns priorities for the testing.

PUBLIC COMMENT

A _Federal Register_ notice is published, which lists the chemicals reviewed by the CEC and the recommended types of testing. The notice also solicits comments from interested parties and information on completed, current, and planned testing in the private sector. The list of chemicals is also published in the _NTP Technical Bulletin_, with a request for comments. These steps are taken to enable other individuals and groups to provide data useful to the chemical evaluation process.

PEER REVIEW

The revised executive summaries and public comments on the nominated chemicals are forwarded to the NTP Board of Scientific Counselors, which meets to evaluate the data and to make recommendations.

CHEMICAL SELECTION

The chemical-review staff then incorporates the board's ratings and pertinent public input into final executive summaries, which are submitted to the NTP Executive Committee. That committee decides whether to test, defer, or delete each of the nominated chemicals for the various types of testing. Its decisions are published in the _NTP Technical Bulletin_. The committee also recommends priorities for testing, test development, and test validation to NTP.

After Executive Committee action, the NTP Steering Committee refers the chemicals to one or more of the three organizational units participating in NTP: NIEHS, NIOSH, and NCTR. A chemical manager is then assigned to evaluate the data developed during the NTP chemical evaluation process and other information retrieved from detailed searches of the published literature and from industry. The manager presents a proposal to the Toxicology Design Committee (TDC) either to perform appropriate testing or to delete the chemical from consideration for testing. The TDC, which consists of research scientists from NIEHS and NTP, assesses the proposal and either develops a final protocol for testing or recommends no further testing; the latter recommendation is based on technical difficulties in testing, budgetary reasons, or the existence of adequate outside testing.

All chemicals selected through this process are then tested as time and resources permit.

The results of testing are reviewed routinely, to determine whether further types of testing are appropriate, and candidates for additional testing are submitted to the NTP chemical nomination and selection process for evaluation and decision-making.

SEQUENTIAL TESTING FOR CHEMICAL RISK ASSESSMENT
(ASTILL ET AL., 1981)

This scoring system was developed by the Eastman Kodak Company to determine the extent of toxicity testing required for production chemicals. Four categories of information are used to derive a total score, on the basis of which one of four testing levels is recommended. Available health and environmental data are compiled and rated independently, composite health-effects scores are computed, and the appropriate tests are selected and performed. Results of these tests are then used to revise the ratings. New scores are obtained, and the testing level is revised. This process is repeated until testing information is complete. Thus, the system is dynamic, in that it incorporates a feedback mechanism that allows for continuing review of the testing needs of a specific chemical. This system provides a basis for a multistage screening system.

Four categories of information are used: magnitude of human exposure, magnitude of environmental exposure, effects on human health, and effects on the environment. The two magnitude categories have four components each, and the two effects categories have three components each.

The four components considered in the rating of the magnitude of human exposure are production volume, number of people exposed, hours per year exposed, and number of population types exposed. Scores for the four components are added to yield a value for the magnitude of exposure. The assessment of health effects considers the LD_{50}, acute effects (reversible and irreversible), and chronic effects (reversible and irreversible).

Each of the 14 components of the four categories is scored from 1 to 3, with 3 indicating the most severe or hazardous score. The scores for the two human categories (health effects and magnitude of human exposure) are summed, as are the scores for the two environmental categories. The resulting scores range from 7 to 21 and are associated with specific testing levels, as follows:

Health (or Environmental) Score	Testing Level
7-9	I
10-13	II
14-17	III
18-21	IV

The level of testing becomes increasingly specific and sophisticated with increasing score. Level I testing is based on the use of physicochemical evaluation and health screening, as well as acute-toxicity studies. Although it is not specifically stated, with respect to human data, Level I might include surveillance of morbidity, mortality, and fertility patterns of exposed human populations. Level II testing consists of toxicity tests that are intermediate between acute tests and subchronic feeding studies, whereas Level III testing includes subacute exposure studies. Long-term (or chronic) health effects are evaluated through Level IV testing.

The health-effects criteria are not very specific, but readily quantified in an objective and replicable manner. The health-effects criteria and ratings are as follows:

		Rating
LD_{50}, mg/kg	> 500	1
	50-500	2
	< 50	3
Immediate effects	None	1
	Reversible	2
	Irreversible	3
Prolonged effects	None	1
	Reversible	2
	Irreversible	3

This system appears to be efficient, in that it uses a minimum of subjective input (expert opinion or judgment), although such judgment may be used in the review and rating of health effects.

This system appears to be practical, in that it facilitates decision-making in an efficient and objective manner. Any compound can be evaluated; in the absence of available data, baseline information is compiled before any testing is done. The baseline information compiled consists of:

- Quantities manufactured and disposed of.

- Exposure estimates.

- Product function and application.

- Structure-activity correlation.

- Literature search.

- Cancer hazard evaluation.

Such baseline information may be sufficiently complete for hazard assessment, particularly if previously published toxicity studies are available.

This scheme has been evaluated by the authors with a wide range of industrial chemicals, although the specifics of evaluation are not provided.

A RANKING ALGORITHM FOR EEC WATER POlLUTANTS
(BROWN ET AL., 1980)

The purpose of this scheme is to rank, for possible regulatory action, water pollutants as potential hazards to humans and to aquatic organisms. The scheme considers about 1,500 compounds used in countries of the European Economic Community and suspected of entering rivers.

The algorithm is based on a simplified mathematical model relating production and use of a chemical to its occurrence in drinking water and in food of freshwater origin. Standard assumptions are made as to intake of fish and water; daily maximal and annual average intakes through ingestion are calculated.

The amount of a chemical estimated to reach the water is calculated by multiplying production by the fraction that reaches the water; the fraction is estimated on the basis of manufacturing practices and the chemical's use. A typical dilution volume of the chemical is estimated from its half-life in water and from river-flow data. Estimated concentrations are used to calculate human exposure from consuming drinking water and freshwater fish. A concentration factor is used to calculate ingestion from consumption of fish, assuming typical diets.

The list of 1,500 chemicals was reduced to about 1,400 when mercury and cadmium compounds were eliminated because they were already controlled by the EEC and persistent synthetic substances (mainly plastic materials) were eliminated because, although objectionable in water, they are not toxic.

For the remaining 1,400 compounds, production and consumption data are obtained, and all those estimated to be produced at under 100 metric tons per year are eliminated. The remaining 426 compounds are then processed through a screening algorithm based on production, environmental half-life, and acute-toxicity factors.

Some elements of toxicity testing for human health are applied in this scheme. The acute-mammalian-effect dose is represented by the lowest reported lethal oral dose for humans. If this information is not available, the lowest oral LD_{50} value for other mammalian species is used. If no oral LD_{50} value is available, the lowest LD_{50} value for the dermal or inhalation route is applied. If no LD_{50} values have been reported at all, the lowest lethal dose for the oral, dermal, or inhalation route is used. If no acute-lethality data are available, an estimate is devised on the basis of comparison with other compounds in the same chemical class. If a reasonable estimate cannot be made this way, the default entry "unknown" is used in the program.

Chronic mammalian effects are also used when available. If the data file indicates that carcinogenicity, mutagenicity, or teratogenicity information is available, it is factored into the algorithm. If a chemical exhibits all three effects, only one is entered, preferably carcinogenicity. The chronic-mammalian-effect dose is the lowest dose that caused the reported effect.

ESTIMATION OF TOXIC HAZARD--A DECISION TREE APPROACH (CRAMER ET AL., 1978)

This scheme ranks food chemicals in three classes of concern for toxicity testing on the basis of chemical structure and oral-toxicity data. It is applied to structurally defined organic and organometallic compounds. Polymers and inorganic compounds are excluded.

By answering a series of questions about chemical structure, the operator of the system follows a decision tree until the chemical considered falls into Class I (low concern), Class II (moderate concern), or Class III (serious concern). In each class, chemicals are ranked by comparison witn no-observed-effect doses. The data on no-effect doses were derived from literature values based on short-term or chronic studies.

Class I substances are those whose structures and toxicity data, when combined with low human exposure, suggest low priority for investigation. Class III substances are those whose structures and toxicity data would not permit presumptions of safety and which thus require the highest priority for investigation. Class II substances are intermediate between Classes I and III. High exposures to substances in any class would increase the priority for investigation or testing. The number of chemicals found to be in Class II is not large.

The tabulation of compounds in classes, with the exception of compounds with no-effect exposures above 500 mg/kg of body weight per day, is restricted to toxicity tests in which the next higher feeding exposure above the no-effect exposure is no more than 5 times the no-effect exposure. It was the general intent of the authors that the most toxic substances in Class I (low concern) should have a no-effect exposure in animal tests at or above 50 mg/kg of body weight per day. This exposure, subjected to a safety factor of 100, corresponds to human exposure at approximately 25 mg/day.

Use of this procedure requires knowledge of chemical structure and reasonably accurate estimates of human intake. The authors made it clear that chemical structure is to be used only as a guideline for testing decisions and that such use of structure-activity analysis is intended as a guide to the acquisition of data, not as a substitute for data.

AN AUTOMATIC PROCEDURE FOR ASSESSING POSSIBLE CARCINOGENIC ACTIVITY OF CHEMICALS PRIOR TO TESTING (DEHN AND HELMES, 1974)

This scheme uses structure-activity relationships to predict carcinogenesis. There is no exposure element. The corresponding exposure element has been described by Gori (1977).

The procedure incorporates the collective knowledge of a panel of experts and attempts to automate the key features of that knowledge to select candidate compounds for carcinogenicity testing. The basis of the procedure is an activity tree constructed so that more specific details of chemical structure (as related to carcinogenicity) are applied at each decision point in the tree. This subdivision of structures continues until an end group (called a node) containing compounds of closely related chemical structure is identified. An estimate is then made of the probability that the chemicals in a node are carcinogenic and of the relative potency of each. Reflecting the expertise of the panel, construction of the tree concentrates on the following groups of chemicals: naturally occurring substances; nitroso, hydrazino, and azo compounds; polycyclic aromatic hydrocarbons; aromatic amines; and inorganic compounds.

Although structure-activity relationships can be useful in setting priorities for carcinogenicity testing, the accuracy of analysis of such relationships in predicting carcinogenicity has not been verified. If the decision tree could be compared with test data generated since the scheme was completed, its utility could be better assessed. Exceptions within a given node (i.e., negative compounds within a carcinogenic chemical class) are extremely instructive and should serve as a cautionary guide when one attempts to apply analysis of structure-activity relationships in too broad a manner.

TOXICOLOGICAL PRINCIPLES FOR THE SAFETY OF DIRECT FOOD ADDITIVES AND COLOR ADDITIVES USED IN FOOD (U.S. DEPARTMENT OF HEALTH AND HUMAN SERVICES, 1982)

This scheme was developed to establish priorities (and extent) for toxicity testing of direct food additives.

Chemicals are divided into three categories of suspicion based on structure-activity considerations by following a short decision tree. A suspicion category is combined with exposure information to define a level of concern (I, II, or III). Once the level of concern is determined, tests may be required. Results of tests already done are placed in three categories (well done; not well enough done, but usable to some degree as a "core" test; and unusable). On the basis of this further information, additional testing may be required.

Toxicity is not estimated quantitatively, so there is no quantitative assessment of uncertainty for it. There is judgmental consideration of uncertainty (specification error) in the evaluation of toxicity tests in the literature.

There is a discussion of tests for each level of concern and for various combinations of concern and test information.

RANKING OF ENVIRONMENTAL CONTAMINANTS FOR BIOASSAY PRIORITY (GORI, 1977)

The purpose of this scheme is to establish, on the basis of exposure, a priority ranking for chemicals to be tested in a carcinogenicity bioassay. All chemicals in commerce are considered by the scheme. Total intake of a chemical by a given route is estimated for all members of a population group with similar exposures, and intake is then summed over population groups and sources of exposure. Intake by route is combined with probability of carcinogenicity and expected potency to produce a ranking index that, in theory, reflects the expected annual number of cancer cases.

The scheme depends on the quantitative prediction of carcinogenic activity from structure-activity comparisons (see Dehn and Helmes, 1974). This requires the identification of substructures, derived from known carcinogens, to which activity indexes can be attached--a process that requires expert opinion. A chemical of unknown carcinogenic potential is then inspected for such substructures, and an activity value is ascertained on the basis of their presence.

Exposure assessment takes account of chemical production and use, but not disposal or discharges explicitly.

Although it may not be clear from the text, the scheme estimates an uncertainty factor or confidence range for every variable. One notes and keeps track of the route of exposure and maintains an "audit trail" to the information in the data base.

Deriving an exposure estimate for a chemical might require up to a person-day of effort, on the average. Considerable subjective input is required.

PRIORITY-SETTING OF TOXIC SUBSTANCES FOR GUIDING MONITORING PROGRAMS (KORNREICH ET AL., 1979)

This system, prepared for the Office of Technology Assessment by Clement Associates, is designed to compile a priority list for selecting potentially toxic chemicals for monitoring in food.

309

The criteria used in developing 32 existing priority lists of toxic chemicals are examined, and criteria developed for ranking chemicals on the basis of their likelihood of endangering human health through contamination of the food supply. Three preliminary lists of possible food contaminants (organic substances, inorganic substances, and radionuclides) are compiled. Data on each chemical on these lists are assembled and used to assign scores to each chemical for various factors. Scores for the factors are combined, and the combined scores are used for ranking the chemicals on the three lists.

Selection criteria include both exposure and toxicity factors. Weights are assigned to reflect the relative importance of each criterion and to allow the total score to be a measure of the overall propensity of a chemical to contaminate foods. The individual score for each factor is multiplied by the assigned weight, and the weighted scores are added. The total exposure score and the total biologic score are each adjusted to a maximum of 50 points and summed to allow for a possible total of 100 points.

This system is designed to use quantitative information, with considerable reliance on expert opinion for the assigning of scores. For toxicity factors, a score of 0 is assigned for negative results and for absence of data.

No cost estimates are given for this system, which was intended for one-time, rather than repeated, use.

RANKING CHEMICALS FOR TESTING: A PRIORITY-SETTING EXERCISE UNDER THE TOXIC SUBSTANCES CONTROL ACT (NISBET, 1979)

This scoring system was developed to set priorities for testing chemicals under the authority of the Toxic Substances Control Act. The scheme is intended for application to chemicals in commerce that are not covered by other statutes. Drugs, cosmetics, food additives, and pesticides are excluded, unless they also have other uses. Also excluded are chemicals with an annual production volume of 1,000 lb or less. The system is intended for chemicals already in commerce at the time of compilation of the TSCA Inventory, which now defines "old" chemicals for the purposes of the statute. Because the Inventory did not exist when the first testing recommendations were required by the statute, the system was originally applied to a list of chemicals derived from lists of chemicals of high production volume or previously reported toxicity. Thus, the initial "universe" of chemicals was limited to chemicals already identified as of potential concern or nominated for inclusion by Interagency Testing Committee (ITC) members or other experts.

Of 24 priority lists reviewed, 19 were used as a basis for the initial compilation of compounds. Noncommercial chemicals were then eliminated.

Chemicals that were not on the U.S. ITC list were designated to be eliminated from the list, but were screened initially and were included if nominated by the expert panel. Later screening evaluated use and eliminated substances already regulated under some statute other than the TSCA.

These initial screening steps resulted in a list of approximately 900 chemicals for scoring. ITC divided the scoring process into two discrete phases—potential exposure and biologic effects. Screening and scoring of biologic effects were postponed until potential exposure was evaluated. The following factors were used in the first stage of exposure scoring:

● General population exposure—number of people exposed, frequency of exposure, exposure intensity, and penetrability.

● Quantity released into and persistence in the environment.

● Production volume.

● Occupational exposure.

Some 330 chemicals were then selected from the list for biologic scoring. The TSCA requires that ITC give priority to compounds that are known or thought to cause or contribute to cancer, gene mutations, or birth defects. Seven factors were selected for scoring on biologic activity:

● Carcinogenicity.

● Mutagenicity.

● Teratogenicity.

● Acute toxicity.

● Other toxic effects.

● Ecologic effects.*

● Bioaccumulation.

Because ITC seeks to identify chemicals that require testing, rather than simply scoring compounds for known biologic activity, it was decided that the biologic scoring system should have two independent components—a measure of known biologic activity and a measure of the need for further testing. These components provided the basis for the biologic scoring system, as follows:

_____

*Note that this scheme and its variants (Nisbet, 1979; Ross and Lu, 1980) are designed to set priorities among chemicals for potential effects on the environment, as well as on human health.

311

<u>Positive numerical score 1 to 3</u>:

- Substance does not need further testing.

- The higher the number, the more positive the results.

<u>Zero score</u>:

- Negative test results.

- Biologically inactive compound.

- Low index of suspicion.

<u>Negative numerical score -1 to -3</u>:

- Lack of data--substance should be tested further.

- The more negative the number, the greater the need for testing (as judged by other data on biologic activity or data on structural analogues).

Early in 1979, ITC sponsored a workshop to review the ITC system and to make recommendations for improvements. The proceedings of the workshop (Nisbet, 1979) includes a number of papers on priority-setting systems and reports by 11 subgroups that reviewed different elements of the ITC scoring system and recommended changes in scoring methods for individual exposure and toxicity elements. The workshop did not propose a comprehensive alternative scheme and did not produce a synthesis of the recommendations of the subgroups.

CHEMICAL SCORING SYSTEM DEVELOPMENT
(ROSS AND LU, 1980)

This draft scheme was designed to screen relatively large numbers of chemicals and to identify those with the greatest need for control or testing. The scheme considers subsets of the TSCA Inventory, including chemicals on which EPA expects to receive additional production- and exposure-related information under Section 8(a) of TSCA.

The scheme consists of several screening processes grouped into five components: biologic toxicity I, biologic toxicity II, environmental fate, production and release, and human exposure. There are several criteria for each component. Each criterion is assigned a numerical score from 0 or 1 to 9 or 10.

The screening system is applied to chemicals on the TSCA Inventory in two phases. The first phase screens chemicals into groups of low, moderate, and high concern on the basis of exposure characteristics (production volume, environmental fate, potential environmental release, and potential human exposure). For chemicals that have similar scores on these major exposure criteria, scores on a group of modifier criteria can be applied to determine which compounds have the greater exposure potential. These modifier criteria can receive a maximal score of 9 and are to be used only in case of ties in the scores on the primary exposure criteria.

The second phase separates chemicals into groups of low, moderate, or high concern on the basis of potential toxic effects. Chemicals that are identified as being of high concern in the first phase are to be considered first in the second phase.

The biologic-effects criteria are divided into two categories: biologic toxicity I includes carcinogenicity, mutagenicity, embryotoxicity and fetotoxicity, and reproductive effects; biologic toxicity II includes all other criteria for biologic effects and contains effects on plants, bacteria, fungi, and aquatic organisms. The authors stated:

Biological toxicity is divided into 2 components because the areas of health effects in the biological toxicity I component are of particular societal and regulatory agency interest and therefore warrant consideration separate from other aspects of toxicity. Another difference between the biological toxicity I and biological toxicity II components is that the scoring systems in the biological toxicity I component are not dose dependence [sic] but are based on expressions of confidence, whereas the scoring systems in the biological II component are either dose or concentration dependent.

In the carcinogenicity scoring process, a precursor is defined as "a chemical which in itself is not carcinogenic but which is responsible for the formation of a chemical which is carcinogenic, e.g., a metabolite." However, the precursor is assigned a score of 4, rather than a potentially higher one.

This scoring process is strictly qualitative and does not deal with the potency of a carcinogen. It appears that absence of data is considered to imply low priority; "no data but suspect" is given a score of 3; "no data but not considered suspect" is given a score equal to that for "no data available, no estimate made."

The mutagenicity scoring procedure considers the potential for genetic impairment at both the somatic cell and germinal cell levels. Like the carcinogenicity scoring procedure, it is strictly qualitative, and a suspect chemical on which no data are available will score low (2 or 3).

Several types of prenatal effects are combined under the broad terms of "embryotoxicity" and "fetotoxicity." Whether other reproductive effects are distinguished from true teratogenic action is unclear.

The chronic-toxicity scoring procedure has two notable components: first, it scores on the basis of quantitative dosage criteria; second, it scores on the basis of the severity of an effect. No guidelines are given to indicate what specific effects would be examined or called for. Again, suspect chemicals with no data get low scores.

The acute-toxicity scoring system considers lethal end points, but not functional impairment. Several opportunities for scoring are possible, because data from any route are considered. When several routes have been studied, the data that provided the highest score are used in the final priority-setting. Chemicals "suspected to have a score of 8 to 10" are assigned a score of 3 when there are no data to confirm the suspicion. Again, suspect chemicals with no data get low scores.

The first phase of the screening program uses the exposure component and subcomponent scores to screen and set testing priorities for chemicals on which additional biologic-effects data are needed. The actual priority-setting treats the data as a set of component scores (for either exposure or biologic effects) that are made up of combinations of subcomponents. Each component has a maximal score of 10. The ratio of the assigned score to the maximal score is displayed. If any subcomponent receives a score of 10, it is automatically placed in a rank of high concern. Otherwise, the accumulated subcomponent ratios within a component are assigned scores, and a hazard index is calculated.

Subcomponent scores are added and form the numerator of a fraction whose denominator is the sum of possible scores for each of the subcomponents within the component. A hazard index is the expression of the ratios as a percentage. With the exception that a score of 10 in any subcomponent automatically places that chemical in a category of high concern, the hazard indexes for each component are to be used to place the chemicals in categories of high, moderate, or low concern.

SELECTING PRIORITIES FROM LARGE SETS OF ALTERNATIVES: THE CASE OF TOXIC SUBSTANCES REGULATION (WILHELM, 1981)

Although it is not explicitly stated, this scheme seems designed to rank the TSCA Inventory list of chemicals for further toxicity testing.

Seventeen scores are developed per chemical. The author argued against using a single aggregation function for these scores. Instead, he suggested nine aggregation functions, each designed for a special purpose (picking out regulatory targets, establishing testing priorities by ranking chemicals on the basis of volume and suspicion of toxicity, possible environmental problems, possible occupational problems, and suspicion of toxicity based on chemical structure). These aggregation functions are defined in terms of inequality constraints on the summary scores.

A score for exposure potential is derived from a simply calculated function of production volume. Factors for exposure potential are production volume, number of chemical-plant sites, and estimated number of workers exposed. The data are to be read, and processing performed, by computer.

Indicators of suspicion are expressed as a series of 10 scores that are reduced to three summary scores. Each score refers to the number of lines on the RTECS file on an item of interest--total number of toxic-dose lines, number of reviews (one each line), number of toxic-dose lines that deal with teratogenic, carcinogenic, and mutagenic studies, etc.

Further indicators are developed for closely related chemicals, and searches are made for toxic-element components for the chemical in question.

The summary scores appear to depend heavily on quantity of information, as contrasted with quality of information. For example, the human-toxicity score is 1 if there is one line in the RTECS file on human toxicity and 5 if there are five lines. Scoring by the number of lines in the RTECS file ignores both the nature and the quality of the published data.

In defense of this approach, it is hard to imagine schemes capable of processing the 55,000 TSCA Inventory chemicals without severe simplifications. Examining the whole list of chemicals requires the use of simple indicators that almost inevitably treat some unequal things as equal.

Because of the simple and mechanical nature of the scheme, it might be most useful as part of a larger scheme. Its role would be to scan the entire universe of chemicals and to put those most in need of testing on a series of (relatively) short lists. Each list could be augmented or reduced by other methods.

The author believed expert judgment to be essential. Experts are to make decisions from the shorter lists generated by the aggregation functions, working on the summary of scores from the entire universe of chemicals. The scheme does not describe how the experts are to perform this role.

The scheme is designed to use quantitative information. Qualifications come at the level of expert judgment, once the lists are obtained, and at the level of discussion that motivates the particular scores and summaries. These qualifications would be more convincing if the scheme were placed in the context of a larger scheme of priority-setting that explained how expert judgments were to be used and how the short lists could be augmented by other means that might compensate for possible weaknesses due to the simplifications inherent in this scheme.

The principal virtue of this scheme is its moderate use of resources. It would be useful to have some estimates of what it would cost in time, money, and personnel to implement the scheme for all 55,000 chemicals.

The scheme appears to be well designed for a narrow, but highly important, role in a larger priority-setting system.

REFERENCES

Administrative Conference of the United States. 1982. Federal regulation of cancer-causing chemicals (Recommendation No. 82-5). Fed. Reg. 47:30710-30715.

Astill, B. D., H. B. Lockhart, Jr., J. B. Moses, A. N. M. Nasr, R. L. Raleigh, and C. J. Terhaar. 1981. Sequential testing for chemical risk assessment. In R. Conway, Ed. Environmental Risk Analysis of Chemicals. Second International Congress of Toxicology, Brussels, 1978. New York: Van Nostrand-Reinhold Company. (in press)

Brown, S. L., R. L. Cofer, T. Eger, D. H. W. Liu, W. R. Mabey, K. Suttinger, and D. Tuse. 1980. A Ranking Algorithm for EEC Water Pollutants. CRESS Report No. 136. Menlo Park, Calif.: SRI International. 8 pp.

Cramer, G. M., R. A. Ford, and R. L. Hall. 1978. Estimation of toxic hazard--A decision tree approach. Food Cosmet. Toxicol. 16:255-276.

Dehn, R. L., and C. T. Helmes. 1974. An Automatic Procedure for Assessing Possible Carcinogenic Activity of Chemicals Prior to Testing. Menlo Park, Calif.: Stanford Research Institute. [128] pp.

Gori, G. B. 1977. Ranking of environmental contaminants for bioassay priority, pp. 99-111. In U. Mohr, D. Schmahl, L. Tomatis, and W. Davis, Eds. Air Pollution and Cancer in Man. IARC Scientific Publications No. 16. Lyon, France: International Agency for Research on Cancer.

Kornreich, M. R., I. C. T. Nisbet, R. Fensterheim, M. Beroza, M. Shah, D. Bradley, J. Turim, A. Pinkney, and D. Smith. 1979. Priority Setting of Toxic Substances for Guiding Monitoring Programs. Report to Office of Technology Assessment. Washington, D.C.: Clement Associates, Inc. [194] pp.

Nisbet, I. C. T. 1979. Ranking chemicals for testing: A priority-setting exercise under the Toxic Substances Control Act, pp. B-41--B-54. In Scoring Chemicals for Health and Ecological Effects Testing. TSCA-ITC Workshop. Rockville, Md.: Enviro Control, Inc.

Ross, R. H., and P. Lu. 1980. Chemical Scoring System Development. Draft report to U.S. Environmental Protection Agency, Office of Toxic Substances. Oak Ridge, Tenn.: U.S. Department of Energy, Oak Ridge National Laboratory. 121 pp.

U.S. Department of Health and Human Services. 1983. National Toxicology Program Fiscal Year 1983 Annual Plan. NTP-82-119. Washington, D.C.: U.S. Department of Health and Human Services, Public Health Service, National Toxicology Program.

U.S. Department of Health and Human Services, Food and Drug Administration, Bureau of Foods. 1982. Toxicological Principles for the Safety Assessment of Direct Food Additives and Color Additives Used in Food. Washington, D.C.: U.S. Department of Health and Human Services, Food and Drug Administration. 240 pp.

U.S. Environmental Protection Agency, Toxic Integration Information System. 1980. Chemical Selection Methods: An Annotated Bibliography. Washington, D.C.: U.S. Environmental Protection Agency.

Wilhelm, S. 1981. Selecting Priorities from Large Sets of Alternatives: The Case of Toxic Substances Regulation. Ph.D. dissertation. Providence, R.I.: Brown University, Department of Chemistry. 259 pp.

APPENDIX B

MATHEMATICAL MODELING OF THE PRIORITY-SETTING PROCESS AND RESULTING DECISION RULES

THE FORMAL MODEL

Mathematical models of the priority-setting process were constructed so that the process could be examined in a systematic way, including an examination of the following questions: Does one system have a lower misclassification cost than another? How is the system affected by changes in one or more of the design characteristics, such as prevalence rates, effectiveness of the tests, or cost of gathering information? Calculation of misclassification cost is particularly important, because it is the main criterion for selecting one priority-setting system over another (see Appendix E).

The entire collection of chemicals to be considered for priority-setting is divided into N categories; categories are defined by ranges of toxicity and exposure. For example, if we were considering three degrees of toxicity of a specific type (low, medium, and high) and three degrees of exposure (low, medium, and high), there would be nine categories in all (low exposure, low toxicity; low exposure, medium toxicity; and so on). In general, there will be several end points, and the ranges may be divided more finely than into low, medium, and high; so there will be more than nine categories in all. The main limitation on defining a large number of categories is the availability of information on exposure and toxicity. A category is denoted s_i, and the entire collection of categories is denoted as the set

$$S = \left\{ s_1, s_2, \ldots, s_N \right\}.$$

Priority-setting systems can be regarded as decision trees on which each node or decision point corresponds to a test. In this context, a test is not limited to a laboratory test, but includes other information-gathering activities and their interpretation.

A single test is denoted t_i, and the entire collection of tests available to the priority-setting process is denoted as the set

$$T = \left\{ t_1, t_2, \ldots, t_k \right\}$$

where k is the total number of tests available. Each test is associated with a set of possible results

$$R_k = \left\{ r_{k_1}, r_{k_2}, \ldots, r_{k_{m(k)}} \right\}$$

319

where r_{ki} is the ith result of test k, and there are m(k) possible results of tests k.

A path π_j through a priority-setting process is defined by a sequence of tests and test results. In the example shown in Figure B-1, path π_i is the sequence of (test, result) pairs $(t_1 r_1)$, $(t_2 r_2)$, $(t_3 r_3)$, and $(t_4 r_4)$. The collection of possible paths is denoted $\pi = \left[\pi_i\right]$.

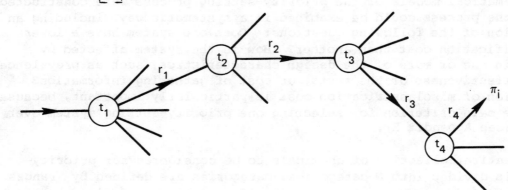

FIGURE B-1 Decision tree with example path π_i, consisting of $t_1 r_1$, $t_2 r_2$, $t_3 r_3$, and $t_4 r_4$.

The performance of a test t_k may be described by the conditional probabilities $P(r_{ki}|s_j)$, where $i = 1, \ldots, m(k)$ and $j = 1, \ldots, N$. Probability $P(r_{ki}|s_j)$ is the probability that test k will have result r_{ki} when the chemical tested belongs to category s_j. In other words, each test, t_k, is associated with an array of conditional probabilities, as shown in Table B-1.

TABLE B-1 Performance Characteristics of Test t_k Expressed as Conditional Probabilities

Test Results	Category		
	s_1	s_2	s_N
r_{k1}	$P(r_{k1}|s_1)$	$P(r_{k1}|s_2) \ldots$	$P(r_{k1}|s_N)$
r_{k2}	$P(r_{k2}|s_1)$	$P(r_{k2}|s_2) \ldots$	$P(r_{k2}|s_N)$
$\vdots$			
$r_{k,m(k)}$	$P(r_{k,m(k)}|s_1)$	$P(r_{k,m(k)}|s_2) \ldots$	$P(r_{k,m(k)}|s_N)$

The fraction of all the chemicals in the universe that belongs to category s_j is denoted $P(s_j)$. The probability that a chemical randomly drawn from the chemical universe will be of type s_j is $P(s_j)$.

The probability that a chemical chosen at random from the initial collection (the universe) will be put into the ith result category by the kth test is:

$$\sum_{j=1}^{N} P(r_{ki}|s_j)P(s_j).$$

A test t_k transforms pretest distribution $P(s_j)$ into a set of posttest distributions $P(s_j|r_{ki})$, where $i = 1, \ldots, m(k)$. The improvement in our understanding about the chemicals is reflected in changes in the posttest distributions relative to the pretest distribution. If tests are effective, the posttest distributions are changed from the pretest distribution. Repeated tests will produce a proliferation of outcomes whose distributions should become narrower and narrower as testing proceeds. Eventually, if there are many tests and each is effective, the multitude of distributions would converge until they had a zero standard deviation. Every chemical would then be assigned to its true category. However, because of budget constraints, lack of information, and limitations of existing tests, an actual priority-setting system will not achieve perfect classification.

MEASURING PERFORMANCE

Different priority-setting processes will result in different kinds of misclassification of chemicals. One requirement for comparing the performance of priority-setting processes or recommending one over another is specification of the severity of misclassification. None of the designs of the priority-setting systems reviewed by the Committee on Priority Mechanisms included a means of measuring performance and choosing among systems.

To compare priority-setting systems, a concept of misclassification cost has been developed (Appendix E). At the end of the priority-setting and testing process, each chemical is classified into some category. If a chemical is assigned to category s_i when it belongs in category s_j, a misclassification cost (distinct from a monetary cost) is incurred. The magnitude of the misclassification cost depends on how much the assigned category differs from the actual category. For example, if a highly toxic chemical with high exposure is erroneously classified as having low exposure and low toxicity, the associated misclassification cost is greater than it would have been if the chemical had been classified as having low exposure and medium toxicity. $C(\hat{s}_i|s_j)$ is the cost of classifying a chemical in category s_i when it belongs in category s_j.

In principle, the above concept requires a penalty to be specified for each type and severity of misclassification. In practice, costs can be assigned for the extreme cases, and then costs can be assigned for the intermediate cases by interpolation. Also in principle, one type of true classification--$C(\hat{s}_i|s_i)$--might be treated as more important than another type of true classification--$C(\hat{s}_j|s_j)$--and different costs assigned to each. However, sufficient information to make use of this refinement appears not to be available, and in the current model all true classifications are treated as incurring zero cost.

BUDGET COST

A second requirement for comparability is definition of the budget cost of a priority-setting process. As stated above, a priority-setting process is characterized by a set of paths $\{\pi_i; \text{ where } i = 1, M^\pi\}$, where M^π is the number of paths. If the priority-setting scheme is regarded as one big test, the paths can be viewed as defining the M^π test outcomes. As in the case of a simple test, the priority-setting scheme is characterized by the conditional probabilities $P(\pi_i|s_j)$.

If path π_i is the sequence (t_1r_1), (t_2r_2), . . ., (t_nr_n) of tests and results, then

$$P(t_1r_1|s_j)\ P(t_2r_2|s_j. \ . \ . \ P(t_nr_n|sj)$$

The dollar testing cost, C_{π_i}, associated with path π_i is simply the sum of the costs of the tests along the path, that is:

$$C_{\pi_i} = C_{t_1} + C_{t_2} + \ . \ . \ . + C_{t_n}$$

where C_k is the cost of test t_k.

Rather than a single testing cost, the priority-setting scheme has a vector of expected costs, E_j^π, where $j = 1, \ . \ . \ . ., N$, because chemicals of different types follow the various paths with different probabilities:

$$E_j^\pi = \sum_{i=i}^{M^\pi} P(\pi_i|s_j)C_{\pi_i} = \sum_{j=1}^{M^\pi} P(\pi_i|s_j)\ (C_{t_1} + C_{t_2} + \ . \ . \ . + C_{t_n}).$$

The total expected cost (TCT^π) of process π per chemical is

$$TCT^\pi = \sum_j^N P(s_j)E_j^\pi$$

MINIMIZING MISCLASSIFICATION COST

For an initial probability distribution $P(s_j)$, the probabilities associated with outcome categories of a process are given by $P(\pi_i|s_j)$, where $i = 1, \ldots, M^\pi$ and $j = 1, \ldots, N^\pi$.

The best final classification of chemicals in the outcome category is that which minimizes expected misclassification cost. The minimal misclassification cost for path π_i is

$$CM_i^\pi = \underset{h=1\ldots N}{Min} \ \sum_j \ P(\pi_i|s_j)P(s_j)C(\hat{s}_h|s_j).$$

and h is chosen to minimize the cost.

The total expected misclassification cost for a priority-setting process is

$$TCM^\pi = \sum_{i,j} \ CM_i^\pi P(\pi_i|s_j)P(sj)$$

per chemical.

The foregoing shows how a priority-setting process acting on an initial collection of chemicals described by a probability distribution $P(s_j)$ incurs a misclassification cost TCM^π and a testing cost TCT^π.

The optimal priority-setting process minimizes TCM^π while satisfying the budget constraint $TCT^\pi \leq B$, where B is the budget per chemical.

THE PILOT MODEL

In designing a priority-setting process, many interacting factors must be considered. To sharpen the committee's judgment as to the interaction of these factors, a small-scale or pilot version of the formal model was quantified to permit sensitivity analysis. The main value of the pilot model is in providing a systematic framework for the many judgments on these factors and in making it possible to sort out important implications of the interactions.

The model is a bookkeeping device; it does not tell us what numbers should be entered on the books--that is a matter of judgment for toxicologists, epidemiologists, chemists, chemical engineers, and other scientists with empirical knowledge about exposure, toxicity, availability of data, and properties of tests. Nor does the model select the best priority-setting scheme--that selection is based on judgment of the predictive value of data elements and tests (and other judgments).

In setting priorities for testing chemicals, many judgments must be made. The pilot model reduces the number of judgments required to approximately 25 estimated probabilities related to three data elements for Stage 1, six data elements in Stage 2, and two data elements in Stage 3 (one dossier on exposure and one on toxicity). One short-term test (the Ames Salmonella/microsome test) and a bioassay could be recommended at the end of the priority-setting process. In the model's present form, the only human health effect being considered is cancer. On a mainframe computer, the pilot model takes about 1 min to run.

Questions that can be investigated with the aid of the quantified model include the following:

● What allocation of expenditures between gathering exposure data and gathering toxicity data minimizes costs of each stage of the process? For example, what proportions of the resources should be devoted to writing exposure and toxicity sections of a dossier?

● What effort is sufficient for writing a dossier? More generally, what amount of data is sufficient to gather in each stage of the priority-setting process?

● How many chemicals should be winnowed out of the process in each stage?

● What is the best allocation of resources between short-term tests and long-term tests? What allocation among the many possible candidate short-term tests minimizes testing costs?

Empirical investigation of these questions requires empirical judgments on the factors that are included in the pilot model. Some of these factors have already been estimated, and some are being assessed. The pilot model is available in the NAS open file.

COMPARISON WITH WEINSTEIN'S MODEL

The formal model decribed above and quantified in the pilot model is similar to the one developed by Weinstein (1979), in that both embody the value-of-information concept and the minimization of expected costs. Four principal differences between Weinstein's approach and the approach used by the Committee on Priority Mechanisms should be pointed out.

● The committee's approach focused on the design of a priority-setting process, whereas Weinstein stressed the application of a given process to a list of chemicals. In Weinstein's model, priorities are based on minimizing cost for a given priority-setting process.

• Weinstein's model does not consider the cost of information acquisition incurred during the gathering of data elements and the performance of toxicity tests. For the committee's task, these costs were important design considerations; thus, they were included in the model. This difference has important implications for the design of the priority-setting process.

• The costs of misclassification were defined by Weinstein as the costs of expected final actions. This was in accord with the decision-theory approach developed by Raiffa (1968); however, it requires two conditions that do not apply to the committee's task of designing a priority-setting system. First, it must be possible to estimate, at low cost and with some accuracy, the costs of likely regulatory, market, court, and individual actions expected to result from each of the possible outcomes of the priority-setting process. Second, one must assume that regulatory and other action will be based on the minimization of expected cost.

• Weinstein's model was not developed to the point of being operational. The committee attempted to take the next step by doing two things. First, it expressed quantitatively the empirical information required to design the system. Second, it wrote a program for the pilot model to enable sensitivity analysis and further exploration of the practical limitations of the modeling approach.

RULES FOR CHOOSING CHEMICALS GENERATED BY THE MODEL

This example is designed to illustrate the rules produced by the optimization model for selecting chemicals. The example is illustrative, in that it depends on the accuracy of the estimates of the characteristics of the population of chemicals and estimates of the accuracy of the tests and on the results of the model changes as these estimates change. This model is for cancer testing, but it appears that the model can be used to design a priority-setting system for other health effects or groups of effects. A computer program of the model is available from the National Academy of Sciences.

Data elements used in the model have been developed by performing the following steps: choose information; define a small number of mutually exclusive outcomes that include all chemicals considered; arrange the outcomes in order of estimated degree of exposure or toxicity; estimate the degree of probability of given degrees of toxicity of or exposure to chemicals with each outcome.

Three data elements are used in Stage 1--one for exposure and two for toxicity.

The exposure data element is based on intended use and production. The 15 possible outcomes (the product of 5 use categories and 3 production ranges) have been grouped into 13 mutually exclusive outcomes (see Table 3).

The first toxicity data element is a crude application of structure-activity relationships based on membership in a chemical group associated with a human health effect (see Table 5). A nonstructurable chemical is considered to be slightly less likely than one randomly drawn from the universe to be a carcinogen. A structurable chemical not a member of a chemical group associated with a health effect is considered to be least likely to be a carcinogen. A chemical that is a member of a chemical group associated with some health effect other than cancer is considered slightly less likely than a randomly drawn chemical to be a carcinogen. Chemical groups associated with cancer are divided into groups of low, medium, and high suspicion.

The second toxicity data element used in Stage 1 is listing in RTECS. The outcomes and the estimated probabilities associated with them are shown in Table 7. A chemical not listed in RTECS at all is considered to be slightly less likely than a chemical randomly drawn from the universe to be a carcinogen. If a chemical is listed in RTECS, but without mention of CAR or MUT, it is considered a little more suspect, but still not as suspect as a randomly drawn chemical. If MUT is mentioned, but not CAR, suspicion increases substantially; if CAR is also mentioned, suspicion increases still more.

The data elements for Stages 2 and 3 are less precisely defined, and there is less information on which to base an estimate of accuracy than in the case of data elements for Stages 1 and 4. The reasoning used to estimate the accuracy for data elements in Stages 2 and 3 is as follows. The cost of the data elements is assumed to be about $30-300 per chemical for Stage 2 and $1,000-4,000 for Stage 3 (otherwise, the budget constraint chosen for the model cannot be met if thousands of chemicals are considered in Stage 2 and hundreds in Stage 3). We assume that accuracy of Stages 2 and 3 will be intermediate between that of Stages 1 and 4. In terms of increasing accuracy, we can probably arrange the stages in the order 1, 2, 3, and 4. The procedure for estimating the accuracy of Stages 2 and 3 differs from that for Stages 1 and 4. For the latter two, expert judgment and quantitative information were used. For Stages 2 and 3, a sensitivity analysis determined the accuracy compatible with data searches costing $30-300 per chemical in Stage 2 and with writing dossiers costing $1,000-4,000 in Stage 3. (If the accuracy of Stages 2 and 3 is too high, the model of the priority-setting system will not specify short-term or long-term testing. If the accuracy is too low, the model will not specify using Stages 2 and 3, but will specify going directly from Stage 1 to testing in Stage 4.) The model is used to estimate how accurate Stages 2 and 3 must be for them to be used moderately. Once the priority-setting system is operating, data may be obtained to validate the estimated accuracy of all four stages.

The data elements in Stage 4 are a short-term test and a long-term test. The short-term test is a battery consisting of an Ames Salmonella/microsome test and a cell-transformation test, which is assumed to be slightly more accurate than the Salmonella/microsome test alone. The test accuracy is based on data reported by McCann and Ames, Purchase et al., and Lave et al. The long-term test is a rodent bioassay. Its accuracy has been analyzed by Tarone, Fears, and Chu, among others. The long-term test is expected to identify 95% of the noncarcinogens correctly as negative and 93% of carcinogens correctly as positive.

There are 6 possible outcomes of the exposure data element, 6 possible outcomes of the chemical-group toxicity data element, and 4 possible outcomes of the RTECS toxicity data element. Thus, there are (6 x 6 x 4 =) 144 possible outcomes of the combined data elements in Stage 1. Similar outcomes were combined to simplify calculations. Outcomes are "similar" if they lead to similar estimates of the 9 possible degrees of public-health concern (each formed from one of the 3 degrees of toxicity and one of the 3 degrees of exposure). After examination of the estimated means and standard deviation of the estimated probability distributions, the 144 possible outcomes of Stage 1 were grouped into 9 outcome categories; each of these 9 "branches" leads from Stage 1 to the later stages. Branch 1 has the lowest standard deviation and the lowest mean; Branch 9 has the highest standard deviation and the highest mean.

Most of the chemicals use Branch 1, because it has a low mean for public-health concern and most chemicals are considered to have low exposure and low toxicity. Chemicals in Branch 1 are removed from further consideration. Dossiers are not written in Stage 2; however, a limited search for exposure data is provided for in Stage 2.

Fewer chemicals use Branches 2-9. These branches have higher means and higher standard deviations for public-health concern. Note that Branch 2 calls for the larger search for exposure data and the smaller search for toxicity data in Stage 2. Branch 2 provides for dossiers, except when the results of the search for toxicity data suggest low toxicity and the results of the search for exposure data suggest low or medium exposure.

Branches 3-9 all provide for dossiers (except Branch 3 when results of Stage 2 data searches suggest both low exposure and low toxicity), and Branches 7, 8, and 9 often lead to recommendations for a long-term test, depending on the results of the Stage 2 data searches and the assessments based on the dossiers.

On the basis of the assumption that 95% of the chemicals considered by the priority-setting system are noncarcinogens, 4% are moderate carcinogens, and 1% are highly potent carcinogens and on the basis of the estimated accuracy of the three selection stages (Stages 1-3) and the tests in Stage 4, the selection rules are expected to lead to recommendations for 62 Stage 4 long-term tests, 690 Stage 4 short-term tests, and 417 Stage 3 dossiers for each 10,000 chemicals.

To demonstrate how the selection rules are used, we assume that a chemical is a pesticide with unknown production, that it is a member of a chemical group associated with a low suspicion of cancer, and that it is listed in RTECS with no mention of CAR or MUT. We go to the breakdown for the Stage 1 exposure data elements "Intended Use and Production"; the fourth entry includes pesticides with unknown production. That entry leads us to Stage 1 toxicity data element Part 4, where we seek a member of a chemical group with a low suspicion of cancer (the fourth entry). There, we find "in RTECS, no mention of MUT or CAR"; that leads to Branch 2 and completes Stage 1. Via Branch 2, we are led to Stage 2 information-gathering activities consisting of a larger search for exposure information and a limited search for toxicity information. Assume that analysis of the information gathered in Stage 2 leads to the assessment that the chemical has medium exposure and is highly toxic. This leads us to prepare a dossier in Stage 3. If an expert committee reviews the dossier and concurs in the assessment of high toxicity, then the decision rules recommend a long-term test in Stage 4.

SELECTION RULES FOR AN ILLUSTRATIVE
PRIORITY-SETTING SYSTEM FOR CANCER

STAGE 1: EXPOSURE DATA ELEMENT

	Go to Stage 1 Toxicity Data Elements Part
Intended Use and Production (based on Table 3)	
General commerce (TSCA), other, or unclassified with production less than 10^4 lb/yr	1
Cosmetic with production less than 10^4 lb/yr; or general commerce, other, or unclassified with production between 10^4 and 10^6 lb/yr	2
Drug, food chemical, pesticide or unknown with production less than 10^4 lb/yr; or cosmetic, general commerce, other, or unclassified with unknown production	3
Food chemical, drug, or pesticide with unknown production	4
General commerce, other, or unclassified with production between 10^6 and 10^8 lb/yr; or cosmetic with production between 10^4 and 10^5 lb/yr	5
Drug, food chemical, or pesticide with production equal to or greater than 10^4 lb/yr; cosmetic with production greater than 10^5 lb/yr; or general commerce or other with production greater than 10^8 lb/yr	6

STAGE 1:   TOXICITY DATA ELEMENTS

Part 1

Go to Stages 2-4
via Branch

Not a member of a chemical group (Table 5)

    not in RTECS   1

    in RTECS, no mention of MUT or CAR.............   1

    in RTECS, mention of MUT, but not CAR.........   2

    in RTECS, mention of CAR......................   2

Member of a chemical group, but not one
associated with cancer

    not in RTECS..................................   1

    in RTECS, no mention of MUT or CAR............   1

    in RTECS, mention of MUT, but not CAR.........   2

    in RTECS, mention of CAR......................   2

Not structurable

    not in RTECS..................................   1

    in RTECS, no mention of MUT or CAR............   1

    in RTECS, mention of MUT, but not CAR.........   2

    in RTECS, mention of CAR......................   2

Member of a chemical group associated with cancer
with "low suspicion"

    not in RTECS..................................   1

    in RTECS, no mention of MUT or CAR............   1

    in RTECS, mention of MUT, but not CAR   4

    in RTECS, mention of CAR.....................   4

Member of a chemical group associated with cancer
with "moderate suspicion"

    not in RTECS..................................   1

    in RTECS, no mention of MUT or CAR.........   2

    in RTECS, mention of MUT, but not CAR......   4

    in RTECS, mention of CAR...................   4

Member of a chemical group associated with cancer
with "high suspicion"

    not in RTECS..................................   1

    in RTECS, no mention of MUT or CAR..........   2

    in RTECS, mention of MUT, but not CAR.......   4

    in RTECS, mention of CAR...................   4

Part 2

Go to Stages 2-4
via Branch

Not a member of a chemical group (Table 5)

 not in RTECS 1

 in RTECS, no mention of MUT or CAR........... 1

 in RTECS, mention of MUT, but not CAR........ 2

 in RTECS, mention of CAR..................... 2

Member of a chemical group, but not one
associated with cancer

 not in RTECS................................. 1

 in RTECS, no mention of MUT or CAR........... 1

 in RTECS, mention of MUT, but not CAR........ 2

 in RTECS, mention of CAR..................... 2

Not structurable

 not in RTECS................................. 1

 in RTECS, no mention of MUT or CAR........... 1

 in RTECS, mention of MUT, but not CAR........ 3

 in RTECS, mention of CAR..................... 3

Member of a chemical group associated with cancer
with "low suspicion"

 not in RTECS................................. 1

 in RTECS, no mention of MUT or CAR.......... 1

 in RTECS, mention of MUT, but not CAR 4

 in RTECS, mention of CAR.................... 4

Member of a chemical group associated with cancer
with "moderate suspicion"

 not in RTECS................................. 1

 in RTECS, no mention of MUT or CAR.......... 2

 in RTECS, mention of MUT, but not CAR....... 6

 in RTECS, mention of CAR.................... 4

Member of a chemical group associated with cancer
with "high suspicion"

 not in RTECS................................. 1

 in RTECS, no mention of MUT or CAR.......... 2

 in RTECS, mention of MUT, but not CAR....... 6

 in RTECS, mention of CAR.................... 4

Part 3

Not a member of a chemical group (Table 5)
    not in RTECS 1
    in RTECS, no mention of MUT or CAR.......... 1
    in RTECS, mention of MUT, but not CAR....... 2
    in RTECS, mention of CAR.................... 2

Member of a chemical group, but not one
associated with cancer
    not in RTECS................................ 1
    in RTECS, no mention of MUT or CAR.......... 1
    in RTECS, mention of MUT, but not CAR....... 2
    in RTECS, mention of CAR.................... 2

Not structurable
    not in RTECS................................ 1
    in RTECS, no mention of MUT or CAR.......... 1
    in RTECS, mention of MUT, but not CAR....... 4
    in RTECS, mention of CAR.................... 2

Member of a chemical group associated with cancer
with "low suspicion"
    not in RTECS................................ 1
    in RTECS, no mention of MUT or CAR.......... 1
    in RTECS, mention of MUT, but not CAR 6
    in RTECS, mention of CAR.................... 4

Member of a chemical group associated with cancer
with "moderate suspicion"
    not in RTECS................................ 1
    in RTECS, no mention of MUT or CAR.......... 2
    in RTECS, mention of MUT, but not CAR....... 6
    in RTECS, mention of CAR.................... 4

Member of a chemical group associated with cancer
with "high suspicion"
    not in RTECS................................ 2
    in RTECS, no mention of MUT or CAR.......... 2
    in RTECS, mention of MUT, but not CAR....... 6
    in RTECS, mention of CAR.................... 5

Part 4

Go to Stages 2-4
via Branch

Not a member of a chemical group (Table 5)
        not in RTECS 1
        in RTECS, no mention of MUT or CAR.......... 1
        in RTECS, mention of MUT, but not CAR....... 3
        in RTECS, mention of CAR.................... 2

Member of a chemical group, but not one
associated with cancer
        not in RTECS............................... 1
        in RTECS, no mention of MUT or CAR.......... 1
        in RTECS, mention of MUT, but not CAR....... 4
        in RTECS, mention of CAR.................... 2

Not structurable
        not in RTECS............................... 1
        in RTECS, no mention of MUT or CAR.......... 1
        in RTECS, mention of MUT, but not CAR....... 4
        in RTECS, mention of CAR.................... 4

Member of a chemical group associated with cancer
with "low suspicion"
        not in RTECS............................... 1
        in RTECS, no mention of MUT or CAR.......... 2
        in RTECS, mention of MUT, but not CAR 6
        in RTECS, mention of CAR.................... 6

Member of a chemical group associated with cancer
with "moderate suspicion"
        not in RTECS............................... 1
        in RTECS, no mention of MUT or CAR.......... 2
        in RTECS, mention of MUT, but not CAR....... 6
        in RTECS, mention of CAR.................... 6

Member of a chemical group associated with cancer
with "high suspicion"
        not in RTECS............................... 1
        in RTECS, no mention of MUT or CAR.......... 2
        in RTECS, mention of MUT, but not CAR....... 7
        in RTECS, mention of CAR.................... 6

Part 5

Not a member of a chemical group (Table 5)
     not in RTECS 1
     in RTECS, no mention of MUT or CAR.......... 1
     in RTECS, mention of MUT, but not CAR....... 4
     in RTECS, mention of CAR.................... 3

Member of a chemical group, but not one
associated with cancer
     not in RTECS................................ 1
     in RTECS, no mention of MUT or CAR.......... 1
     in RTECS, mention of MUT, but not CAR....... 4
     in RTECS, mention of CAR.................... 4

Not structurable
     not in RTECS................................ 1
     in RTECS, no mention of MUT or CAR.......... 1
     in RTECS, mention of MUT, but not CAR....... 4
     in RTECS, mention of CAR.................... 4

Member of a chemical group associated with cancer
with "low suspicion"
     not in RTECS................................ 1
     in RTECS, no mention of MUT or CAR.......... 2
     in RTECS, mention of MUT, but not CAR 8
     in RTECS, mention of CAR.................... 6

Member of a chemical group associated with cancer
with "moderate suspicion"
     not in RTECS................................ 1
     in RTECS, no mention of MUT or CAR.......... 2
     in RTECS, mention of MUT, but not CAR....... 8
     in RTECS, mention of CAR.................... 7

Member of a chemical group associated with cancer
with "high suspicion"
     not in RTECS................................ 1
     in RTECS, no mention of MUT or CAR.......... 2
     in RTECS, mention of MUT, but not CAR....... 8
     in RTECS, mention of CAR.................... 8

Part 6

Go to Stages 2-4
via Branch

Not a member of a chemical group (Table 5)
     not in RTECS 1
     in RTECS, no mention of MUT or CAR........... 1
     in RTECS, mention of MUT, but not CAR........ 4
     in RTECS, mention of CAR..................... 4

Member of a chemical group, but not one
associated with cancer
     not in RTECS.................................. 1
     in RTECS, no mention of MUT or CAR........... 1
     in RTECS, mention of MUT, but not CAR........ 4
     in RTECS, mention of CAR..................... 4

Not structurable
     not in RTECS.................................. 1
     in RTECS, no mention of MUT or CAR........... 1
     in RTECS, mention of MUT, but not CAR........ 6
     in RTECS, mention of CAR..................... 4

Member of a chemical group associated with cancer
with "low suspicion"
     not in RTECS.................................. 1
     in RTECS, no mention of MUT or CAR........... 2
     in RTECS, mention of MUT, but not CAR 9
     in RTECS, mention of CAR..................... 8

Member of a chemical group associated with cancer
with "moderate suspicion"
     not in RTECS.................................. 2
     in RTECS, no mention of MUT or CAR.......... 3
     in RTECS, mention of MUT, but not CAR....... 9
     in RTECS, mention of CAR.................... 9

Member of a chemical group associated with cancer
with "high suspicion"
     not in RTECS.................................. 2
     in RTECS, no mention of MUT or CAR.......... 4
     in RTECS, mention of MUT, but not CAR....... 9
     in RTECS, mention of CAR.................... 9

STAGES 2-4

Branch 1

Stage 2: Perform a limited search for exposure data. Do not consider
          chemical further.

Branch 2

Stage 2: Perform search for exposure data and limited search for
          toxicity data.

Assessments Based on Data Gathered in Stage 2		Stage 3:	Assessments Based on Data Gathered	Stage 4:
Exposure	Toxicity	Data Gathering	in Stage 3	Recommended Testing[a]
Low	Low	No dossier	None	ST
Low	Medium	Dossier	None	ST
Low	High	Dossier	High toxicity	LT
			Otherwise	ST
Medium	Low	No dossier	None	ST
Medium	Medium or high	Dossier	High toxicity and high exposure	LT
			Otherwise	ST
High	Low, medium, or high	Dossier	High exposure	LT

----

[a] ST = short-term test; LT = long-term test.

336

Branch 3

Stage 2: Perform search for exposure data and limited search for toxicity data.

Assessments Based on Data Gathered in Stage 2		Stage 3: Data Gathering	Assessments Based on Data Gathered in Stage 3	Stage 4: Recommended Testing[a]
Exposure	Toxicity			
Low	Low	No dossier	High toxicity and high exposure	LT
			Otherwise	ST
Low	Medium	Dossier	High toxicity	LT
			Otherwise	ST
Low	High	Dossier	High toxicity; or high exposure and medium toxicity	LT
			Otherwise	ST
Medium	Low	Dossier	High toxicity; or high exposure and medium toxicity	LT
			Otherwise	ST
Medium	Medium	Dossier	High exposure or high toxicity	LT
			Otherwise	ST
Medium	High	Dossier	High exposure or high toxicity; or medium exposure and high toxicity	LT
			Otherwise	ST

High	Low	Dossier	High exposure; or medium exposure and high toxicity	LT
			Otherwise	ST
High	Medium	Dossier	High exposure or high toxicity; or medium exposure and medium toxicity	LT
			Otherwise	ST
High	High	Dossier	Low exposure and low toxicity	ST
			Anything but low exposure and low toxicity	LT

_____

a ST = short-term test; LT = long-term test.

Branch 4

Stage 2: Perform search for exposure data and limited search for toxicity data.

Assessments Based on Data Gathered in Stage 2		Stage 3: Data Gathering	Assessments Based on Data Gathered in Stage 3	Stage 4: Recommended Testing[a]
Exposure	Toxicity			
Low	Low	Dossier	High toxicity and low or medium exposure	LT
			Medium exposure and low or medium toxicity	ST
			Low exposure and low or medium toxicity	None
Low	Medium	Dossier	High toxicity	LT
			Medium toxicity; or low toxicity and medium exposure	ST
			Low toxicity and low exposure	None
Low	High	Dossier	High toxicity	LT
			Otherwise	ST
Medium	Low	Dossier	High toxicity	LT
			Medium toxicity	ST
			Low toxicity and low exposure	None
Medium	Medium or high	Dossier	High toxicity or high exposure	LT
			Otherwise	ST
High	Low, medium, or high	Dossier	High toxicity or high exposure	LT
			Otherwise	ST

---

[a]ST = short-term test; LT = long-term test.

Branch 5

Stage 2: Perform search for exposure data and limited search for toxicity data.

| Assessments Based on Data Gathered in Stage 2 | | Stage 3: | Assessments Based on Data Gathered | Stage 4: Recommended |
Exposure	Toxicity	Data Gathering	in Stage 3	Testing[a]
Low	Low	Dossier	High toxicity and low or medium exposure	LT
			Medium toxicity and low or medium exposure	ST
			Low exposure and low toxicity	None
Low	Medium	Dossier	High toxicity and low or medium exposure	LT
			Otherwise	ST
Low	High	Dossier	High toxicity; or medium toxicity and medium or high exposure	LT
			Otherwise	ST
Medium	Low	Dossier	High toxicity and low or high exposure	LT
			Otherwise	ST
Medium	Medium	Dossier	High toxicity and low or high exposure	LT
			Otherwise	ST
Medium	High	Dossier	High exposure or high toxicity; or medium exposure and medium toxicity	LT
			Otherwise	ST

Branch 5 (continued)

High	Low	Dossier	High or medium exposure or high toxicity	LT
			Otherwise	ST
High	Medium	Dossier	High exposure or high toxicity; or medium toxicity and low or medium exposure	LT
			Otherwise	ST
High	High	Dossier	High or medium toxicity	LT
			Otherwise	ST

---

[a] ST = short-term test; LT = long-term test.

Branch 6

Stage 2:   Perform search for exposure data and limited search for toxicity data.

Assessments Based on Data Gathered in Stage 2		Stage 3: Data Gathering	Assessments Based on Data Gathered in Stage 3	Stage 4: Recommended Testing[a]
Exposure	Toxicity			
Low	Low, medium, or high	Dossier	High toxicity	LT
			Otherwise	ST
Medium	Low, medium, or high	Dossier	High exposure; or high toxicity and low exposure	LT
			Otherwise	ST
High	Low	Dossier	High toxicity and medium or high exposure; or low toxicity and high exposure	LT
			Otherwise	ST
High	Medium	Dossier	High toxicity; or high exposure and low toxicity	LT
			Otherwise	ST
High	High	Dossier	High toxicity and medium or high exposure; or medium toxicity and medium exposure	LT
			Otherwise	ST

---

[a] ST = short-term test; LT = long-term test.

342

Branch 7

Stage 2: Perform search for exposure data and limited search for toxicity data.

Assessments Based on Data Gathered in Stage 2		Stage 3: Data Gathering	Assessments Based on Data Gathered in Stage 3	Stage 4: Recommended Testing[a]
Exposure	Toxicity			
Low	Low	Dossier	High toxicity	LT
			Otherwise	ST
Low	Medium	Dossier	High toxicity or high exposure	LT
			Otherwise	ST
Low	High	Dossier	High toxicity or high exposure	LT
			Otherwise	ST
Medium	Low	Dossier	High exposure; or high toxicity and low exposure	LT
			Otherwise	ST
Medium	Medium	Dossier	High toxicity or high exposure	LT
			Otherwise	ST
Medium	High	Dossier	Low or high exposure and high toxicity; or medium exposure and medium or high toxicity	LT
			Medium exposure and low toxicity	ST
High	High, medium, or low	Dossier	Low or high exposure and high toxicity; or medium exposure and medium or high toxicity	LT
			Medium exposure and low toxicity	ST

[a] ST = short-term test; LT = long-term test.

343

Branch 8

Stage 2: Perform search for exposure data and limited search for toxicity data.

Assessments Based on Data Gathered in Stage 2		Stage 3:	Assessments Based on Data Gathered	Stage 4: Recommended
Exposure	Toxicity	Data Gathering	in Stage 3	Testing[a]
Low	Low	Dossier	High exposure; or high toxicity and low exposure	LT
			Otherwise	ST
Low or medium	Medium	Dossier	High exposure or high toxicity	LT
			Otherwise	ST
Low	High	Dossier	High toxicity; or high exposure; or medium toxicity and medium exposure	LT
			Otherwise	ST
Medium	Low	Dossier	High exposure; or medium exposure and high toxicity	LT
			Otherwise	ST
Medium	High	Dossier	Medium or high toxicity or high exposure	LT
			Otherwise	ST
High	Low	Dossier	Medium or high toxicity or high exposure	LT
			Otherwise	ST
High	Medium	Dossier	Medium or high toxicity; or low toxicity and medium or high exposure	LT
			Low toxicity and low exposure	ST
High	High	Dossier	All assessments	LT

_____
[a] ST = short-term test; LT = long-term test.

344

Branch 9

Stage 2: Perform search for exposure data and limited search for toxicity data.

Assessments Based on Data Gathered in Stage 2		Stage 3: Data Gathering	Assessments Based on Data Gathered in Stage 3	Stage 4: Recommended Testing[a]
Exposure	Toxicity			
Low	Low	Dossier	High exposure; or high toxicity and medium exposure	LT
			Otherwise	ST
Low	Medium	Dossier	High toxicity or high exposure; or medium toxicity and medium exposure	LT
			Otherwise	ST
Low	High	Dossier	High or medium toxicity; or high exposure	LT
			Low toxicity and medium or low exposure	ST
Medium	Low	Dossier	High toxicity or high exposure	LT
			Otherwise	ST
Medium	Medium	Dossier	High or medium exposure or high toxicity	LT
			Otherwise	ST
Medium	High	Dossier	Low toxicity and low exposure	ST
			Otherwise	LT

Branch 9 (continued)

High	Low	Dossier	Low or medium exposure; or high exposure and high toxicity	LT
			Otherwise	ST
High	Medium	Dossier	Low toxicity and low exposure	ST
			Otherwise	LT
High	High	Dossier	All assessments	LT

[a] ST = short-term test; LT = long-term test.

REFERENCES

Raiffa, H. 1968. Decision Analysis: Introductory Lectures on Choices under Uncertainty. Reading, Mass.: Addison-Wesley. 309 pp.

Weinstein, M.C. 1979. Decision making for toxic substances control: Cost-effective information for the control of environmental carcinogens. Public Policy 27:333-383.

APPENDIX C

EXPERT JUDGMENT AND THE TREATMENT OF UNCERTAINTY

Priority-setting systems inescapably involve uncertainty and the exercise of judgment. The data on which such systems operate--the toxic properties of candidate chemicals and the circumstances in which humans are exposed to them--can be known only within a range of uncertainty. A method is needed for taking this uncertainty into account and for registering how the acquisition of more knowledge will affect it.

Even with perfect knowledge of the toxicity of all chemicals, it would be necessary to rank the various forms of toxicity by reference to agreed standards of severity, which are based on judgments. Scientific comparisons of forms of toxicity are, however, only part of an overall evaluation of the impact of these toxicities on society. The potential severity of various health effects may be greatly increased or reduced by factors that affect the rate, characteristics, and consequences of exposure and actual incidence. These factors include the genetic and demographic characteristics of populations; patterns of housing and transportation; social organization, in general and in relation to work; the organization and adequacy of public-health measures, health care, and social welfare (both private--including familial--and public); and the amenability of these and other relevant factors to changes made to reduce or mitigate exposure, incidence, and their consequences. Thus, the valuation of the social impacts of toxicities requires a broad array of judgments.

Finally, the assignment of priorities for testing is, in itself, an allocation of scarce resources that may materially affect the interests of social groups competing for those resources. Such allocations involve issues of social justice and, in a democracy, politics. Their resolution requires additional kinds of judgment concerning matters of procedure, as well as substance.

In its first report, the Committee on Priority Mechanisms recommended that the priority-setting process to be developed for NTP be explicit about the use of uncertainty and judgment, provide a rationale and means for taking them into account, and address their use in a deliberate manner that is documented and articulated, but also preserve a record for later review and evaluation (National Research Council, 1981).

PROBABILITY JUDGMENTS

Existing priority-setting systems fail to offer a persuasive treatment of the uncertainty with which the health hazards of chemicals are evaluated (National Research Council, 1981). Often, information is summarized in a single number representing a best guess at toxicity or exposure. However accurate such point estimates may be, they are relatively uninformative in the setting of priorities in the face of two key questions: How much uncertainty surrounds the estimate? What are the opportunities for reducing that uncertainty through testing? Moreover, existing schemes do not characterize the potential of the tests for reducing uncertainty. Without that characterization, one has no basis for arguing that a particular test offers the best value for money spent--nor can the policy-makers who must rely on the results of testing know the likely false-negative and false-positive rates for the data at which they are looking.

Where uncertainty is considered, the treatment is often unsatisfactory. For example, in one system examined by this committee, default values for missing data are combined with estimates, without attention to the differing degrees of uncertainty about each. In another system, "strong" evidence is scored with positive numbers and "weak" evidence with negative numbers, without explanation of how the two types of information should be treated in the priority-setting process. Such treatment of uncertainty not only obscures the logic of a scheme, but also precludes any systematic test of its predictive accuracy.

The committee has attempted to develop a scheme that will deal with uncertainty more adequately. In its current configuration, the scheme requires experts in exposure and toxicity to judge a number of probabilities for simple events, such as a finding that a chemical selected from a particular group will prove carcinogenic or a finding that a chemical that appeared carcinogenic in a particular test is actually not carcinogenic. These judgments can be regarded as estimates of prevalence and of false-positive rates.

This innovation makes explicit some judgments that scientists ordinarily make implicitly when selecting tests. This allows the proposed scheme to provide a number of attractive features. When supplemented with even crude estimates of the costs of tests and of different kinds of errors, these probability judgments provide the basis of value-of-information analyses, which allow one to identify in a logically defensible manner which chemicals to test and which tests to use. Although the identification of chemicals and tests depends on probability judgments, as well as cost estimates, the values used are open to public scrutiny, and objective analysis of the process is facilitated.

The use of a mathematical model enables investigators to analyze the sensitivity of the data elements used and of the values attached to them. When supplemented with test data, probability judgments can be made explicit, thereby enabling investigators to evaluate the performance of the entire scheme.

Probability judgments are constrained primarily by the requirement that the "judges" have substantive expertise, but not necessarily in making probability assessments per se. It must be possible to explain probability assessment procedures to these experts simply, effectively, and unintrusively.

USE OF PROBABILITIES IN DECISION-MAKING

The studies of Ramsay, von Neumann and Morgenstern, Wald, and others have provoked rapid development in the theory of decision-making over the last 30 years (see reviews by Howard, 1975; Mishan, 1976). Their work has shown that coherent decision-making methods for situations involving uncertainty require an explicit treatment of the uncertainty. Failing to provide such treatment leads to schemes that are confusing, in that untested and often false values are implicitly assigned to uncertainties (Fischhoff et al., 1981).

On a practical level, decision theorists have developed a sophisticated repertoire of frameworks for characterizing decision-making situations and aids for resolving them. Decision analysis, value-of-information analysis, probabilistic risk assessment, and probabilistic information-processing systems are some of the procedures that have been developed. Where appropriate, each procedure involves the use of judgmentally assessed probabilities.

Probably the greatest successes with these procedures have been achieved in management and business administration. Other applications of explicit probability judgments include meteorology (Murphy, 1973), atmospheric-pollution studies (Moreau, 1980), and nuclear-power projects (U.S. Nuclear Regulatory Commission, 1975). These examples indicate that others in both government and business have found judgmental probabilities to be useful and practical. Fields in which such probabilities are used in tasks akin to value-of-information analysis include petroleum geology and clinical diagnosis.

One of the earliest practical uses of decision analysis was in the allocation of resources for oil exploration. Reliance was placed on expert geologists to assess the probabilities of various outcomes of different possible test drills (Raiffa, 1968). To improve the allocation of scarce medical resources, several projects have been undertaken to study the ability of physicians to predict the outcomes and value of various diagnostic procedures. In radiology, for example, such projects have resulted in recommendations that some procedures be discontinued (Christensen-Szalanski et al., 1982).

ELICITATION OF PROBABILITY JUDGMENTS

As the preceding applications suggest, it has proved possible to obtain explicit probability judgments from a wide variety of persons dealing with diverse topics. It is reasonable to ask the price of these demonstrations and what profits were realized.

349

The evidence from both laboratory experiments and field applications (Lichtenstein et al., 1982) indicates that, once people have accepted the idea of providing probability judgments, instruction in the actual procedures is straightforward (Howell and Burnett, 1978). Some response modes require no explanation at all. People know what it means to say that "the probability of rain tomorrow is 0.30" or "the odds that this drilling site will not yield gas or oil above 8,000 feet are 4:1." For these situations, there has been only a negligible improvement in responses after the meaning of probabilities was explained.

With other, less familiar response modes, some direct instruction may be necessary (Lichtenstein and Newman, 1967). For example, when expressing incomplete knowledge about a continuous quantity, one may assess a cumulative probability distribution (e.g., there is a 0.05 chance that the rate is less than 0.003, a 0.10 chance that it is less than 0.005, etc.) or use the fractile method described by Raiffa (1968) (e.g., there is a 0.05 chance that the rate is less than 0.003, and a 0.05 chance that the rate is greater than 0.012, and the underlying distribution is log-normal). The former method has been used in studies of atmospheric oxidation rates, the latter in probabilistic risk assessments for nuclear-power stations.

An occasional source of difficulty is confusion about the event in question. The recent EPA effort to use probabilistic methods to set ambient-air quality standards, although thoughtfully conceived, was hampered by the poorly defined and unfamiliar units whose likelihood experts were required to judge. For example, one key question asked scientists to estimate the probability that severe health effects would result under specified conditions. This required the scientists to assign some value to a term (severe health effects) that should be defined by policy-makers (Moreau, 1980).

The design of assessment procedures requires the exercise of good sense. It cannot be assumed that the asking of a question ensures an articulate response. Part of the art of analysis is the ability to divide a complex problem into portions encompassing subjects on which people have the required expertise. The recommended scheme asks for rather straightforward judgments about data that are familiar to respondents.

A potential danger is that good probability assessment will be achieved at the expense of attention to the scientific issues being judged. One might imagine respondents being fascinated by the newness of probability judgments or eager to please the investigator with their assessments. Here, too, good sense may be sufficient to allay most fears. Indeed, if an expert in probabilities is playing an active role as the questioner in the proceedings, he or she should be evaluating the respondents' performance with a scoring rule that rewards them for responding with their true beliefs (Murphy, 1973). Evidence suggests that the procedures for organizing one's knowledge in a manner that will produce the best appraisal of what is known are the same procedures that produce the best probability assessment (i.e., a careful review of all that is known with an emphasis on evidence that seems to contradict the dominant opinion). Better substantive judgment and better probability assessment should go hand in hand.

This discussion is based on the assumption that experts in pertinent substantive matters are willing to participate in the assessment procedure. The product of the elicitation is likely to suffer if the respondents do not have confidence in the procedure. There are a number of reasons for hesitancy. Some resistance comes from mistrust, not of probability assessments, but of the analytic schemes in which they are embedded. Analysts have, at times, promised more than they can deliver, namely, a value-free, definitive solution to a difficult social problem.

Scientists may also be unwilling to engage in a procedure based on an unfamiliar concept to which their experience cannot be applied easily. Addressing this reluctance requires a spirit of compromise. The assessor needs to be willing to make an attempt.

Finally, scientists may simply resist being explicit about their uncertainties. A common ploy of analysts in such situations is to ask different people for the numerical equivalent of verbal expressions of uncertainty and then to show the range of responses elicited. If, for example, the probabilities associated with "further fighting in Cyprus is quite likely this year" vary from 0.25 to 0.80, then one can argue that the use of verbal expressions constitutes a barrier to communication. One hopes that scientists would also accept such demonstrations as persuasive and would be less likely to seek the shelter of vagueness when expressing their uncertainty.

QUALITY OF PROBABILITY JUDGMENTS

A variety of judges, topics, instructions, and response modes have been used in several hundred empirical studies of the quality of probability assessments (Lichtenstein et al., 1982). Two major conclusions resulted from these studies:

● Expressed confidence is correlated with justified confidence. Events that are judged to be more probable are, in fact, more likely to occur. People typically know more when they express great certainty than when they express uncertainty.

● Expressed confidence is not an infallible guide to the absolute extent of people's knowledge. Although confidence is sensitive to knowledge, it is not sensitive enough. The most commonly observed overall "bias" is a tendency toward overconfidence, except when judgments are very easy (Fischhoff et al., 1977).

These conclusions seem to be independent of such factors as the instructions used, the amount of practice in making such judgments, the stakes riding on good performance, familiarity with the subject in question (except insofar as it reduces uncertainty to nil), and the response mode used (provided that it has been properly explained).

IMPROVING THE QUALITY OF PROBABILITY JUDGMENTS

Two procedures have been found to improve the quality of probability judgments: providing personalized feedback after a round of practice questions (Lichtenstein and Fischhoff, 1980) and requiring assessors to be explicit about the set of reasons supporting and (especially) contradicting their beliefs (Koriat et al., 1980).

A number of theoretically based procedures have been developed for aggregating the probability assessments of a set of judges (Lindley et al., 1979). These procedures require one to make some assumptions about the quality of the probabilities that the different judges provide. Empirical evidence suggests that people who express more confidence tend to know more, other things being equal. The identity and power of those "other things" are not altogether clear. One factor that obviously needs to be considered is how well the respondents are motivated toward candor.

To determine whether the assessment that can be expected in specific situations is "good enough," it is necessary to make a comparative analysis of the consequences associated with potential biases and those associated with forgoing the probability assessment altogether. In setting priorities for testing chemicals for health effects, the utility of estimating explicit probabilities is likely to be substantial. They will lead to better testing decisions, and they will facilitate evaluation of the scheme's efficacy. If included from the beginning of the process, even rudimentary assessments will facilitate the inclusion of more sophisticated assessment procedures as they become available (and if evaluation shows the need for them). A modest amount of modeling might also provide some indications of the effect of biases on estimates of probabilities. Many estimates will prove to be only slightly affected by even fairly large biases; greater effects may be caused by errors originating in other sources.

Instruction in probability assessment is straightforward and can be given in a short time. Care is needed to explain explicitly the particular method used (e.g., probabilities, odds, or fractiles). Performance of an expert may be improved considerably by providing a little training, including practice questions whose answers are compared with known values.

The consequences of inaccurate probability estimates depend on the testing decision associated with them. Many testing decisions are likely to be unaffected by errors in those estimates.

REFERENCES

Christensen-Szalanski, J., P. Diehr, and R. Wood. 1982. Phased trial of a proven algorithm at a new primary care clinic. J. Public Health 21:16-22.

Fischhoff, B. 1980. Clinical decision analysis. Oper. Res. 28:28-43.

Fischhoff, B., P. Slovic, and S. Lichtenstein. 1977. Knowing with certainty: The appropriateness of extreme confidence. J. Exp. Psychol. Hum. Percept. Perform. 3:552-564.

Fischhoff, B., S. Lichtenstein, P. Slovic, S. L. Derby, and R. L. Keeney. 1981. Acceptable Risk. New York: Cambridge University Press. 185 pp.

Howard, R. 1975. Social decision analysis. Proc. Inst. Electr. Electron. Eng. 63:359-371.

Howell, W. C., and S. A. Burnett. 1978. Uncertainty measurement: A cognitive taxonomy. Organ. Behav. Hum. Perform. 22:45-68.

Koriat, A., S. Lichtenstein, and B. Fischhoff. 1980. Reasons for confidence. J. Exp. Psychol. [Hum. Learn.] 6:107-118.

Lichtenstein, S., and B. Fischhoff. 1980. Training for calibration. Organ. Behav. Hum. Perform. 26:149-171.

Lichtenstein, S., and J. R. Newman. 1967. Empirical scaling of common verbal phrases associated with numerical probabilities. Psychon. Sci. 9:563-564.

Lichtenstein, S., B. Fischhoff, and L. D. Phillips. 1982. Calibration of probabilities: The state of the art to 1980. Pp. 306-334 in D. Kahneman, P. Slovic, and A. Tversky, Eds. Judgment under Uncertainty: Heuristics and Biases. New York: Cambridge University Press.

Lindley, D. V., A. Tversky, and R. V. Brown. 1979. On the reconciliation of probability assessments. J. R. Stat. Soc., Ser. A 142:146-180.

Mishan, E. J. 1976. Cost-Benefit Analysis. New York: Praeger. 478 pp.

Moreau, D. H. 1980. Quantitative Risk Assessment of Non-carcinogenic Ambient Air Quality Standards: A Discussion of Conceptual Approaches, Input Information, and Output Measures. Research Triangle Park, N.C.: U.S. Environmental Protection Agency, Office of Air Quality Planning and Standards. 35 pp.

Murphy, A. H. 1973. A new vector partition of the probability score. J. Appl. Meteorol. 12:595-600.

National Research Council, Steering Committee on Identification of Toxic and Potentially Toxic Chemicals for Consideration by the National Toxicology Program. 1981. Strategies to Determine Needs and Priorities for Toxicity Testing. Vol. 1. Design. Washington, D.C.: National Academy Press. 143 pp.

Raiffa, H. 1968. Decision Analysis: Introductory Lectures on Choices under Uncertainty. Reading, Mass.: Addison-Wesley. 309 pp.

U.S. Nuclear Regulatory Commission. 1975. Reactor Safety Study. (Wash 1400). Washington, D.C.: U.S. Nuclear Regulatory Commission.

APPENDIX D

THE ANALYSIS OF STRUCTURE-ACTIVITY RELATIONSHIPS IN SELECTING POTENTIALLY TOXIC COMPOUNDS FOR TESTING

Information on many chemicals now in commercial use is sparse or lacking, except for their structural formulas and a few physical constants. Because essentially nothing may be known about their toxicity to various forms of life, it would be highly desirable to use our knowledge of the relationships between chemical structure and biologic activity--i.e., structure-activity relationships (SARs)--to formulate algorithms for selecting the potentially most toxic compounds for extensive biologic tests. This would make it possible to predict a specific kind of toxicity simply from a chemical structure of a substance and a few physical properties.

It is difficult to use SARs to predict the biologic activity of two types of compounds:

● Congeners with a common pharmacophoric or other biologic function presumed to have the same mechanisms of action.

● Heterogeneous compounds that produce qualitatively the same biologic response, but do not have a common function or a common mechanism of action.

During the last 20 years, considerable progress has been made in predicting the activity of congeners by using the techniques of quantitative SAR analysis (Martin, 1978). Few published studies have been devoted to theoretical solutions to the problem of predicting biologic activity of heterogeneous compounds, which is of interest when using SARs to set priorities. In one empirical approach, results on a large number of organic compounds tested in a standard system (e.g., the Ames _Salmonella_/microsome test and tests to determine LD_{50}) are used to derive a correlation equation that relates chemical features and physical properties of the molecules to the observed biologic response. Free and Wilson (1964) developed the first general approach to computerized SARs, which was based on chemical structure. Using regression analysis, they assigned values _de novo_ to various substituents (molecular fragments that could be used to predict the biologic activity of untested compounds). Their effort was developed further by Cramer _et al._ (1974), Hodes _et al._ (1977), and Tinker (1981).

In effect, these models postulate that

$$BA = aA + bB + cC +, \ldots, + nN, \qquad (1)$$

where BA is biologic activity, and the "best" values of the coefficients a, b, c, . . ., n are selected by a computer program to weight the contributions of the chemical's structural features or molecular fragments, A, B, C, . . ., N. Equation 1 can often be used to correlate activity and structure when the biologic activity is produced by the same mechanism for all compounds in the data set, even though the mechanisms of action at the molecular level are not known. For unknown reasons, the correlation sometimes fails badly, even for small sets of closely related molecules.

A heterogeneous set of compounds may cause a specific type of toxicity through a variety of mechanisms at the molecular level. For example, DNA can be damaged in many ways to initiate cancer. To express this, Equation 1 should be rewritten as:

$$BA_i = a'A + b'B + c'C, \ldots, + n'N, \qquad (2)$$

where each BA has a different mechanism of action and, hence, most likely a different SAR. Thus, there are an unknown number of variables on both sides of the equation. The meaning of such an equation is not clear.

Two approaches have been used to define the molecular fragments used in the algorithm. On the basis of the work of Free and Wilson (1964), Enslein and Craig (1981) use molecular fragments developed by chemists from accumulated experience about the chemical reactivity of clusters of atoms within a molecule. This method of defining molecular fragments allows them to use hundreds, instead of thousands, of fragments for correlation, thereby greatly reducing the redundancy in the independent variables. Hodes (1981) uses arbitrarily defined molecular fragments, such as three contiguous atoms excluding hydrogen, generated by a computer program. For a large data set (results of testing 1,000 compounds), Hodes starts with more than 10,000 variables (fragment constants) and eliminates redundancy as much as possible. The final molecular fragments selected bear almost no relation to those conventionally used by chemists. The advantage of this approach is that combinations of atoms not considered important by chemists are not overlooked.

The accuracy of an algorithm is determined by using it to predict the activity of a portion of the data set. (Some of the data points are set aside and not used in developing the algorithm.) The data of Enslein and Craig suffer in comparison with those of Hodes. Their data were taken from the literature and, hence, are of variable quality, whereas those of Hodes were derived from tests conducted by the National Cancer Institute, which made a concerted effort to achieve uniformity.

NONADDITIVITY

The approaches used by Hodes et al. (1977), Tinker (1981), Enslein and Craig (1981), Cramer et al. (1974), and Free and Wilson (1964) are based on Equations 1 and 2, which assume additivity--the biologic activity of a molecule is assumed to be the sum of the contributions of its parts. That is, a given molecular fragment, such as a $-CH_2CH_2CH_3$ group, makes a constant contribution, either positive or negative, to biologic activity; and other fragments make constant contributions; so one can add the weighting factors of the fragments of an untested compound to estimate its activity. These weighting factors are average values that work best for the set of data under consideration. Such additivity does not apply to a wide range of activity or variety of fragments.

One of the most carefully studied and important properties related to the biologic activity of organic compounds is their hydrophobicity. The most common way of describing that is to measure how a compound distributes between water and a nonpolar solvent, such as octanol. The partition coefficient, P, is the ratio of the concentrations in the two phases, and log P is used as a numerical scale of hydrophobicity. The relationship between log B (biologic activity) and log P is roughly parabolic, if a wide enough range of log P is considered. Activity in a given set of congeners (other factors remaining constant) increases as log P increases until a characteristic point (log P_O) is reached, at which point activity decreases with further increase in log P until it is inevitably lost. This can be understood as follows: As P approaches negative infinity, a compound becomes so water-soluble that it cannot cross a lipid barrier and therefore remains localized in the first aqueous compartment with which it comes into contact. Likewise, as P approaches positive infinity, the compound becomes so lipid-soluble that it remains in the first lipid compartment it encounters. Actually, activity falls to zero far short of infinity. For example, carcinogenic aromatic hydrocarbons, as a class, have one of the highest log P_O values; however, these compounds become inactive when log P reaches approximately 8 (Hansch and Fujita, 1964). The nonlinear relationship between P and activity eliminates the possibility of an additive relationship between hydrophobicity and bioactivity.

Another important effect of fragments is their relative attraction for or release of electrons, which determines the electron distribution in molecules. The relative electron density around functional groups can have an enormous effect on the reactivity of the groups. Some reactions are favored by high electron density; some, by low density. For example, Hansch et al. (1980) demonstrated a linear relationship (over a wide range) between the electron-withdrawing effect of substituents and the mutagenicity of cis-platinum ammines; the stronger the electron withdrawal, the more potent the ammine was in the Ames Salmonella/microsome test. In an earlier study, Venger et al. (1979) found exactly the opposite effect in triazines in the same test: strong electron withdrawal by substituents linearly reduced mutagenicity. The two classes of compounds induce mutagenesis in different ways; hence, one cannot assign a single number to a three-atom fragment, such as a nitro group, to account for its activity in both these classes of compounds in the test. If one had a set of compounds of which half were triazines and half cis-platinum ammines, the average contribution of the nitro group might be close to zero; however, if the set were composed only of compounds behaving like the cis-platinum ammines, a nitro group would increase mutagenicity in the Ames Salmonella/microsome test by almost 3 orders of magnitude.

This illustrates a most serious problem in trying to formulate an algorithm for estimating toxicity of large, miscellaneous sets of compounds. Each set will be composed of subgroups of congeners, each of which acts via its own mechanism. Twenty years of work in quantitative SAR analysis has shown that the SAR will generally be different for each mechanism of toxicity.

The examples provided above illustrate problems of additivity with hydrophobic and electronic effects in bioactive compounds. Additional problems are encountered when steric factors are considered. For example, an isopropyl group on a benzene ring para to the active group might increase activity, whereas the same group ortho to the active group could easily destroy it.

Even more dramatic is the difference in activity of stereoisomers that contain the same fragments, but in such arrangements that the compounds are mirror images of each other. One often finds huge differences in activity between such compounds; however, it has been shown that one can even do quantitative SAR work with stereoisomers when the mechanism of action is uniform throughout the set (Hansch et al., 1977). The concept of lock-and-key fit of biologically active compounds and bioreceptors is still valid. The only change in recent times is to view them as being more flexible.

The problems encountered when using chemical structures to predict biologic activity are similar to those found when translating a language by computer. Errors are made in translation, because the context in which a word is used is not considered by the computer. Assigning a toxicity weight to the hydroxyl (OH) group illustrates the necessity of considering the group within the context of the molecule. The parent compound in this series is water (HOH)--the least toxic compound known. Simple alcohols (ROH) containing 1-10 carbon atoms are toxic to all forms of life from enzymes to humans. The toxicity of alcohols containing more than 10 carbon atoms, depending on the test system, decreases rapidly with increasing chain length, and simple alcohols having more than 14 carbon atoms are not toxic in the usual tests. Lipophilic alcohols are highly toxic, but, in the context of superoptimal hydrophobicity or hydrophilicity, the OH is nontoxic. If OH is assigned a weighting factor for toxicity on the basis of experience with a number of hydrophobic alcohols, the factor is useless when applied to the OH group in sugars, starches, or some antibiotics. If one uses a large and evenly balanced set of data to form a correlation equation, the weighting factors of each of the fragments may approach zero as the size of the data set increases.

The examples described above illustrate several difficulties with the additivity principle and the limitation inherent in this approach when it is used to develop a global algorithm for correlating a specific biologic activity with chemical structure. Hence, it is reasonable to inquire about the effectiveness of these models and their verification. Hodes (1981) has published results of a critical evaluation of this type of SAR. He used an algorithm developed from test data on leukemia to estimate the relative probability of antitumor activity of 988 compounds that had not yet been tested against P358 mouse lymphocytic leukemia cells. In the same study, an experienced chemist also classified the 988 compounds and selected 298 as likely to have antileukemia activity, 14 as having novel structures, and 676 as likely to be inactive. In later National Cancer Institute tests for antileukemia activity, 26 compounds were found to be active; 33 were presumed to be active, but to require further testing; 10 were judged to have potential activity; and 27 were found to be too toxic to test. The chemist's selected 298 contained 11 of the 26 active substances, 8 of the 33 presumed to be active, 5 of the 10 with doubtful activity, and 19 of the 27 toxic compounds. The 298 compounds ranked most active by Hodes's algorithm contained 13 of the 26 active substances, 10 of the 33 presumed to be active, 4 of the 10 doubtful chemicals, and 14 of the 27 toxic compounds. Hodes and the chemist each identified about one-third of the active compounds. Hodes indicated that this is better than chance, but not by much.

CHEMICAL CLASSES RELATED TO HEALTH EFFECTS IN HUMANS

Given the paucity of toxicity data on most chemicals in commerce, it is impossible to formulate general correlation equations that relate chemical structure to biologic activity. The best we can do is make guesses about toxicity that are little more than expressions of suspicion. An organized approach to such guessing might be the development of a matrix of structural classes and health effects in humans.

The list of structural elements could be kept arbitrarily low (fewer than 100), and the toxicity classes could be made as large as necessary. Expert toxicologists could be asked to make some kind of simple estimate of the end points--i.e., health effects in humans associated with the structural class--and an estimate of their confidence in the association. These "guesses" would be based on a combination of experience and intuition. Given enough estimates, one should obtain an indication of which compounds may be toxic. However, would such a selection system be better than having the same "guesses" made directly without an attempt to classify the structural features?

A choice must be made between using a computer to sort chemicals into broad structural classes and using experienced persons to rate the potential toxicity of chemicals directly. Using structural classes, toxicologists must make judgments about only a few classes. The alternative approach requires experienced persons to make judgments about tens of thousands of compounds. The advantage of judging the compounds directly is that functional groups can be judged within the context of the entire molecule. Moreover, the experienced persons would be expected to make more accurate judgments than would be required if the broad structural classes were used; however, the use of these experts may be more expensive.

Computer sorting into structural classes is only a crude screening device. The effectiveness of the screen depends on the number of classes and on the accuracy of the computer searching procedures. An increase in the number of classes increases accuracy and search costs. If the number of classes were increased sufficiently, one would have a model similar to that developed by Enslein and Craig.

Sorting by structural classes can be expected to fail in some cases. If the number of classes is kept low, many chemicals will not be included in a class, and the procedure will not yield an estimate. Complicated molecules may be placed in two or more classes, so that intrepetation of the results is not straightforward. Furthermore, sorting by the use of these classes will produce many false-positives, because many members of a class have little toxicity.

SAR modeling seems to be a necessity, in that current testing resources do not permit testing all the compounds in commerce. Some method of generalizing toxicity data is required, and current SAR models are a modest beginning. Development of such models has suffered from the lack of an adequate data base and from the theoretical difficulties outlined above. There appear to be few alternative methods for making inexpensive judgments about the toxicity of tens of thousands of compounds on which there is little information other than chemical structure.

Future progress in building SAR models depends on the acquisition of data on a carefully selected group of organic compounds. This group should contain a few representative compounds from each of the important classes of industrial chemicals, and there should be some diversity among their physical properties. In addition, a useful data base would contain information on a few subsets containing 20-30 compounds each, all appearing to have the same toxic function, which has been modified by structural changes and which can be correlated quantitatively through the use of SARs. Such correlations could probably be made and would probably provide information that would be useful in developing a global algorithm.

REFERENCES

Craig, P. N., and K. Enslein. 1980. Application of structure-activity studies to develop models for estimation of toxicity, pp. 411-419. In D. B. Walters, ed. Safe Handling of Chemical Carcinogens, Mutagens, Teratogens and Highly Toxic Substances. Vol. 2. Ann Arbor, Mich.: Ann Arbor Science Publishers, Inc.

Cramer, R. D., III, G. Redl, and C. E. Berkoff. 1974. Substructural analysis. A novel approach to the problem of drug design. J. Med. Chem. 17:533-535.

Enslein, K., and P. Craig. 1981. Structure-activity in hazard assessment, pp. 389-420. In J. Saxena and F. Fisher, Eds. Hazard Assessment of Chemicals. Vol. 1. Current Developments. Academic Press, New York.

Enslein, K., and P. Craig. Carcinogens: A statistical structure activity model. J. Toxicol. Environ. Health. (in press)

Free, S. M., and J. W. Wilson. 1964. A mathematical contribution to structure-activity studies. J. Med. Chem. 7:395-399.

Hansch, C., and T. Fujita. 1964. Analysis. A method for correlation of biological activity and chemical structure. J. Am. Chem. Soc. 86:1616-1626.

Hansch, C., C. Grieco, C. Silipo, and A. Vittoria. 1977. Quantitative structure-activity relationship of chymotrypsin-ligand interactions. J. Med. Chem. 20:1420-1435.

Hansch, C., B. H. Venger, and A. Panthananickal. 1980. Mutagenicity of substituted (o-phenylenediamine) platinum dichloride in the Ames test. A quantitative structure-activity analysis. J. Med. Chem. 23:459-461.

Hodes, L. 1981. Computer-aided selection of compounds for antitumor screening: Validation of a statistical-heuristic method. J. Chem. Inf. Comput. Sci. 21:128-132.

Hodes, L., G. F. Hazard, R. I. Geran, and S. Richman. 1977. A statistical-heuristic method for automated selection of drugs for screening. J. Med. Chem. 20:469-475.

Martin, Y. C. 1978. Quantitative Drug Design: A Critical Introduction. Marcel Dekker, Inc., New York. 425 pp.

Tinker, J. 1981. Relating mutagenicity to chemical structure. J. Chem. Inf. Comput. Sci. 21:3-7.

Venger, B. H., C. Hansch, G. J. Hatheway, and Y. U. Amrein. 1979. Ames test of 1-(x phenyl)-3,3-dialkyltriazenes. A quantitative structure-activity study. J. Med. Chem. 22:473-476.

APPENDIX E

COSTS OF MISCLASSIFICATION

This appendix describes a plausible approach--based on economic theory--to the assignment of costs to errors in classification. The concepts of net social benefit and health costs are introduced and related to production. For this discussion, a chemical is assumed to be regulated on the basis of classification of its hazard. The social costs of regulatory errors caused by misclassification of the hazard are calculated from health costs and social benefits.

Supply and demand curves for a chemical are shown in Figure E-1. If the chemical is unregulated, it is produced at the market-determined level q_m, where marginal supply cost equals the price supported by demand. The net social benefit (excluding health costs) is defined as the area between the supply and demand curves up to the point of production. When the chemical is regulated to production q_r, the net benefit (excluding health costs), $b(q_r)$, is a function of production, q. Note that unregulated production and marketing are carried to the point where b(q) becomes flat (Figure E-2). This is also the point where supply and demand curves cross at q_m. Note also that the curve in Figure E-2 is concave downward; that is an important property of the typical structure of supply and demand relationships.

If health cost is assumed to be proportional to the product of exposure and toxicity and exposure is assumed to be proportional to production it is possible to express health costs as a function of the product of production and toxicity. Therefore, health cost is a linear function of production with a slope related to toxicity. Toxicity does not vary with production, but its magnitude is uncertain; the implications of this uncertainty will be considered later.

According to the criterion of net benefit, which includes health costs, optimal regulatory control leads to the production, q_r, that maximizes b(q) − c(q) or b(q) − tq. Figure E-3 shows social benefits, b(q), and health costs, c(q), plotted as functions of production, q. Optimal regulatory control leads to the production, q_r, that maximizes the vertical distance, y, between the curves b(q) and c(q). Inasmuch as c(q) is a straight line with slope t, this distance is maximized when the slope of b(q) is equal to t.

In the simplest case, only toxicity is uncertain. Exposure and production are assumed to be known, and exposure is assumed proportional to production. The supply and demand for the chemical in question are known (hence the net benefits). The harm of the chemical is the product of its (known) exposure and (unknown) toxicity. The correct classification of the chemical into a category of harm or hazard is uncertain.

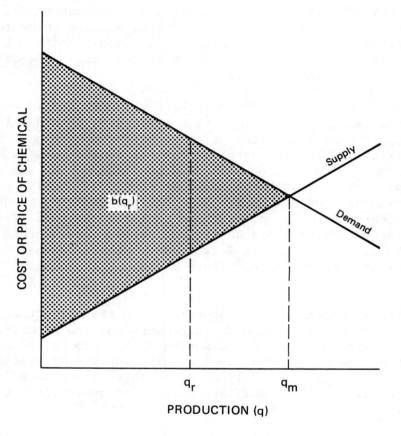

FIGURE E-1 Variation in supply and demand caused by changes in production, q. q_m, market-determined production; $b(q_r)$, net benefit of regulation of production.

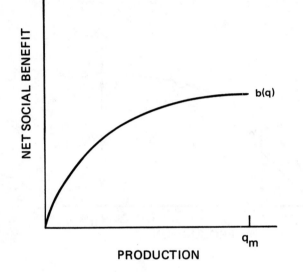

FIGURE E-2 Variation in net social benefit, $b(q)$, with production q. q_m, market-determined production.

365

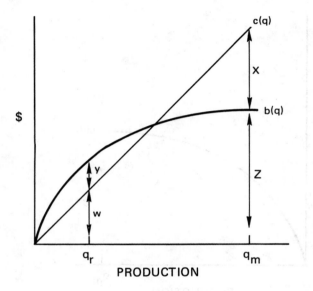

FIGURE E-3 Health costs, c(q), and benefits, b(q).

Errors in assessing toxicity are assumed to lead to errors in regulation. Define $\underline{t}$ as the best estimate of toxic potency. Suppose a chemical is classified as nontoxic ($\underline{t} = 0$) when in fact it is toxic (t 0). If society acts on this erroneous information, it does not regulate the chemical. Production then is set by the market at q_m, with a net benefit (excluding health costs) of $b(q_m)$. The net social benefit is net economic benefit minus health costs. This is shown as the vertical distance x in Figure E-3. With perfect knowledge, the optimal control is q_r, with a total net social <u>benefit</u>, y, in Figure E-3. The cost of a regulatory false-negative is the difference in benefit betwen what society obtains with erroneous information and what it could obtain with perfect information. In Figure E-3, the cost of a regulatory false-negative is the benefit that could have been obtained with regulation (y) minus the benefit actually obtained. This net benefit is negative because $c(q_m)$ is less than $b(q_m)$, so that the cost of the false-negative is x + y.

In the opposite error, suppose we classify the chemical as toxic ($\underline{t} > 0$) when in fact it is nontoxic (t = 0). If society acts on this erroneous information, it regulates to a production of q_r. It thus obtains a net benefit of y + w (Figure E-3) when it could have obtained a benefit of z, if it had known that the chemical was nontoxic and did not warrant regulation. The cost of a regulatory false-positive is z - y - w (the difference between what it could have obtained with perfect information and what it actually obtained with erroneous information).

Clearly, the costs of the two types of mistakes usually are not equal. The cost of a mistake depends on how much the estimated toxicity, $\underline{t}$, departs from the true toxicity, t. The cost of a regulatory error, $R(\underline{t}, t)$, is a function of the estimated toxicity and the true toxicity. If $\underline{t}$ is greater than t, the mistake is a false-positive; and if $\underline{t}$ is less than t, it is a false-negative. It is easy to check that when $\underline{t} = t$, $R(\underline{t}, t) = O$. Also, $R(\underline{t}, t)$ is nonnegative.

As measured by TD_{50}, carcinogenic potency of chemicals may range over 7 orders of magnitude. Three categories of carcinogenic potency are defined: low (noncarcinogenic), with a TD_{50} above 10^8 μg/kg per day; medium with a TD_{50} between 10^8 and 10^4 μg/kg per day; and high, with a TD_{50} below 10^4 μg/kg per day.

A carcinogen of medium potency, such as chloroform, has an associated health cost as represented by line $c_2(q)$ in Figure E-4. Chloroform is subject to some control, so there is some reduction in use with special care in handling, which lowers the amount of exposure for a given level of production. Thus, Figure E-4 shows a regulated production, q_r, that is about half the market-determined production, q_m, because at q_r the largest vertical distance occurs between $b(q)$ and $c_2(q)$.

The health cost as a function of exposure for a highly potent carcinogen is shown by $c_3(q)$, which has a slope 1,000 times greater than that for $c_2(q)$. The health cost for a noncarcinogen is virtually the horizontal axis, $c_1(q)$--its slope is 0.0001 that of $c_2(q)$.

The structure of relative health costs shown above will be used to consider various costs of regulatory mistakes, $R(\underline{t}, t)$.

First, consider the case of a chemical classified as of medium potency when it is actually nontoxic (t_1). The error cost is the benefit that society could have achieved minus the benefit actually achieved, assuming that society acts on its best available information, $\underline{t}_2$. As shown in Figure E-4, society, acting on the true information, would not regulate and hence would obtain benefit $b(q_m)$. However, acting on the best available information, society regulates to a production of q_r, with benefit $b(q_r)$. The cost of this regulatory false-positive is $b(q_m) - b(q_r)$.

Now consider the case of a chemical classified as nontoxic ($\underline{t}_1$) when it is actually of medium potency (t_2). With perfect information, society would regulate to a production of q_r, with benefit $b(q_r) - c_2(q_r)$. With available information, society does not regulate, with benefit $b(q_m)$. The cost of this regulatory false-negative is $b(q_r) - c_2(q_r) - [b(q_m) - c_2(q_m)]$.

In the case of a chemical classified as nontoxic ($\underline{t}_1$) when it is actually highly potent (t_3) and with correct information and optimal regulation, society would ban this chemical entirely (see slopes in Figure E-4), with zero benefit. With actual information, society does not control the chemical at all, with benefit $b(q_m) - c_3(q_m)$, where $c_3(q_m)$ is the health cost with no control for highly potent carcinogenic chemical. Note that the net benefit is negative and large, because $c_3(q_m)$ is much larger than $b(q_m)$. The cost of this regulatory false-negative is $[0 - b(q_m)] - [0 - c_3(q_m)] = c_3(q_m) - b(q_m)$.

Suppose we classify a chemical as highly potent ($\underline{t}_3$) when it is actually nontoxic (t_1). Acting on the erroneous information, we ban the chemical, with zero benefit. With correct information, there would be no regulation. The cost of this false-positive is $b(q_m) - 0 = b(q_m)$.

If we classify a chemical as highly potent ($\underline{t}_3$) when it is actually of medium potency, the cost of the false-positive is $b(q_r) - c_2(q_r) - 0 = b(q_r) - c_2(q_r)$.

The above cases are shown together in Table E-1. When the toxicity of a chemical is classified correctly ($\underline{t}_i = t_i$), there is no regulatory mistake; therefore, the regulatory error, $R(\underline{t}, t)$, is zero.

Returning to Figure E-4, we can attempt to assess some relative magnitudes for the costs of regulatory mistakes. The benefit, $b(q_m)$, may be considered as a unit. Assume that this unit is about $1 million.

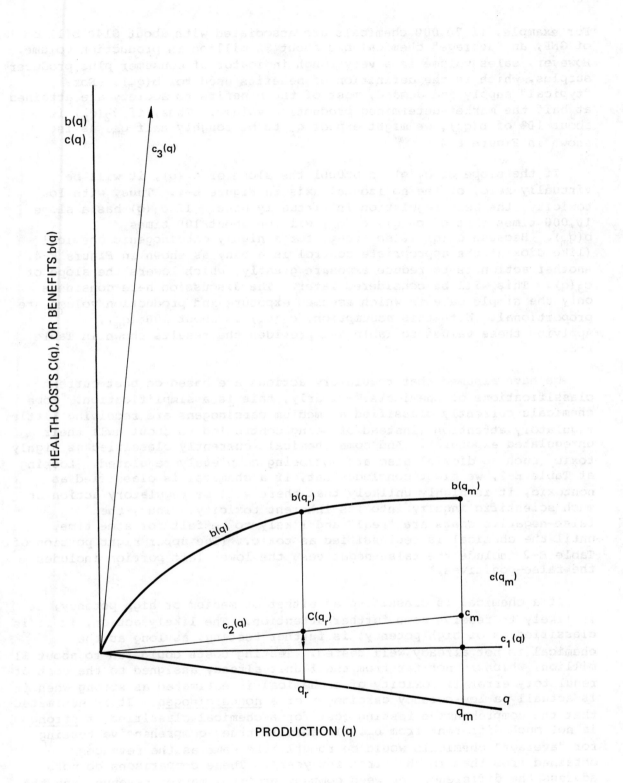

FIGURE E-4   Health costs c(q) and benefits b(q).

369

For example, if 70,000 chemicals are associated with about $140 billion of GNP, an "average" chemical has about $2 million in production volume. However, sales volume is a very rough indicator of consumer plus producer surplus, which is the definition of benefits used for $b(q_m)$. For "typical" supply and demand, most of the benefits to society are attained at half the market-determined production volume. Thus, if $c_2(q_m)$ is about 10% of $b(q_m)$, we might expect q_r to be roughly half q_m, as is shown in Figure E-4.

If the slope of $c_1(q)$ is 0.0001 the slope of $c_2(q)$, it will be virtually zero, or the horizontal axis in Figure E-4. Thus, with low toxicity, the best regulation is virtually none. If $c_3(q)$ has a slope 10,000 times that of $c_2(q)$, $c_3(q_m)$ will be about 100 times $b(q_m)$. Because $c_3(q)$ is so steep, for a highly carcinogenic chemical (like dioxin) the appropriate control is a ban, as shown in Figure E-4. Another action is to reduce exposure greatly, which lowers the slope of $c_3(q)$. This will be considered later. The discussion here considers only the simple case in which assumed exposure and production volume are proportional. With this assumption, $c_3(q_m)$ is about $100b(q_m)$, Applying these values to Table E-1 provides the results shown in Table E-2.

We have assumed that regulatory actions are based on best current classifications of chemicals. Clearly, this is a simplification. Some chemicals currently classified as medium carcinogens are receiving little regulatory attention (instead of being controlled to about half their unregulated exposure). And some chemicals currently classified as highly toxic (such as dioxin) also are not being completely regulated. Looking at Table E-2, we might conclude that, if a chemical is classified as nontoxic, it is highly unlikely that there will be regulatory action or much scientific inquiry into its inherent toxicity. Thus, the false-negative costs are "real" and likely to be felt for some time, until the chemical is reclassified as toxic. (The upper right portion of Table E-2 include the false-negatives; the lower left portion includes the false-positives.)

If a chemical is classified as either of medium or high potency, it is likely to receive some further attention. The likely action, if it is classified as of high potency, is further testing, as long as the chemical is not already well tested. Testing costs could run to about $1 million, which is not far from the 1 unit already assigned to the cost of regulatory error if toxicity of a chemical is estimated as strong when it is actually a low-potency carcinogen or a noncarcinogen. It is estimated that the comprehensive testing cost for a chemical classified as strong is not much different from b_m. If that is true, comprehensive testing for "average" chemicals would be roughly the same as the revenues obtained from them in the first few years. These comparisons do not address the differences between company profits, market revenue, and the consumers' plus producers' surplus.

TABLE E-1 Cost of Regulatory Error due to Disagreement Between Estimated Toxicity, $\underline{t}$, and True Toxicity, t

Estimated Toxicity	Cost		
	t_1	t_2	t_3
$\underline{t}_1$	0	$b(q_r) - c_2(q_r)$ $- [b(q_m) - c_2(q_m)]$	$c_3(q_m) - b(q_m)$
$\underline{t}_2$	$b(q_m) - b(q_r)$	0	$c_3(q_r) - b(q_r)$
$\underline{t}_3$	$b(q_m)$	$b(q_r) - c_2(q_r)$	0

TABLE E-2 Cost of Regulatory Error in Terms of Social Benefit per Average Chemical under Market Conditions, $b(q_m)$

Estimated Toxicity	Cost, $b(q_m)$		
	t_1	t_2	t_3
$\underline{t}_1$	0	0.05	99
$\underline{t}_2$	0.1	0	49
$\underline{t}_2$	1.0	0.92	0

It is probable that further scrutiny would follow a classification of a chemical as a medium carcinogen; the assignment of 49 units for the cost of regulatory error may overstate the cost of this misclassification. With further scrutiny, there is some chance that the chemical in question would be correctly classified, with eventually a lower cost to society. The cost of this scrutiny in further testing might be $100,000 - 500,000 (less than if the chemical had been classified as <u>high</u>). Thus, if the chemical is truly nontoxic, but is misclassified as medium, the cost of the mistake might be a little more than 0.1 unit ($100,000), suggested in Table E-2. Thus, we modify the costs of a regulatory error when a chemical is classified as a medium carcinogen, as shown in Table E-3.

Thus far, exposure has been assumed to be known and to be proportional to production. Exposure has been estimated to range over about 6 orders of magnitude. Furthermore, exposure often is fairly independent of production. Taking into account a range of 6 orders of magnitude in exposure and a range of 7 orders of magnitude in toxic potency, a hazard index of the product of toxicity and exposure would range over 13 orders of magnitude. At the same time, the link between regulatory action and hazard classification becomes more complicated and more diffuse.

A more complicated example of misclassification occurs when a chemical is correctly classified as having medium toxicity, but erroneously classified as having high exposure when in fact it has low exposure. Is the chemical likely to be regulated on the basis of this misclassification? Hardly. Before regulation, its exposure is likely to be studied more carefully and (it is hoped) sufficiently well to correct the misclassification of exposure. Because the exposure category is 3 orders of magnitude wide, this discovery is fairly likely.

Thus, the main cost of classifying exposure too highly is the extra cost (ultimately found to be unnecessary) of learning that the chemical belongs in a lower exposure category.

Now consider a chemical, correctly classified as high in toxicity, but erroneously classified as having low exposure when it actually has high exposure. Would such a chemical receive no further attention, on the basis of its erroneous exposure classification? Once a chemical is classified as having high toxic potency, it would receive further attention, even if currently classified as having low exposure. With the further attention, the erroneous classification of exposure would have a good chance of being correctd. The main cost of this misclassification is not the cost of research on exposure, which is warranted in this case, but the cost arising from the chance that the error in exposure would not be corrected through investigation.

Similar reasoning applies in the case of a chemical correctly classified as being medium in toxicity, but misclassified as to exposure. Too high a classification of exposure leads to unnecessary research on exposure, whereas too low a classification leads to a chance that research will not correct the error. Chemicals classified as having medium toxicity are assumed to receive further attention.

TABLE E-3 Cost of Regulatory Error in Terms of Social Benefit
per Average Chemical under Market Conditions, $b(q_m)$

Estimated Toxicity	Cost, $b(q_m)$		
	t_1	t_2	t_3
	0	0.05	99
$\underline{t}_1$			
$\underline{t}_2$	0.4	0	25
$\underline{t}_3$	1.0	0.95	0

Further attention is more likely to be devoted to these chemicals than in the previous case, because less attention will probably be devoted to a chemical classified as having medium than high toxic potency.

Taking these ideas into account, we might arrive at a table of regulatory error, $R(\underline{t},t)$, as shown in Table E-4. The underlined numbers correspond to medium exposure and correspond to the entries of Table E-3, with some modification.

The obvious thing to note in Table E-4 is the large range in the costs of false-negatives, which appear above and to the right of the diagonal whose values are zeroes. This range is due to the range of 13 orders of magnitude in hazard. The range of regulatory false-positives is about 7 orders of magnitude, which is less than the range in hazard because of the way false-positive and false-negative costs are defined, in comparison with benefit, $b(q_m)$. However, the range is still enormous. Moreover, the cost of the largest false-negative (99,999) is enormous, compared with the cost of the largest false-positive (0.8), and probably much larger than most people would think realistic.

What might lead to overstating these ranges? Consideration of the answers to this question leads to examination of two assumptions: exposure is proportional to production, and a classification of medium toxicity leads to some control. These assumptions may decrease the differences between false-negatives and false-positives.

The "average" chemical is assumed to have market-determined production, q_m, with a net benefit, $b(q_m)$. Consider what happens when q_m also ranges over 6 orders of magnitude and, with it, net benefit. In the simplest case, exposure is proportional to production, and regulation follows directly from classification. The matrix of error costs is similar to Tables E-2 and E-3, where the variation in $b(q_m)$ and q_m, from high to low production, scales everything up and down. An overestimate of exposure is also a mistake in overestimating $b(q_m)$, and these mistakes cancel out, as long as costs of mistakes are expressed in terms of $b(q_m)$.

In another simple case, benefit, $b(q_m)$, and production, q_m, are fixed and exposure still varies over 6 orders of magnitude. This case is shown in Table E-4. A more realistic case has costs of regulatory error intermediate between the values for the extreme cases shown in Table E-4.

The main reason for the high relative cost of the largest false-negative in Table E-2 (99) is the assumption that there should be some regulatory control for medium carcinogens. The control considered was a restriction of production to about 50% of that determined by the market. With the typical concavity of the benefit curve (which follows from typical supply-and-demand schedules) and the range of 3 orders of magnitude in toxic potency, this led to the asymmetric structure of Table E-2.

TABLE E-4 Expected Cost of Regulatory Error, as Modified by Expected Reaction to Misclassification[a]

Estimated Classification	True Classification								
	e_1t_1	e_2t_1	e_3t_1	e_1t_2	e_2t_2	e_3t_2	e_1t_3	e_2t_3	e_3t_3
e_1t_1	0	0	0.05	0	0.05	99	0.05	99	99,999
e_2t_1	0	0	0.05	0	0.05	99	0.05	99	99,999
e_3t_1	0	0	0	0	0.05	99	0.05	99	99,999
e_1t_2	0.7	0.7	0.7	0	0.40	40	0.5	40	40,000
e_2t_2	0.7	0.7	0.7	0.2	0	10	0.5	10	10,000
e_3t_2	0.7	0.7	0.7	0.2	0.1	0	0.5	5	5,000
e_1t_3	0.8	0.8	0.8	0.4	0.4	0.2	0	2	200
e_2t_3	0.8	0.8	0.8	0.4	0.4	0.2	0.1	0	100
e_3t_3	0.8	0.8	0.8	0.4	0.4	0.2	0.1	0.1	0

---

[a] e = exposure class; t = toxicity class. Subscripts: 1 = low or non-; 2 = medium; 3 = high. Underlined numbers correspond to medium exposure and correspond to entries of Table E-3, with some modification.

But there is another, probably more realistic, scenario in which chemicals classified as of medium toxicity may be subjected to some control. For the same production, more careful handling and use might halve exposure at a management cost that is low relative to the benefit. To have some control for medium toxicity, it is not necessary to posit $c_2(q)$ with a slope of about 10% of that of $b(q_m)$ (see Figure E-4). Suppose the slope of $c_2(q)$ is only 1% of that of $b(q_m)$, instead of 10%. This assumption implies very little restriction in production, but it could imply large reductions in exposure per unit of production by restricting use and adopting stringent handling requirements. With this change, we can revise Table E-2 to Table E-5. Again, consider the cost of various cases of misclassification:

- $R(\underline{t}_1 t_3)$. Because <u>high</u> toxicity is 1,000 times more than <u>medium</u> toxicity, the slope of $c_3(q)$ is only 10 times that of $b(q_m)$. Classifying a chemical as <u>nontoxic</u> when it has a <u>high</u> toxicity leads to a false-negative regulatory cost of $10b(q_m) - b(q_m) = 9b(q_m)$ (upper right corner of Table E-5 in terms of benefit from production set by market conditions).

- $R(\underline{t}_1 t_2)$. If a chemical is classified as <u>nontoxic</u> when it is actually <u>medium</u>, it is possible to halve the exposure at a small cost (say, 0.1%), relative to $b(q_m)$. The cost of regulatory error is $[b(q_m) = c_2(q_r) - 0.011b(q_m) - b(q_m) - c(q_m)]$. Because $c_2(q_r)$ is half $c_2(q_m)$, the cost of this false-negative is 0.004 unit of $b(q_m)$.

- $R(\underline{t}_2 t_3)$. If society acts directly on a chemical classified as <u>medium</u> when it is actually <u>high</u>, the benefit is $b(q_m) - c_3(q_r) - 0.001b(q_m)$, or $b(q_m) - 5 - 0.001\ b(q_m)$. With correct information, the chemical would be banned, with zero benefit. So the cost of this false-negative is $0 - (-3.999) = 3.999$.

- $R(\underline{t}_2 t_1)$. If a chemical is classified as <u>medium</u> when it is actually <u>nontoxic</u>, the benefit is $b(q_m) - 0.001b(q_m)$; it could have been b_m. The cost of this false-positive is 0.001.

- $R(\underline{t}_3 t_2)$. If a chemical is classified as of <u>high</u> toxicity when it is actually <u>medium</u>, it is banned, with zero benefit, whereas we could have halved the exposure, at production q_m, with a benefit of $b(q_m) - c_2(q_r) - 0.001\ b(q_m) = 1 - 0.005 - 0.001 = 0.994b(q_m)$. The cost of this false-positive is $0.994b(q_m)$.

- $R(\underline{t}_3 t_1)$. If a chemical is classified as <u>highly</u> toxic when it is actually <u>nontoxic</u>, it is banned, with zero benefit, whereas we could have had the benefit, $b(q_m)$, with no health cost. Therefore, the cost of this false-negative is $b(q_m) - 0 = b(q_m)$.

376

TABLE E-5 Cost of Regulatory Error Due to Misclassification of Toxicity in Terms of Social Benefit under Market Conditions, $b(q_m)$

Estimated Toxicity	Cost, $b(q_m)$		
	t_1	t_2	t_3
$\underline{t_1}$	0	0.004	9
$\underline{t_2}$	0.001	0	3.999
$\underline{t_3}$	1	0.994	0

Table E-5 was caculated on the assumption that exposure (unless separately controlled) is proportional to production and $b(q_m)$ is proportional to q_m. Now we can take into account the idea that exposure and production are not always proportional. Suppose that for high-exposure chemicals the proportionality between exposure and production is twice that for medium-exposure chemicals, and suppose further that for medium-exposure chemicals the proportionality between exposure and production is twice that for low-exposure chemicals. In other words, high-production chemicals tend to cost less per pound, and the assumption is that the market benefit per pound of production volume is half that for medium-production chemicals and one-fourth that for low-production chemicals. This assumption leads to Table E-6. The false-negative section (upper right portion) is revised, but the false-positive portion is unchanged. Most of the false-positive costs are associated with research to develop a chemical before it is regulated, and these costs are limited to 1 unit; the total benefit having been foregone, the chemical is erroneously banned.

It is difficult to assess regulatory costs resulting from misclassification of a chemical with regard to exposure or toxicity. But, although the above analysis can be refined, some features of the structure of error costs are discernible. It seems reasonable for the costs in the upper right portion to be asymmetrically larger than those in the lower left portion--perhaps as much as 20 times as large. This indicates that the social cost of underregulating a chemical is much greater than that of overregulation. Exposure clasification is clearly important, but it does not play a role entirely symmetric with that of toxicity classification, as the simple concept, "hazard equals exposure times toxicity," might suggest. The reason for the asymmetry is that exposure tends to be proportional to the benefit of a chemical, whereas toxic potency is an inherent property of a chemical and does not vary with production, benefit, or exposure. An implication of this difference is that, for a given health effect, information on toxic potency is permanent or changes only as science improves, whereas correct information on exposure is more dynamic and changes as production responds to market changes and technologic advances.

378

TABLE E-6 Regulatory Error Due to Misclassification of Exposure and Toxicity[a]

Estimated Classification	True Classification								
	e_1t_1	e_2t_1	e_3t_1	e_1t_2	e_2t_2	e_3t_2	e_1t_3	e_2t_3	e_3t_3
e_1t_1	0	0	0	0.002	0.004	0.008	4.5	9	18
e_2t_1	0	0	0	0.002	0.004	0.008	4.5	9	18
e_3t_1	0	0	0	0.0015	0.003	0.006	4.5	8	16
e_1t_2	0.7	0.7	0.7	0	0.01	0.04	2	4	8
e_2t_2	0.7	0.7	0.7	0.2	0	0.01	2	4	8
e_3t_2	0.7	0.7	0.7	0.2	0.1	0	1.8	3	6
e_1t_3	0.8	0.8	0.8	0.4	0.4	0.2	0	0.01	0.05
e_2t_3	0.8	0.8	0.8	0.4	0.4	0.2	0.1	0	0.01
e_3t_3	0.8	0.8	0.8	0.4	0.4	0.2	0.1	0.1	0

[a] e = exposure class; t = toxicity class. Subscripts: 1 = low or non-; 2 = moderate; 3 = high.

APPENDIX F

DIFFERENCES BETWEEN PART 1 AND PART 2

The stated objectives of the Committee on Sampling Strategies and the Committee on Toxicity Data Elements differed from those of the Committee on Priority Mechanisms. Accordingly, the concepts and practices used by the committees to achieve the two objectives were different:

• The study included an examinination of the extent of toxicity testing of chemicals to which humans are exposed and methods of selecting chemicals for testing. The committees were thus assigned different aspects of the same problem: evaluating toxicity testing for chemicals of concern to NTP. The Committees on Toxicity Data Elements and Sampling Strategies devised a method to assess the current state of toxicity information used to determine health hazard and to estimate additional needs for toxicity testing. The Committee on Priority Mechanisms developed a design approach for priority-setting systems and applied that approach to a demonstration system for the select universe defined by the Committee on Toxicity Data Elements.

• The Committee on Toxicity Data Elements identified a list of test types that served as a basis for examining the adequacy of past toxicity testing and estimating current testing needs. The committees acknowledge that under a variety of conditions it would not be necessary for all such tests to be done. For example, although it is useful to have information on chronic toxicity for a specific substance, the presence of positive data from a well-conducted subchronic study might obviate the development of any further information on chronic exposure. A need would still remain to establish a mechanism for deciding which tests and which substances should be examined and which ones would be given a higher priority. The Committee on Priority Mechanisms examined this issue.

• The select universe defined by the Committee on Toxicity Data Elements was fixed once the sample was taken so that, after sample analysis, useful estimates for the universe could be made. The universe of substances that the Committee on Priority Mechanisms considered, however, by definition is constantly expanding as more substances with a potential for human exposure are identified.

• Working documents developed and used by the Committee on Toxicity Data Elements were designed to assess the status and quality of toxicity-testing information. The dossier concept adopted by the Committee on Priority Mechanisms is intended to provide an assessment of exposure and toxicity. The approach of the Committee on Sampling Strategies and the Committee on Toxicity Data Elements results in an estimation of toxicity-testing frequency, adequacy, and needs based on a retrospective analysis of existing information. The approach of the Committee on Priority Mechanisms results in a priority-setting framework that could be useful in determining which chemicals to test and which tests would yield the most informative data.

● The select universe of chemicals used in this study contained, by design, substances with a potential for human exposure. A sample was drawn from this select universe by the Committee on Sampling Strategies for use by the Committee on Toxicity Data Elements in its determination of toxicity-testing needs. Although the degree of potential human exposure was used in the determination of testing needs, it was not a determinant in selecting the sample of substances. In contrast, the Committee on Priority Mechanisms suggests procedures that use information on the degree of potential human exposure early in the chemical selection process.

● For each substance in its sample, the Committee on Toxicity Data Elements searched comprehensively for and nonselectively used all information that might assist it in determining the testing needs for that substance. The Committee on Priority Mechanisms was selective in applying information to each of the various stages of its priority-setting system.

● Analysis of information was approached differently in each activity. The purpose of the Committee on Toxicity Data Elements was to determine the type and quality of available data, rather than review existing assessments of toxicity. These determinations were based on the expert judgment of committee members. The Committee on Priority Mechanisms devised an approach to making assessments of public-health concern as part of a system to select chemicals for testing. This approach explicitly provides for estimating the degree of uncertainty in the assessment.

● Finally, the Committee on Toxicity Data Elements and the Committee on Sampling Strategies examined available information to determine whether there is enough to conduct at least partial health-hazard assessments. The Committee on Priority Mechanisms provides a framework for using the information to conduct such assessments.